T0205356

Advances in Intelligent Systems and Computing

Volume 827

Series editor

Janusz Kacprzyk, Polish Academy of Sciences, Warsaw, Poland
e-mail: kacprzyk@ibspan.waw.pl

The series "Advances in Intelligent Systems and Computing" contains publications on theory, applications, and design methods of Intelligent Systems and Intelligent Computing. Virtually all disciplines such as engineering, natural sciences, computer and information science, ICT, economics, business, e-commerce, environment, healthcare, life science are covered. The list of topics spans all the areas of modern intelligent systems and computing such as: computational intelligence, soft computing including neural networks, fuzzy systems, evolutionary computing and the fusion of these paradigms, social intelligence, ambient intelligence, computational neuroscience, artificial life, virtual worlds and society, cognitive science and systems, Perception and Vision, DNA and immune based systems, self-organizing and adaptive systems, e-Learning and teaching, human-centered and human-centric computing, recommender systems, intelligent control, robotics and mechatronics including human-machine teaming, knowledge-based paradigms, learning paradigms, machine ethics, intelligent data analysis, knowledge management, intelligent agents, intelligent decision making and support, intelligent network security, trust management, interactive entertainment, Web intelligence and multimedia.

The publications within "Advances in Intelligent Systems and Computing" are primarily proceedings of important conferences, symposia and congresses. They cover significant recent developments in the field, both of a foundational and applicable character. An important characteristic feature of the series is the short publication time and world-wide distribution. This permits a rapid and broad dissemination of research results.

More information about this series at http://www.springer.com/series/11156

Sebastiano Bagnara · Riccardo Tartaglia
Sara Albolino · Thomas Alexander
Yushi Fujita
Editors

Proceedings of the 20th Congress of the International Ergonomics Association (IEA 2018)

Volume X: Auditory and Vocal Ergonomics, Visual Ergonomics, Psychophysiology in Ergonomics, Ergonomics in Advanced Imaging

 Springer

Editors
Sebastiano Bagnara
University of the Republic of San Marino
San Marino, San Marino

Riccardo Tartaglia
Centre for Clinical Risk Management
  and Patient Safety, Tuscany Region
Florence, Italy

Sara Albolino
Centre for Clinical Risk Management
  and Patient Safety, Tuscany Region
Florence, Italy

Thomas Alexander
Fraunhofer FKIE
Bonn, Nordrhein-Westfalen
Germany

Yushi Fujita
International Ergonomics Association
Tokyo, Japan

ISSN 2194-5357   ISSN 2194-5365   (electronic)
Advances in Intelligent Systems and Computing
ISBN 978-3-319-96058-6   ISBN 978-3-319-96059-3   (eBook)
https://doi.org/10.1007/978-3-319-96059-3

Library of Congress Control Number: 2018950646

This Springer imprint is published by the registered company Springer Nature Switzerland AG
The registered company address is: Gewerbestrasse 11, 6330 Cham, Switzerland

Preface

The Triennial Congress of the International Ergonomics Association is where and when a large community of scientists and practitioners interested in the fields of ergonomics/human factors meet to exchange research results and good practices, discuss them, raise questions about the state and the future of the community, and about the context where the community lives: the planet. The ergonomics/human factors community is concerned not only about its own conditions and perspectives, but also with those of people at large and the place we all live, as Neville Moray (Tatcher et al. 2018) taught us in a memorable address at the IEA Congress in Toronto more than twenty years, in 1994.

The Proceedings of an IEA Congress describes, then, the actual state of the art of the field of ergonomics/human factors and its context every three years.

In Florence, where the XX IEA Congress is taking place, there have been more than sixteen hundred (1643) abstract proposals from eighty countries from all the five continents. The accepted proposal has been about one thousand (1010), roughly, half from Europe and half from the other continents, being Asia the most numerous, followed by South America, North America, Oceania, and Africa. This Proceedings is indeed a very detailed and complete state of the art of human factors/ergonomics research and practice in about every place in the world.

All the accepted contributions are collected in the Congress Proceedings, distributed in ten volumes along with the themes in which ergonomics/human factors field is traditionally articulated and IEA Technical Committees are named:

   I. Healthcare Ergonomics (ISBN 978-3-319-96097-5).
  II. Safety and Health and Slips, Trips and Falls (ISBN 978-3-319-96088-3).
 III. Musculoskeletal Disorders (ISBN 978-3-319-96082-1).
 IV. Organizational Design and Management (ODAM), Professional Affairs, Forensic (ISBN 978-3-319-96079-1).
  V. Human Simulation and Virtual Environments, Work with Computing Systems (WWCS), Process control (ISBN 978-3-319-96076-0).

VI. Transport Ergonomics and Human Factors (TEHF), Aerospace Human Factors and Ergonomics (ISBN 978-3-319-96073-9).

VII. Ergonomics in Design, Design for All, Activity Theories for Work Analysis and Design, Affective Design (ISBN 978-3-319-96070-8).

VIII. Ergonomics and Human Factors in Manufacturing, Agriculture, Building and Construction, Sustainable Development and Mining (ISBN 978-3-319-96067-8).

IX. Aging, Gender and Work, Anthropometry, Ergonomics for Children and Educational Environments (ISBN 978-3-319-96064-7).

X. Auditory and Vocal Ergonomics, Visual Ergonomics, Psychophysiology in Ergonomics, Ergonomics in Advanced Imaging (ISBN 978-3-319-96058-6).

Altogether, the contributions make apparent the diversities in culture and in the socioeconomic conditions the authors belong to. The notion of well-being, which the reference value for ergonomics/human factors is not monolithic, instead varies along with the cultural and societal differences each contributor share. Diversity is a necessary condition for a fruitful discussion and exchange of experiences, not to say for creativity, which is the "theme" of the congress.

In an era of profound transformation, called either digital (Zisman & Kenney, 2018) or the second machine age (Bnynjolfsson & McAfee, 2014), when the very notions of work, fatigue, and well-being are changing in depth, ergonomics/human factors need to be creative in order to meet the new, ever-encountered challenges. Not every contribution in the ten volumes of the Proceedings explicitly faces the problem: the need for creativity to be able to confront the new challenges. However, even the more traditional, classical papers are influenced by the new conditions.

The reader of whichever volume enters an atmosphere where there are not many well-established certainties, but instead an abundance of doubts and open questions: again, the conditions for creativity and innovative solutions.

We hope that, notwithstanding the titles of the volumes that mimic the IEA Technical Committees, some of them created about half a century ago, the XX Triennial IEA Congress Proceedings may bring readers into an atmosphere where doubts are more common than certainties, challenge to answer ever-heard questions is continuously present, and creative solutions can be often encountered.

Acknowledgment

A heartfelt thanks to Elena Beleffi, in charge of the organization committee. Her technical and scientific contribution to the organization of the conference was crucial to its success.

References

Brynjolfsson E., A, McAfee A. (2014) The second machine age. New York: Norton.

Tatcher A., Waterson P., Todd A., and Moray N. (2018) State of science: Ergonomics and global issues. Ergonomics, 61 (2), 197–213.

Zisman J., Kenney M. (2018) The next phase in digital revolution: Intelligent tools, platforms, growth, employment. Communications of ACM, 61 (2), 54–63.

Sebastiano Bagnara
Chair of the Scientific Committee, XX IEA Triennial World Congress
Riccardo Tartaglia
Chair XX IEA Triennial World Congress
Sara Albolino
Co-chair XX IEA Triennial World Congress

Organization

Organizing Committee

Riccardo Tartaglia (Chair IEA 2018)	Tuscany Region
Sara Albolino (Co-chair IEA 2018)	Tuscany Region
Giulio Arcangeli	University of Florence
Elena Beleffi	Tuscany Region
Tommaso Bellandi	Tuscany Region
Michele Bellani	Humanfactor[x]
Giuliano Benelli	University of Siena
Lina Bonapace	Macadamian Technologies, Canada
Sergio Bovenga	FNOMCeO
Antonio Chialastri	Alitalia
Vasco Giannotti	Fondazione Sicurezza in Sanità
Nicola Mucci	University of Florence
Enrico Occhipinti	University of Milan
Simone Pozzi	Deep Blue
Stavros Prineas	ErrorMed
Francesco Ranzani	Tuscany Region
Alessandra Rinaldi	University of Florence
Isabella Steffan	Design for all
Fabio Strambi	Etui Advisor for Ergonomics
Michela Tanzini	Tuscany Region
Giulio Toccafondi	Tuscany Region
Antonella Toffetti	CRF, Italy
Francesca Tosi	University of Florence
Andrea Vannucci	Agenzia Regionale di Sanità Toscana
Francesco Venneri	Azienda Sanitaria Centro Firenze

Scientific Committee

Sebastiano Bagnara (President of IEA2018 Scientific Committee)	University of San Marino, San Marino
Thomas Alexander (IEA STPC Chair)	Fraunhofer-FKIE, Germany
Walter Amado	Asociación de Ergonomía Argentina (ADEA), Argentina
Massimo Bergamasco	Scuola Superiore Sant'Anna di Pisa, Italy
Nancy Black	Association of Canadian Ergonomics (ACE), Canada
Guy André Boy	Human Systems Integration Working Group (INCOSE), France
Emilio Cadavid Guzmán	Sociedad Colombiana de Ergonomia (SCE), Colombia
Pascale Carayon	University of Wisconsin-Madison, USA
Daniela Colombini	EPM, Italy
Giovanni Costa	Clinica del Lavoro "L. Devoto," University of Milan, Italy
Teresa Cotrim	Associação Portuguesa de Ergonomia (APERGO), University of Lisbon, Portugal
Marco Depolo	University of Bologna, Italy
Takeshi Ebara	Japan Ergonomics Society (JES)/Nagoya City University Graduate School of Medical Sciences, Japan
Pierre Falzon	CNAM, France
Daniel Gopher	Israel Institute of Technology, Israel
Paulina Hernandez	ULAERGO, Chile/Sud America
Sue Hignett	Loughborough University, Design School, UK
Erik Hollnagel	University of Southern Denmark and Chief Consultant at the Centre for Quality Improvement, Denmark
Sergio Iavicoli	INAIL, Italy
Chiu-Siang Joe Lin	Ergonomics Society of Taiwan (EST), Taiwan
Waldemar Karwowski	University of Central Florida, USA
Peter Lachman	CEO ISQUA, UK
Javier Llaneza Álvarez	Asociación Española de Ergonomia (AEE), Spain
Francisco Octavio Lopez Millán	Sociedad de Ergonomistas de México, Mexico

Donald Norman	University of California, USA
José Orlando Gomes	Federal University of Rio de Janeiro, Brazil
Oronzo Parlangeli	University of Siena, Italy
Janusz Pokorski	Jagiellonian University, Cracovia, Poland
Gustavo Adolfo Rosal Lopez	Asociación Española de Ergonomia (AEE), Spain
John Rosecrance	State University of Colorado, USA
Davide Scotti	SAIPEM, Italy
Stefania Spada	EurErg, FCA, Italy
Helmut Strasser	University of Siegen, Germany
Gyula Szabò	Hungarian Ergonomics Society (MET), Hungary
Andrew Thatcher	University of Witwatersrand, South Africa
Andrew Todd	ERGO Africa, Rhodes University, South Africa
Francesca Tosi	Ergonomics Society of Italy (SIE); University of Florence, Italy
Charles Vincent	University of Oxford, UK
Aleksandar Zunjic	Ergonomics Society of Serbia (ESS), Serbia

Contents

Auditory and Vocal Ergonomics

Auditory and Vocal Harmonie

Effects of Self-selected Music and the Arousal Level of Music on User Experience and Performance in Video Games

Arthur Abia and Loïc Caroux[(✉)]

CLLE (University of Toulouse & CNRS),
5 allées Antonio Machado, 31058 Toulouse, France
arthurabia@gmail.com, loic.caroux@univ-tlse2.fr

Abstract. Music is important in video games. It has to be considered more than just an environmental stimulus. However, the role of music in user experience in video games has received little attention in the literature. In particular, there is still a need for investigation of the effects of self-selected music and the arousal level of music on player experience. An experiment was designed to study the effects of these music characteristics on player performance and player experience in different genres of games. Although no significant effect on performance was observed, results showed that the type of music displayed when playing has an impact on players' affective enjoyment and music perception. Moreover, further analyses revealed significant interactions between the genre of game and the type of music presented on player experience. Overall, these findings suggest that music should be better taken in account in video game design to optimize player experience.

Keywords: Game design · Player experience · Enjoyment
Musical preference · Game genre

1 Introduction

Music has become an important part of everyday life. A growing number of studies has been therefore dedicated to it [1]. In the field of cognitive ergonomics, music study is mainly task-performance oriented. According to the literature [2, 3], music can be considered as an additional distraction, interfering with the completion of a task. The concept of distraction, refers here to the process of attention, as defined by Broadbent [4]. In this model, attention is described as a cognitive filter, selecting and processing the most relevant information as well as inhibiting the other messages in goal-oriented tasks. According to this model, attention is a limited process. Processing stimuli during task execution or in the presence of external stimulus drains from the cognitive system of an operator or a user. Exposition to musical stimuli may then be detrimental performance-wise.

However, in video games, soundtracks are a common part of the player experience. They have to be considered more than ambient stimuli and they are not used just as auditory feedback anymore [5]. The role of music in user experience in video games

© Springer Nature Switzerland AG 2019
S. Bagnara et al. (Eds.): IEA 2018, AISC 827, pp. 3–12, 2019.
https://doi.org/10.1007/978-3-319-96059-3_1

has received little attention in the literature. There is still much to discover about the impact of audio in video games. In particular, there is still a need for investigation of the effects of self-selected music and the arousal level of music on player experience.

The present study aimed to investigate the effects of music, and more specifically the effects of musical preference and the arousal level of music on user performance and user experience in a given task: playing video games. Based on the work of Cassidy and MacDonald [6], an experiment was designed to assess the effects of these music characteristics on player performance and player experience in different types of games.

1.1   Influence of Music on Human Behavior

Musical preference has been widely studied in the literature of psychology of art. According to Berlyne [7], art can be expressed in terms of "aesthetics stimuli". The author suggested a theory in which the subjective experience of art, visually or auditory, may provoke a variation in arousal levels, both physiologically and psychologically. This variation of stimulation would be related to the characteristics of the experienced stimuli, such as their novelty, their complexity, or their tempo for example in the case of music. The preference one feels for any form of aesthetics stimulus would then be linked to the perceived degree of arousal.

Listening to music requires cognitive resources, music being a stimulus needing to be analyzed and processed. Konečni [8] studied the relation between musical preference and the listener's performance in a dual-task context. The author showed that the musical preference is influenced by the arousal provoked by music and also by the amount of cognitive resources available during the dual-task process. The more demanding a task is, the more intrusive a music is perceived, and the less it is appreciated.

The notion of musical preference is as well related to task performance. In a study conducted by Nantais and Schellenberg [9], the participants performed significantly better a given task when they listened to a stimulus they like, may it had been Schubert or even a story narrated by Stephen King.

1.2   Influence of Music in Video Games

Further literature investigated the influence of music in human-computer interaction situations, and in particular in player-video game interaction. With the advances in technology, more and more studies have been published on video games and user experience. While most of them examine the visual aspect of the game interface (see [10] for a review), a number of studies analyzed the auditory dimension of interactive games. For example, Nacke et al. [5] have shown that the presence of sounds and/or music during gameplay was correlated to an increase of positive affects regarding subjective game experience and a decrease of negative ones.

Several studies focused on the topic of musical preference. A study conducted by North and Hargreaves [11] aimed to investigate the relation between music listening, musical preferences and performance in a concurrent task in a context of dual-task while playing a driving game. The authors found that the driving performance was lower when the participants had to listen to the high-arousal music while performing the concurrent task, due to a high cognitive workload. They also found that the driving performance was higher when the participants listened to the low-arousal music with no concurrent task. Finally, the authors reported higher ratings of music appreciation when there was no concurrent task to simultaneously perform. In line with Konečni [8], those results showed that musical preference in the context of task performance is related to the nature of the task, and to the cognitive resources available when listening.

More recent evidence of this phenomenon can be found in a study conducted by Cassidy and MacDonald [6]. The authors asked participants to perform a driving task in a video game. They were separated in 5 groups to compare several auditory conditions: no sound, car sounds alone, car sounds with the addition of self-selected music, car sounds with high-arousal music, and car sounds with low-arousal music. The authors found that the participants who had the opportunity to bring their own songs reported a higher level of appropriateness, lower levels of tension-anxiety, had a higher driving performance, and overall had a better user experience. They also found that the highest driving performance was found in the high-arousal music condition.

These studies showed that music is more than an additional information existing in the auditory dimension. It can be used as a tool to change levels of arousal as well as affects, thus having an effect in goal-oriented contexts. Depending on the task, one could allocate their preferences depending on the cognitive resources available. In the present study, those effects of music were investigated on a different type of material, to appreciate the external validity of previous studies.

1.3   The Present Study

The aim of the present study was to extend current knowledge of the effects of musical preference and arousal level of music on user experience and performance in video games. This investigation, in line with Konečni [8], aimed to measure the user experience and performance as well as taking into account the ecological choice of music self-selection according to the task to perform. Our experimental set up was based on the Cassidy and Macdonald's one [6]. The main differences are that, in the present study, enjoyment of players was measured with a specific questionnaire [12], and that two different types of games were used, namely an action game and a puzzle game. Most of studies tended to focus on driving games [11, 13]. With this in mind, our research tried to examine others contexts of video games to determine the external validity of past studies. The performance of players was measured in two distinct video games, as well as their subjective experience, several times under varying musical conditions for each session. There were three types of musical conditions: high-arousal music, low-arousal music and self-selected music. In this last condition, the music track was brought by the players. The user experience was measured using several existing questionnaires.

The following hypotheses were established on the basis of the previous literature described above. The first hypothesis was that the performance is overall better when participants are exposed to the music they select themselves in comparison to the other music conditions, for both types of games. The second hypothesis was that the subjective experience of players is better when exposed to self-selected music as compared to experimenter-selected music, for both types of games. The third hypothesis was self-selected music is more appreciated and perceived as more appropriate than other music conditions, for both types of games. The fourth hypothesis was that the player performance is better when high-arousal music is displayed during the action game than when low-arousal music is displayed, and vice-versa in the puzzle game. In a similar way, the fifth hypothesis was that high-arousal music is perceived as more appropriate in the action game than low-arousal music, and vice-versa in the puzzle game.

2 Methods

2.1 Participants

A total of 20 participants (10 women) from age 20 to 34 ($M = 23.9$, $SD = 4.1$) were recruited for the study. They were university students from various disciplines, recruited on a voluntary basis, with various gaming expertise.

2.2 Material

Music. The music tracks used in for the low-arousal and high-arousal conditions were selected from the study of Cassidy and MacDonald [2] and prepared in advance. This choice allowed us to use an already empirically validated material. The different musical pieces used were "Distractions" from the album *Simple Things* (2002) of the band Zero 7 and "Attitude" from the album *Roots* (1996) of the band Sepultura. "Distractions" is a down-tempo music with a rhythm of 102 BPM and "Attitude" is a groove metal song with a rhythm of 155 BPM. Regarding the self-selected music, the participants were asked to choose a music before coming to the experimentation. They were asked to "choose a song they like, and that they would like to hear during the game session." Participants knew that the experiment was to play video games under several conditions, but they did not know the type of game in advance. Music was diffused with computer speakers.

Games. The selected video games were *World of Goo* (2D Boy, 2008) and *Geometry Wars 3: Dimensions* (Lucid Games, 2014). They were fairly recent and critically well-received. They were chosen for their typical genre-characteristics. *World of Goo* is a puzzle video game, where the goal is to assemble small creatures called "Goos" into structures, in order to lead them through an exit pipe. This game has a static environment, meaning that the situation on screen only moves when the players enter an input. *Geometry Wars 3: Dimensions* is an action video game, in which the player controls a spaceship and must earn the highest score possible by destroying enemies' spaceships. This game has a dynamic environment, meaning that the situation evolves independently from the players' action, and they have to adapt and react to it. For the action game, the instruction was to "destroy as many ships as possible while dying the least possible during 3 min" and for the puzzle game, it was to "finish the level as fast as possible using as few moves as possible". For this experimentation, only the music was manipulated. The original game sounds (coming from environment, enemies, effects, etc.) were not manipulated. They were always presented during the game sessions.

Questionnaire. The questionnaire was composed of items from the Fang et al.'s "instrument to measure enjoyment of computer game play" [12], and from the study of Cassidy and MacDonald [6], which assesses the players' perception of music in a video game and their global play experience. The Fang et al.'s [12] instrument aims to assess the level of enjoyment, by focusing on 3 dimensions affecting the user experience during gameplay: the affective aspect (related to emotions and feelings), the behavioral aspect (related to the behavior during the game session) and the cognitive aspect (related to the judgment of game elements by the player). We present only the affective dimension in the present paper. The questionnaire included 11 items (Table 1), which had to be rated on a 10-point Likert scale.

Table 1. Items used for the questionnaire.

Study of origin	Dimension	Item
Fang et al. [12]	Affective aspect of enjoyment	1. I feel unhappy when playing this game
		2. I feel worried when playing this game
		3. I feel happy when playing this game
		4. I feel exhausted when playing this game
		5. I feel miserable when playing this game
Cassidy and MacDonald [6]	Global experience	6. I felt distracted while playing
		7. I enjoyed the game experience
	Music perception	8. I liked the music during playing
		9. I felt the music was familiar
		10. I felt the music was appropriate for the task
		11. I found the music arousing

2.3 Design and Procedure

The type of music (low-arousal vs. high-arousal vs. self-selected) and the genre of game (action vs. puzzle) were manipulated as within-participants factors. Once recruited, participants did not know which type of video game they were going to play to. Depending on the sessions' order, the respective video game was presented to the participant, as well as the instructions. The order of the game sessions was counterbalanced. Controls, using keyboard and mouse only, were explained before proceeding. After a training session without any music presented, the experiment began. Each participant completed 6 game sessions. The experiment was segmented in two blocks of 3 sessions for each game, each session corresponding to a different music condition. The order of music conditions was also counter-balanced for each participant. After each session, participants had to fill the subjective experience questionnaires. After 3 sessions with the first game presented, participants were then presented the second game of the experiment, with the same procedure. The mean time of a game session was 2 min, without the time of completing the questionnaire.

2.4 Dependent Variables

Performance indicators used for the action game were the score and the number of destroyed enemies for each session. The elapsed time as well as the number of moves done for each session were used for the puzzle game. The subjective experience of the participants was assessed with the answers given for each item of the questionnaire.

3 Results

3.1 Performance

The performance indicators were analyzed separately for each game because they were not similar and thus not readily comparable. A repeated measures ANOVA was carried out to investigate the differences between the music conditions on defined performance indicators. The ANOVA carried out did not present any significant difference for the score nor for the number of deaths in the action game, and neither for the elapsed time nor for the number of moves in the puzzle game.

3.2 Questionnaire

Items of the questionnaires were analyzed with the two independent variables, namely the type of music and the type of game, as well as their interaction. Repeated measures ANOVAs were performed for all items of the questionnaires.

Affective Aspect of Enjoyment. For Item 3 "I feel happy when playing this game", the main effect of the type of music $F(2,38) = 6.10, p < .01, \eta_p^2 = .24$ on the answers of participants, as well as the interaction between the type of music and the type of game, $F(2,38) = 5.08, p < .01, \eta_p^2 = .21$, were significant. As shown in Fig. 1, participants reported more positive affect when exposed to self-selected music compared to other types of music for both types of game. It was also observed that in the action game, participants reported more positive affect with the high-arousal music than the low-arousal music, and vice-versa for the puzzle game.

For Item 2 "I feel worried when playing this game", there was a significant difference of the type of game $F(1,19) = 4.96, p < .05, \eta_p^2 = .21$ on the answers of participants, in the sense that the participants reported higher scores when playing the action game than the puzzle game. However, there were no significant difference between the types of music $F(2,38) = 1.09, p = .35$. The interaction between these two factors did not reach significance $F(2,38) < 1, p = .85$.

For the other items (1, 4 and 5), none of the observed differences were statistically significant neither for the type of game, nor for the type of music, or their interactions.

Music Perception. For Item 8 "I liked the music during playing", there was a significant difference between the types of music, $F(2,38) = 50.60, p < .001, \eta_p^2 = .73$, and between the types of game, $F(1,19) = 4.43, p < .05, \eta_p^2 = .19$. A significant interaction was observed between the type of music and the type of game $F(2,38) = 4.87, p < .05, \eta_p^2 = .20$. As shown in Fig. 1, participants appreciated the most their self-selected

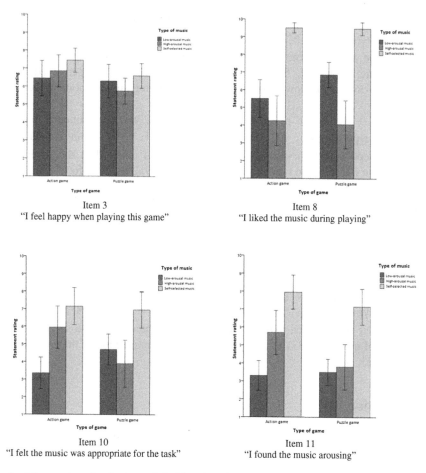

Fig. 1. Mean ratings for the items 3, 8, 10 and 11 for each type of game and type of music. The error bars represent standard errors.

music for both types of game. It was also observed that in the action game, participants appreciated more the low-arousal music in the puzzle game than in the action game.

For Item 10, "I felt the music was appropriate for the task", there was a significant difference between the two types of music $F(2,38) = 15.52, p < .001, \eta_p^2 = .45$ as well as an interaction between the type of music and the type of game $F(2,38) = 9.54, p < .001, \eta_p^2 = .33$. Figure 1 shows that self-selected music was perceived as more appropriate than the other types of music, independently from the type of game. Moreover, high-arousal music was perceived as more appropriate in the action game than in the puzzle game, and vice-versa for the low-arousal music. The analysis did not reveal any significant differences between the types of games $F(1,19) = .87, p = .36$.

The results for Item 11 "I found the music arousing" revealed a significant difference between the type of music $F(2,38) = 39.32, p < .001, \eta_p^2 = .67$ and the type of game $F(1,19) = 8.16$, $p < .01$, $\eta_p^2 = .30$. There was also a significant interaction between the two variables $F(2,38) = 3.57$, $p < .05$, $\eta_p^2 = .16$. As shown in Fig. 1, self-selected music was perceived as more arousing than the other types of music for both types of game. Moreover, there was a significant difference of perceived arousal between high-arousal and low-arousal music in the action game, but none in the puzzle game.

For Item 9 "I felt the music was familiar", there was a significant difference between the type of music $F(2,38) = 70.54$, $p < .001$, $\eta_p^2 = .79$, in the sense that participants reported higher scores for the self-selected music than the other music conditions, but none for the type of game $F(1,19) = 1.49$, $p = .24$ or for the interaction of the two variables $F(2,38) = 1.14$, $p = .33$.

Global Experience. Item 6 "I felt distracted while playing" showed a significant difference between the types of games $F(1,19) = 10.70$, $p < .01$, $\eta_p^2 = .36$, in the sense that players reported higher scores when playing the puzzle game than the action game, but none for the type of music $F(2,38) = 2.10$, $p = .14$ or for the interaction of the two variables $F(2,38) = 2.32$, $p = .112$.

Item 7 "I enjoyed the game experience" also showed a significant difference between the types of games $F(1,19) = 4.84$, $p < .05$, $\eta_p^2 = .20$, in the sense that players reported higher scores when playing the action game than the puzzle game, but none for the type of music $F(2,38) = 2.20$, $p = .13$ or for the interaction of the two variables $F(2,38) = 1.23$, $p = .30$.

4   Discussion

The first hypothesis was not supported. The results did not show any significant difference in the performance indicators defined in both video games. Those findings differ from those found by Cassidy and MacDonald [6]. Such results could be explained by the level of expertise of participants and the genre of games. Many studies only examined the effects of music in driving games [6, 11, 13]. The difference could be linked to the nature of the task. Playing a driving game requires skills acquired in real life, as compared to the skills required for playing a game like a FPS. Because of that, even if some participants are not considered as video game experts, their driving experience could play a role for the task.

The second hypothesis was partially supported. The item 3 "Felt happy" of the questionnaire provided significant differences between self-selected music and the other conditions, in the sense that participants reported more positive affect when exposed to self-selected music, as compared to the other music conditions. Those results are in line with Cassidy and MacDonald's [6] findings. However, the analysis did not confirm any significant differences for the items related to the anxiety, distraction or overall user experience, namely items 2, 6 and 7. Even though those findings are not in line with previous work [6, 11], it can be supposed that only the affective aspect of the enjoyment was affected by the musical preference of participants. It is plausible that the length of our games sessions, being in average two minutes each, which was relatively short, played a role in those results. It can be then assumed that duration of gameplay has an effect on attentional and cognitive factors involved in the user experience.

The third hypothesis was supported. The musical preference effect found by Cassidy and MacDonald [6] was also present in this study, in the sense that self-selected music was the most favorite for participants. This reinforce the external validity of this effect.

The fourth hypothesis was not supported. Contrary to findings of the previous literature about driving games [6], player performance was not improved by the music that would be appropriate for the game genre.

The fifth hypothesis was supported. Interactions between the type of music and the type of game presented were observed regarding music perception (items 8 and 10) and the positive affect reported by the participants (item 3). The low-arousal music condition was linked to better perception and higher levels of positive affect than the high-arousal music condition in the puzzle game, and the opposite for the action game. One could assume that what was observed was merely an aesthetics preference regarding the arousal levels of our material. Nevertheless, self-selected music and high-arousal music were both perceived as more arousing in the action game than in the puzzle game (item 11). This provides further evidence for an interaction between the musical stimuli and the environment presented to the participant. It may mean than the perception of an auditory environment could vary in function of a person's user experience. Such results underline the importance of investigating the effects of music on video games experience, to better understand the underlying mechanisms.

5 Conclusions and Outlook

In this paper, the effects of self-selected music and the arousal level of music on performance and subjective experience in video games were investigated. The findings suggested that participants rate self-selected music as more appropriate, more agreeable to listen to, more arousing, and linked to more positive affect, in both the action and the puzzle games. Our results share a number of similarities with Cassidy and MacDonald's [6] findings, notably about the musical preference effect and the link between self-selected music and positive affect. Although our study did not confirm previous results on the effects of music in video game performance, it highlighted interactions between the types of music presented and the types of game in user experience.

The difference observed with the two types of games underline the importance of keeping in mind that video games do not follow a standard model when it comes to game experience, and each game will solicit different cognitive resources and trigger different behaviors depending on the players. This is part of what makes this field interesting to study, but also what raises many questions about the measurement of game experience [12, 14]. The hypotheses presented to explain the difference in findings from other studies [6] should be taken into account for future studies, in order to determine the interactions between audio and game experience. Not much is known about the auditory dimension in video games [10].

With advances in technology, several video games now offer the possibility to customize the soundtrack, such as the *Grand Theft Auto* series (Rockstar Games) or *Audiosurf 2* (Invisible Handbar, 2015). It is then worthwhile to consider that aspect of user experience during game development. Extending this practice to other types of games, such as sports games or fighting games for example, would allow both developers to manipulate the level of arousal depending of the experience intended, and players to benefit from a more tailored game experience.

References

1. Krumhansl CL (2000) Rhythm and pitch in music cognition. Psychol Bull 126:159–179
2. Cassidy G, MacDonald RAR (2007) The effect of background music and background noise on the task performance of introverts and extraverts. Psychol Music 35:517–537
3. Furnham A, Allass K (1999) The influence of musical distraction of varying complexity on the cognitive performance of extroverts and introverts. Eur J Personal 13:27–38
4. Broadbent D (1958) Perception and communications. Pergamon, New York
5. Nacke LE, Grimshaw MN, Lindley CA (2010) More than a feeling: measurement of sonic user experience and psychophysiology in a first-person shooter game. Interact Comput 22:336–343
6. Cassidy G, Macdonald R (2009) The effects of music choice on task performance: a study of the impact of self-selected and experimenter-selected music on driving game performance and experience. Music Sci 13:357–386
7. Berlyne DE (1971) Aesthetics and psychobiology. Appleton-Century-Crofts, New York
8. Konečni V (1982) Social interaction and musical preference. In: Deutsch D (ed) The psychology of music. Academic Press, New York, pp 497–516
9. Nantais KM, Schellenberg EG (1999) The Mozart effect: an artifact of preference. Psychol Sci 10:370–373
10. Caroux L, Isbister K, Le Bigot L, Vibert N (2015) Player–video game interaction: a systematic review of current concepts. Comput Hum Behav 48:366–381
11. North AC, Hargreaves DJ (1999) Music and driving game performance. Scand J Psychol 40:285–292
12. Fang X, Chan S, Brzezinski J, Nair C (2010) Development of an instrument to measure enjoyment of computer game play. Int J Hum Comput Interact 26:868–886
13. Brodsky W (2001) The effects of music tempo on simulated driving performance and vehicular control. Transp Res Part F Traffic Psychol Behav 4:219–241
14. Jennett C, Cox AL, Cairns P, Dhoparee S, Epps A, Tijs T, Walton A (2008) Measuring and defining the experience of immersion in games. Int J Hum Comput Stud 66:641–661

Development of Fuzzy Data Envelopment Risk Analysis Applied on Auditory Ergonomics for Call Center Agents in the Philippines

Erika Mae Go, Karl Benedict Ong, Jayne Lois San Juan$^{(\boxtimes)}$ ⓘ,
Wendy Gail Sia, Rendell Heindrick Tiu, and Richard Li

Industrial Engineering Department, De La Salle University,
2401 Taft Avenue, 1004 Manila, Philippines
{erika_go, karl_ong, jayne_sanjuan, wendy_sia,
rendell_tiu, richard.li}@dlsu.edu.ph

Abstract. Most contact center employees experience work related injuries, leading to decreased productivity and performance. The increased risk is due to poor workstation design, such as indecent noise level and duration exposure experienced by call center agents daily, while receiving and making calls on a headset or telephone. Some illnesses caused by prolonged exposure to unfavorable acoustics are headaches, increased anxiety levels, tinnitus, and noise-induced hearing loss. Mathematical modelling has only been applied in optimizing systems considering musculoskeletal disorders and only in other industries. Thus, it is important to consider the auditory ergonomic risks faced by call center agents mathematically. This study proposes Fuzzy Data Envelopment Risk Analysis (FDERA), a DEA-based risk analysis tool that considers the presence of imprecise data. The validity of the model is demonstrated using a case example. The results of the proposed tool are relative risk efficiency scores for each agent. Guidelines for interventions to improve risk efficiencies are presented using a matrix that provides possible preventive and corrective measures to address risks.

Keywords: Fuzzy Data Envelopment Risk Analysis · Auditory ergonomics
Call center agents

1 Introduction

One of the fastest growing industries in the Philippines is the business process outsourcing (BPO) industry, employing thousands of Filipinos. In 2014, there were 460,518 Filipinos employed in the BPO industry, majority of which were in the contact center sector. This will continue to grow with the growth of the industry in the Philippines [1]. Around 57% of men and 72% of women in the contact center industry experience work related injuries caused by poor workstation design [2]. It is important to have a well-designed workstation to increase productivity, decrease absences, decrease turnover of employees, among some [3].

The primary role of call center agents is to receive and make calls. Prolonged activities can cause ergonomic risks in the workplace, discomfort and injuries lead to

© Springer Nature Switzerland AG 2019
S. Bagnara et al. (Eds.): IEA 2018, AISC 827, pp. 13–22, 2019.
https://doi.org/10.1007/978-3-319-96059-3_2

reduced performance and productivity [4]. However, most if not all studies do not consider a risk that most call center agents are exposed to, which is the noise level and duration exposure that agents experience on a daily basis. Acoustic design is a key factor on the condition of a work environment and facilities [5]. Noise is considered an occupational hazard, which is best reduced by dealing with the source or through protective gear. Challenges with speech communication, decreased productivity, stress, and annoyance are some of the major concerns for noisy environments; workers experience reduced stress and fatigue under more favorable acoustic environments [6]. Work in call centers are regarded as potentially hazardous activities according to the Ministry of Labor and Social Security in Turkey [7].

Mathematical modeling is only sometimes used to optimize systems and processes based on ergonomic concerns. Typically, the types of ergonomic risks considered are linked only with musculoskeletal disorders [8–10].

Data envelopment analysis (DEA), which is a model that considers multiple inputs and outputs, allows benchmarking and the measure of relative efficiency, was extended by Chang and Chen [11] to discriminate relative workload level among employees treated as a decision-making units (DMUs). The results assigned workload scores to each employee. DEA was also used to analyze the impact of macroergonomic factors in a medicine supply chain [12], and for comparative risk assessment to explore which among several energy supply technologies would have the best benefit-risk ratio or performance [13]. DEA may also be used to identify an ergonomically friendly best practice that provides the least risk but the most productiveness.

However, this is limited because uncertainties and probabilistic conditions are not considered. This is important because the occurrence of negative health cases are uncertain in nature [14]. Azadeh and Alem [15] address this through the fuzzy data envelopment analysis (FDEA) and the chance-constrained data envelopment analysis (CCDEA). With FDEA, imprecise data are represented by fuzzy sets, while data are treated as random variables with known probability functions in the CCDEA method. The results include the average relative efficiency of each DMU, the variance of efficiency scores, and a 95% confidence interval for the average. The use of either tool assumes that data for a certain system is only exclusively uncertain or probabilistic, but real-world situations usually involve a mix of these. Developing a model that is able to handle both crisp and insecure, fuzzy data towards the application of risk assessment with regards to auditory ergonomics would be useful to apply in real world situations, such as in a call center environment.

This study aims to propose a fuzzy data envelopment risk analysis (FDERA) that deals with risk efficiency of call center agents. Risk efficiency is "the minimum risk decision choice for a given level of expected performance" [16]. The results identify the most risk efficient employees and suggest interventions for the inefficient agents.

2 Proposed FDERA Model

The use of the traditional DEA model requires crisp input and output data to evaluate the performance of a DMU. However, these types of data, especially those related to health risks and concerns, are rarely exact or available. Furthermore, measuring the severity of illnesses, such as discomfort and pain, are complex and costly; thus, they rely greatly on the perception of the affected party and are fuzzy in nature. The data values are taken as positive triangular fuzzy numbers (TFN). Because this study focuses on risk efficiency, the outputs are mostly undesirable to represent negative health outcomes. Undesirable outputs are treated similarly with as inputs in the model [17]; DMUs are considered more efficient if it produces less undesirable outcomes [18].

This section develops the FDERA, which uses imprecise data represented by fuzzy sets. This model is developed by adapting and modifying the FDEA model proposed by Azadeh and Alam [23]. Assume that there are n DMUs (DMU_i; $i = 1,...,n$), each using the same m inputs ($j = 1,...,m$) and yielding s undesirable outputs ($k = 1,...,s$) and v desirable outputs ($q = 1,...,v$). The relative fuzzy risk efficiency score of the pth DMU (E_p) may be obtained using the model shown below. The initial model is modified to allow the fuzzy inputs and outputs to be considered as TFN. Azadeh and Alem [15] apply an alternative α-cut approach to transform the fuzzy model into a crisp linear programming problem. This methodology is likewise applied on the model, resulting model is an interval problem. The approach of Puri and Yadav [19] is used to transform the interval programming model into a crisp linear model. This approach is preferred because it provides a single number for the relative efficiency scores, instead of a range. This results in a nonlinear programming problem is transformed into a crisp linear programming model shown below.

$$Max \; E_\alpha^p = \sum\nolimits_{q=1}^{v} \bar{y}_{pq}^g - \sum\nolimits_{k=1}^{s} \bar{y}_{pk}^b \tag{1}$$

$$s.t. \quad \sum\nolimits_{j=1}^{m} \bar{x}_{pj} = 1 \tag{2}$$

$$\sum\nolimits_{q=1}^{v} \bar{y}_{iq}^g - \sum\nolimits_{k=1}^{s} \bar{y}_{ik}^b - \sum\nolimits_{j=1}^{m} \bar{x}_{ij} \leq 0 \quad \forall i \tag{3}$$

$$\sum\nolimits_{q=1}^{v} \bar{y}_{iq}^g - \sum\nolimits_{k=1}^{s} \bar{y}_{ik}^b \geq 0 \quad \forall i \tag{4}$$

$$v_{pj}\left(\alpha x_{ij}^m + (1-\alpha)x_{ij}^l\right) \leq \bar{x}_{ij} \leq v_{pj}(\alpha x_{ij}^m + (1-\alpha)x_{ij}^u) \quad \forall i,j \tag{5}$$

$$u_{pq}^g\left(\alpha y_{iq}^{gm} + (1-\alpha)y_{iq}^{gl}\right) \leq \hat{y}_{iq}^g \leq u_{pq}^g\left(\alpha y_{iq}^{gm} + (1-\alpha)y_{iq}^{gu}\right) \quad \forall i,q \tag{6}$$

$$u_{pk}^b(\alpha y_{ik}^{bm} + (1-\alpha)y_{ik}^{bl}) \leq \hat{y}_{ik}^b \leq u_{pk}^b(\alpha y_{ik}^{bm} + (1-\alpha)y_{ik}^{bu}) \quad \forall i,k \tag{7}$$

$$u_{pq}^g \geq \varepsilon \;\; \forall q, \quad u_{pk}^b \geq \varepsilon \;\; \forall k, \quad v_{pj} \geq \varepsilon \;\; \forall j, \quad \varepsilon > 0, \tag{8}$$

y_{ik}^b is the value of undesirable output k of DMU i, y_{iq}^g is the value of desirable output q of DMU i, x_{ij} is the amount of input j utilized by DMU i, u_{ik}^b is the weight assigned to undesirable output k of DMU i, u_{iq}^g is the weight assigned to desirable output q of DMU i, and v_{ij} is the weight given to input j used by DMU i. The model should be run n-times to calculate the fuzzy risk efficiency of n DMUs. A linear programming model is shown above, where α represents a parameter with values found in the continuous set $(0, 1]$.

For each value of α, an optimal solution, specifically relative risk efficiency scores for each DMU, may be obtained. Similar to the traditional DEA model, a relative risk efficiency score of 1 indicates that a DMU experiences the least risk for the most input among the given set of DMUs. The model may also be applied to deal with data that is entirely deterministic, or a combination of crisp and fuzzy data. For crisp data, the TFN that represent a fuzzy set of a parameter should be assigned equal values.

3   Case Application to Philippine Contact Centers

The risk efficiencies of call center agents working in the inbound department of contact center in Philippines were analyzed. Ten agents were chosen to complete the DMUs for this study to comply with the 2*(input+output factors) DEA requirement. The DMUs chosen were within the same contact center, but far enough from each other to ensure independence, because DMUs that are close to each other may impact the other's noise environment. The DMUs are similar to ensure comparability.

Typically, inputs considered are critical resources that lead to the negative or positive effects of exposure to risk, such as the exposure level and duration. Meanwhile, outputs considered are those significantly impacted by the inputs chosen, these include possible injuries and illnesses, as well as effects on desirable outputs like performance or productivity. In a noise exposure assessment involving an individual, the inputs include average noise level and duration of noise exposure each shift [20]. Noise level is measured in sound pressure level with units of dBA. A higher sound pressure level means that the noise is louder. Another variable needed is the duration of noise exposure each shift, which is measured in time (minutes or hours). The occupational noise exposure administration proposed standards for permissible noise exposure that relates average noise level to the duration of the individual's exposure (Fig. 1).

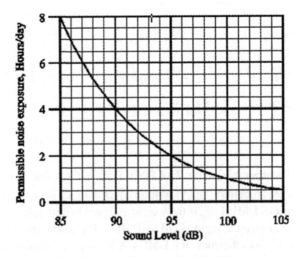

Fig. 1. OSHA Standards for Permissible Noise Exposure

The outputs considered in the assessment are productivity of agent, anxiety level and tinnitus (or ringing in the ear) [21]. Two of the inputs and three of the outputs are presented as fuzzy values that will be placed in the model formulated to compute for the efficiency levels of each DMU. The inputs of average noise level and exposure time are directly taken from the Table 1. On the other hand, for the output, anxiety level can be computed by multiplying headache pain and the frequency of pain. While productivity is directly taken from the table. Tinnitus was not defined as a fuzzy variable because are obtained from a tinnitus hearing-test given by DisMark Tinnitus [22]. Table 2 shows the corresponding values for the tinnitus hearing-test.

Table 1. Input and Output variables with associated fuzzy values

System's variables	Linguistic	Linguistic values	Fuzzy intervals
Inputs	Average Noise Level	Low	65–75
		Medium	72–85
		High	80–90
		Very High	85–95
		Extra High	90–100
	Duration of Noise Exposure each Shift	Short	1–3
		Average	2–6
		Long	5–8
Outputs	Headache Pain	Minimal Pain	1–2
		Mild Pain	1–4
		Moderate Pain	3–7
		Severe Pain	6–10
	Frequency of Pain	Monthly	0.10–0.45
		Thrice a Month	0.40–0.60
		Weekly	0.55–0.75
		Thrice a Week	0.70–0.90
		Daily	0.86–1.00
	Productivity	Very Not Productive	0.10–0.75
		Not Productive	0.72–0.85
		Productive	0.80–0.90
		Very Productive	0.85–0.95
		Extra Productive	0.90–1.00

The intervention matrix that can be seen below was created in order to be able to address the different inputs and outputs given the weight and gap of that input or output. In evaluating the different inputs such as average noise level duration of noise exposure, anxiety and tinnitus, there are certain interventions that could be applied to decrease these inputs and undesirable outputs, and increase the desirable outputs. Interventions are detailed in Table 4, and are supported by existing studies or interview results (Table 3).

Table 2. Tinnitus hearing-test values

Frequency ('00 Hz)	1	3	6	9	16	28	40	58	70	80	90	110
Model input		0.01	0.03	0.05	0.08	0.15	0.25	0.36	0.53	0.64	0.73	0.82	1

Table 3. Intervention matrix for inputs

Input	Weight	Gap		
		High	Medium	Low
Average noise level	High	A	B	B
	Medium	A	A	B
	Low	A	A	A
Duration of Noise Exposure	High	D	C	C
	Medium	D	D	C
	Low	D	D	D

Table 4. Intervention matrix for outputs

Output	Weight	Gap		
		High	Medium	Low
Anxiety	High	D, E	D, E	D, E
	Medium	D, F	D, F	D, E
	Low	D, F	D, F	D, E
Tinnitus	High	F	C	C
	Medium	F	F	C
	Low	F	F	F

   The intervention matrix that can be seen below can be used by identifying the gap and the weight for each of the inputs and outputs presented. In the intervention matrix, it can be seen that the most severe combination is the high gap and the low weight. The gap, which is the difference between the efficiencies of the most and least efficient DMU, can be categorized as high, medium or low depending on the results of the model. Thus, the gap represents the desired increase in the efficiency score of an inefficient DMU. Meanwhile, the weight of an input or output factor represents the amount of improvement in the efficiency score applying an intervention in that factor can bring. The gaps and the weights can be categorized as low if the results range from 0–0.3333, as medium if it ranges from 0.3334–0.6666 and as high if it ranges from 0.6667–0.9999 (Table 5).

   Ten respondents were chosen based on the previously discussed. A survey questionnaire is developed and deployed to the chosen DMUs aimed to determine the noise environment of the agents, specifically tackling their exposure intensity and duration, the type of equipment of used, their performance level, experiences of anxiety and tinnitus, their performance level, and intervention suggestions. After the data was transformed into TFN, it was run through the program LINGO.

Table 5. Intervention code

Intervention code	Definition
A	Active Noise Canceling Headphones (Zhang & Qiu 2014)
B	Prompt operators to lower noise levels (Patel & Broughton, 2002)
C	Short break from work (Patel & Broughton, 2002)
D	Temporary leave of absence
E	Medication (e.g. paracetamol, painkillers)
F	Seek Medical Attention

3.1 Discussion of Results and Proposed Interventions

The results of the study are summarized in Table 6 below. The table shows the individual results or the relative risk efficiency scores of the 10 respondents when the value of the alpha is varied. DMUs attaining a relative risk efficiency score of 1 indicates that the DMU is found to be the most optimal solution, yielding the most efficient results or holding the least amount of risks for the most amount of input of a given set of DMUs.

Table 6. Resulting values of DMUs

DMU	$\alpha = 0.0$	$\alpha = 0.25$	$\alpha = 0.5$	$\alpha = 0.75$	$\alpha = 1$
1	0.8666667	0.8983051	0.9310345	0.9649123	1
2	0.9	0.9240506	0.9487179	0.974026	1
3	0.8666667	0.8983051	0.9310345	0.9649123	1
4	0.8888889	0.915493	0.9428571	0.9710145	1
5	0.8470588	0.8830585	0.9204893	0.9594384	1
6	0.8666667	0.8983051	0.9310345	0.9649123	1
7	0.9	0.9240506	0.9487179	0.974026	1
8	0.8947368	0.92	0.9459459	0.9726027	1
9	0.8666667	0.8983051	0.9310345	0.9649123	1
10	0.8888889	0.915493	0.9428571	0.9710145	1

It could be observed that DMUs 2 and 7 have the highest relative risk efficiency score regardless of the value of α, revealing that the working conditions of agent 2 and 7 are the conditions where agents experience the least amount of health risks and hazards. This is intuitive because despite the higher levels of noise exposure experienced, their anxiety level is lower than the other DMUs. On the other hand, DMU 5 has the worst risk efficiency score, which means that it experiences the most risk. Based on the data in Appendix, it had relatively high exposure to noise in terms of severity and duration, it also experienced higher anxiety level and a lower performance level. Table 7 below shows the rank of the DMUs with 1 being the least risky and 10 being the riskiest.

Interventions are proposed to improve the performance of DMU 5. This demonstrates how the proposed intervention matrix can be applied.

Table 7. Rank of DMUs

DMU	1	2	3	4	5	6	7	8	9	10
Rank	6	1	6	4	10	6	1	3	6	4

The objective of proposing interventions is to improve the risk of DMUs 5 to match that DMUs 2 and 7. There are two approaches that may be done, reducing the uses of the noise level and duration exposure, as well as decrease the occurrence of anxiety and tinnitus. The values used below are taken from when the model is ran with alpha equal to 0.0. Table 8 also shows the appropriate intervention to be applied to the corresponding input and undesirable output factors.

Table 8. Application of proposed intervention matrix for DMU 5

			Equivalent rating	
	Efficiency Gap	0.0529	Low	
		Weight		Intervention
Inputs	Average Noise Level	0.00332	Low	A
	Duration of Noise Exposure	0.01117	Low	D
Outputs	Anxiety	0.05195	Low	D, E
	Tinnitus	0.58205	Medium	C

Table 8 shows that the efficiency gap between the least efficient DMU and DMUs 2 and 7 is 0.0529, this is considered a low efficiency gap. The weights obtained for the input factors average noise level and noise exposure duration are low. In the same way, the weight of anxiety is low, while the weight obtained for tinnitus is medium. Thus, the appropriate interventions to improve the risk efficiency of DMU 5 are to use active noise cancelling headphones, to take a short break or a temporary leave from work, and to take necessary medication. The proposed interventions rest on the assumption of minimum additional effort or costs to implement. To be able to more effectively address the sources of risk experienced by a DMU, combining interventions or implementing an intervention designed for worse cases can be done. For example, to improve average noise exposure level the DMU or call center can implement both active noise cancelling headphones together with an automatic prompt to the operators to lower their volume when an unsafe level is reached. This will ensure the minimization of negative noise environment exposure. In addition, a call center agent can also seek the help of a licensed doctor to treat either their anxiety or tinnitus, instead of just taking medication, which may not be as effective.

4   Conclusions and Recommendations for Future Studies

Call center agents experience work related injuries and illnesses due to poor workstation design and conditions. Auditory ergonomics is an area of call center ergonomics study is only rarely looked into. In the same way, mathematical modelling is rarely applied in the field of ergonomics.

The case analysis demonstrated the validity of the proposed fuzzy data envelopment risk analysis (FDERA) tool, which determines the relative risk efficiency of decision making units based on its inputs and outputs. It is able to consider the uncertainties that exist in real life data, especially those considered in health issues, through TFN. The model is applied to 10 contact center agents in the Philippines. The riskiest among the set of contact center agents was determined. Through the use of the proposed intervention matrix, possible points of intervention are suggested for the agents to undertake to improve their risk efficiency. However, these proposed interventions are not exact, and may be combined with interventions for other cases.

In the interest of expanding the research, other factors relevant to the health and performance of an agent could be explored. For example, both musculoskeletal and auditory risks of the contact center agents may be explored. Other than ergonomic practices, human factors conditions, such as communication and teamwork, in a work area may also be considered. These considerations are able to impact the performance, health, and safety of a call center agent in the long term. Thus, it may be beneficial for any contact center to consider the use of a risk benchmarking tool to determine the best combination practice that ensures the highest desirable outputs, with the least undesirable outputs and inputs.

Appendix

Appendix A. Translated Triangular Fuzzy Numbers

Low/Mid/High

DMU	Noise Level	Noise Duration	Anxiety Level	Tinnitus	Performance
1	65/70/75	2/4/6	0.55/1.625/3	0.03	0.85/0.9/0.95
2	90/95/100	1/2/3	0.1/0.42/0.9	0.53	0.72/0.79/0.85
3	65/70/75	1/2/3	1.2/2.5/4.2	0.01	0.9/0.95/1
4	80/85/90	2/4/6	0.86/2.325/4	0.25	0.72/0.79/0.85
5	72/78.5/85	5/6.5/8	1.65/3.25/5.25	0.05	0.8/0.85/0.9
6	65/70/75	2/4/6	0.7/1.2/1.8	0.01	0.86/0.9/0.95
7	90/95/100	2/4/6	4.2/6.4/9	0.64	0.1/0.43/0.75
8	85/90/95	1/2/3	1.65/3.25/5.25	0.08	0.8/0.85/0.9
9	65/70/75	5/6.5/8	0.1/0.42/0.9	0.03	0.9/0.95/1
10	80/85/90	5/6.5/8	5.167.44/10	0.15	0.72/0.79/0.85

References

1. Errighi, Khatiwada, Bodwell (2016) Business process outsourcing in the Philippines: Challenges for decent work. ILO Asia- Pacific Working Paper Series
2. Norman: Call centre work – characteristics, physical, and psychosocial exposure, and health related outcomes (2005)

3. OSHA (2000) Ergonomics: The Study of Work. https://www.osha.gov/Publications/osha3125.pdf
4. Lan L, Wargocki P, Lian Z (2011) Quantitative measurement of productivity loss due to thermal discomfort. Energy Build 43(5):1057–1062
5. Kurakata K, Mizunami T, Matsushita K (2007) Accessible design in auditory ergonomics, p 1–7
6. Tiesler G, Oberdoerster M (2008) Noise - a stress factor? acoustic ergonomics at schools. J Acoustical Soc Am 123(5):3918
7. Beyan A, Demiral Y, Cimrin A, Ergor A (2016) Call centers and noise-induced hearing loss. Noise Health 18(81):113
8. Choi G (2009) A goal programming mixed-model line balancing for processing time and physical workload. Comput Ind Eng 57(1):395–400
9. Xu Z (2010) Design of Assembly Lines with the Concurrent Consideration of Productivity and Upper Extremity Musculoskeletal Disorders using Linear Models (Master's thesis)
10. Mummolo G, Mossa G, Boenzi F, Digiesi S (2015) Productivity and ergonomic risk in human based production systems: a job-rotation scheduling model. Int J Prod Econ 171:471–477
11. Chang S, Chen T (2006) Discriminating relative workload level by data envelopment analysis. Int J Ind Ergon 36(9):773–778
12. Azadeh A, Motevali Haghighi S, Gaeini Z, Shabanpour N (2016) Optimization of healthcare supply chain in context of macro-ergonomics factors by a unique mathematical programming approach. Appl. Ergon 55:46–55
13. Ramanathan R (2001) Comparative risk assessment of energy supply technologies: a data envelopment analysis approach. Energy 26(2):197–203
14. Rougier J, Rougier J, Hill LJ, Sparks RS (2013) Risk and uncertainty assessment for natural hazards. Cambridge University Press, New York
15. Azadeh A, Alem S (2010) A flexible deterministic, stochastic and fuzzy Data Envelopment Analysis approach for supply chain risk and vendor selection problem: Simulation analysis. Expert Syst Appl 37(12):7438–7448
16. Project Management Institute Inc.: A guide to the project management body of knowledge (PMBOK® guide) (2000)
17. Chapman C, Ward S (2004) Why risk efficiency is a key aspect of best practice projects. Int J Project Manag 22(8):619–632
18. Yang H, Pollitt M (2009) Incorporating both undesirable outputs and uncontrollable variables into DEA: the performance of Chinese coal-fired power plants. Eur J Oper Res 197 (3):1095–1105
19. Puri J, Yadav SP (2014) A fuzzy DEA model with undesirable fuzzy outputs and its application to the banking sector in India. Expert Syst Appl 41(14):6419–6432
20. Golmohammadi R, Eshaghi M, Khoram MR (2011) Fuzzy Logic Method for Assessment of Noise Exposure Risk in an Industrial Workplace 3(2):49–55
21. Pawlaczyk-Łuszczyńska M, Dudarewicz A, Zamojska-Daniszewska M, Rutkowska-Kaczmarek P, Zaborowski K (2017) Noise exposure and hearing threshold levels in call center operators. ICBEN
22. DisMark Tinnitus (2017) Tinnitus hearing-test - Sound therapy helps against tinnitus, ringing in the ears. http://www.tinnitool.com/en/tinnitus_analyse/hoertest.php. Accessed 2017/12/11

Ergonomics and Acoustics in Music Education

Orlando Maria Patrizia[1], Lo Castro Fabio[2], Iarossi Sergio[2],
Mariconte Raffaele[3], Longo Lucia[1], and Giliberti Claudia[3(✉)]

[1] Dipartimento Organi di Senso, Università Sapienza Rome, Rome, Italy
Mariapatrizia.orlando@uniromal.it
[2] CNR I.N.S.E.A.N., sezione Acustica e Sensoristica Orso Mario Corbino,
Rome, Italy
{fabio.locastro,sergio.iarossi}@cnr.it
[3] INAIL Dipartimento Innovazioni Tecnologiche e Sicurezza degli Impianti,
Prodotti ed Insediamenti Antropici, Rome, Italy
{r.mariconte,c.giliberti}@inail.it

Abstract. In music education, teachers and students often complain about acoustical discomfort. The facilities dedicated to study and musical training play an important role in risk assessment and have to respect standards on architectural acoustics. In particular, bad sound insulation of the classrooms, sound level emitted by musical and/or vocal instruments, high level of background noise and an unsuitable reverberation time, can compromise learning and performances and could cause pathologies (especially to auditory system and vocal apparatus) due to excessive noise levels and acoustical discomfort. The work aims to estimate the potential risk for music teachers and students in an Italian conservatory, identifying critical issues by collecting objective and subjective information. Measurement campaigns, medical evaluations, questionnaires have been performed to identify critical issues related to ergonomic discomfort. 129 workers were examined. The measurement campaign has detected that insulation values and reverberation times were often not adequate for the purpose of the classrooms. Moreover, the social survey has pointed out a high discomfort both among teachers and students The analysis of the singers' voices in relation to the posture has allowed to identify possible solutions to remedy the pathologies related to the relationship between vocal effort and posture.

Keywords: Noise · Acoustics · Voice

1 Introduction

A working environment not ergonomically designed could affect productivity, safety, comfort, concentration, job satisfaction and morale of workers, producing adverse health consequences (injury, illness, stress, etc.), reducing job performance and satisfaction (discomfort) [1, 2]. Significant ergonomic factors include building design and age, workplace layout, workstation set-up, furniture and equipment design and quality, space, temperature, ventilation, lighting, noise, vibration, radiation, air quality.

In music education, musicians teachers and students are at particular noise risk because they rehearse and perform daily, often for many consecutive years, in loud

© Springer Nature Switzerland AG 2019
S. Bagnara et al. (Eds.): IEA 2018, AISC 827, pp. 23–32, 2019.
https://doi.org/10.1007/978-3-319-96059-3_3

environments, experiencing sound pressure levels ranging from 86 dBA to 99 dBA and peak noise levels of 105 dBA [3, 4]. These conditions can have a severe impact on a worker's physical health and psychological well-being, causing pathologies to auditory system and vocal apparatus, acoustical discomfort (irritation, stress), and even work injuries [5–7]. The representative noise levels and noise peaks from single musicians and singers are reported in Table 1 [8].

Table 1. Representative noise levels in the music sector

Noise source	dB	Peak
Single musicians		
Violin/viola (near left ear)	85–105	116
Violin/viola	80–90*	104
Cello	80–104*	112
Acoustic bass	70–94*	98
Clarinet	68–82*	112
Oboe	74–102*	116
Saxophone	75–110*	113
Flute	92–105*	109
Flute	98–114	118
Piccolo	96–112*	120
Piccolo (near right ear)	102–118*	126
French horn	92–104*	107
Trombone	90–106*	109
Trumpet	88–108*	113
Harp	90	111
Timpani and bass drum	74–94*	106
Percussion (high-hat near left ear)	68–94	125
Percussion	90–105	123–134
Singer	70–85	94
Soprano	105–110	118
Choir	86	No data
Normal piano practice	60–90*	105
Loud piano	70–105*	110
Keyboards (electric)	60–110*	118
Several musicians		
Chamber music (classical)	70–92*	99
Symphonic music	86–102*	120–137

*Measured at 3 m from source

They give an indication of the variety of levels that musicians can receive from specific instruments. It could cause greater declines in hearing than the general population, showing a hazard ratio of 1.45 for hearing loss, 3.61 for noise-induced hearing loss (NIHL), and 1.57 for tinnitus [9].

In student musicians, a prevalence of NIHL in at least one ear was found to be within the range of 33% – 50%, bilateral and NIHL was served in 11.5% of all tested subjects [3].

The risk of hearing damage depends on sound level, exposure time and specific environment. As a result, listening to loud sounds over many hours per day entails a similar risk as listening to an even louder sound for a shorter period per day. For long-term exposure, time periods of 8 h per day or weekly sound exposure (8 h per day for 5 days per week), are typically considered in order to set protections standards.

New stringent action levels were introduced in the EU with the Directive 2003/10/EC on the minimum health and safety requirements regarding the exposure of workers to the risks arising from physical agents (noise). It recommends three protection levels at the workplace, depending on equivalent noise level for an 8-h working day or C-weighted peak sound pressure level: (1) less than 80 dBA or 135 dBC_{peak}, employers shall make available hearing protectors and below this limit, the risk to hearing is assumed to be negligible; (2) less than 85 dBA or 137 dBC_{peak}: protection of workers is mandatory; (3) 87 dBA or 140 dBC_{peak} are the maximum exposure limit values, that must never be exceeded.

In Italy, the National Regulation (Testo unico sulla sicurezza nei luoghi di lavoro, dlgs 81/08) has implemented EU Directive and in the article n. 198 forces entertainment, music and call centers environments to carry out noise exposure assessment.

Moreover, in music education, acoustic design of the facilities, dedicated to study and musical training, play an important role in risk assessment; in fact, bad sound insulation of the classrooms, sound level emitted by musical and/or vocal instruments, high level of background noise and an unsuitable reverberation time, can compromise learning and performances and could cause pathologies to auditory system and vocal apparatus and acoustical discomfort, for both students and teachers.

The design standards for environments in music study and education generally specify the minimum acceptable requirements for acoustic parameters; unfortunately, these standards may differ at national (regional) and international level.

The Italian regulatory framework for minimum requirements in the building of classrooms for music education includes a regulation dates back to 1997 [10], implementing by Law 447/95 [11] and a more recent regional one [12].

In particular, national regulation sets limits for the following parameters: R'_w that is the apparent sound reduction index of a partition that measures the ability to limit the passage of airborne noise; $D_{2m, n\ Tw}$ that is the standardized level difference for façade sound insulation, which characterizes the partition capacity to reduce airborne noise from outside to inside; L_{eq} that measure background noise level and T_r that is the reverberation time and represents the persistence of sound in a room after the sound stimulus is ended (Table 2).

All the parameters listed in Table 2 are less restrictive than those reported in technical standards or in international regulations [13–19]. For example, different international standards recommend a reverberation time much less than 1 s (around 0,6 s), a background noise less than 35 dBA and higher values for sound reduction index.

Finally, it must also be considered that, playing and singing are static and strenuous muscular exercises, physically tiresome and very demanding on the body. Musicians

Table 2. Synoptic framework of the main mandatory national standards

Parameter	Value	Law source
Apparent sound reduction index R'$_w$ between adjacent classrooms	≥ 40 dB	D.M. 18/12/1975
Apparent sound reduction index R'$_w$ between overlaid classrooms	≥ 42 dB	D.M. 18/12/1975
Standardized impact sound pressure level L'$_{nw}$ between overlaid classrooms	≤ 68 dB	D.M. 18/12/1975
Reverberation time T$_r$ in classroom	≤ 1,2 s	D.M. 18/12/1975
Apparent sound reduction index R'$_w$ between adjacent and overlaid classrooms	≥ 50 dB	D.P.C.M. 5/12/1997
Standardized impact sound pressure level L'$_{nw}$ between overlaid classrooms	≤ 58 dB	D.P.C.M. 5/12/1997
standardized level difference for façade sound insulation D$_{2m, n Tw}$	≥ 48 dB	D.P.C.M. 5/12/1997
Apparent sound reduction index R'$_w$ for partition walls between noisy music classrooms	≥ 62 dB	Decree of the President of the Province of Bolzano 7 July 2008, n. 26
Apparent sound reduction index R'$_w$ for partition walls between music classrooms	≥ 53 dB	Decree of the President of the Province of Bolzano 7 July 2008, n. 26
Apparent sound reduction index R'$_w$ for corridors, offices and non-musical classrooms	≥ 47 dB	Decree of the President of the Province of Bolzano 7 July 2008, n. 26

and vocalists should have good awareness of their body, joints and muscles, and they should find a natural basic posture that causes the least strain. Voice and auditory systems are needfool tools for professional musicians, so the best way to take good care of them is be aware of the potential risks and prevent them.

The present work aims to evaluate the music education workplace environment, estimating the potential risk for music teachers and students in an Italian music school, identifying critical issues by using quantitative and qualitative instrument (measurements and surveys), implementing appropriate measures to prevent or minimize the risk.

2 Materials and Methods

Measurement campaigns (background noise analyzed according to psychoacoustic parameters and sound quality metrics and noise exposure levels), medical evaluations (voice analysis in relation to posture, audiological examination and audiometric evaluation), surveys (lesson duration, average number of students, non-working factors, assessment of noise exposure, etc.) have been performed to identify critical issues related to ergonomic discomfort in a national conservatory.

About noise exposure, 129 subjects between teachers and students, were examined and the evaluation was performed, according the standard ISO 9612 [20].

Three measurement strategies for the determination of workplace noise exposure are offered by ISO 9612: (1) *task-based measurement*: the work performed during the day is analysed and split up into a number of representative tasks, and for each task separate measurements of sound pressure level are taken; (2) *job-based measurement*: a number of random samples of sound pressure level are taken during the performance of particular jobs; (3) *full-day measurement*: sound pressure level is measured continuously over complete working days.

After work analysis, *task based measurement* have been performed, grouping teachers and student in the following homogeneous noise exposure groups: for instruments players (Harp, Bass tuba, Voice, baroque voice, baroque vocal repertoire practice, Guitar, Clarinet, Harpsichord and historical keyboards, Double bass, Horn, Bassoon, Accordion, Flute, transverse flute, Lute, Electronic music, Oboe, Organ, Piano, Saxophone, Percussion, Trumpet, Trombone - low trombone, Viola, Viola da gamba, Violin, baroque violin, Cello); for ensembles players (Piano Accompaniment, Wind instruments ensemble, Bowed string instrument ensemble, Chamber music, Band Instruments, Orchestral practice, Choral Music); no instruments (Offices, Theory, Doorman/usher, Lute manufacturing).

After measurements of sound pressure level, A-weighted noise exposure level normalized to an 8 h working day ($L_{EX,8h}$) was calculated for each group.

In order to evaluate the acoustic comfort, a specific survey has been produced to acquire a subjective judgment on the perceived sound, classifying it as unacceptable, disturbing or acceptable in relation to the activity that is taking place. Moreover, the physical and psychophysical objective data on the interfering noise were obtained from the audio samples. It was used the metrics of "sound quality", widely used in industrial product design and automotive field, in terms of loudness, sharpness, roughness and fluctuation strength [21].

The perception of acoustic comfort was assessed for both teachers and learners. 10 teachers and 25 students were interviewed. The questionnaire consisted of a series of personal initial questions about sex, age, smoking habits, drugs, then musical discipline and instrument used, number of hours of lesson and exercise and average number of instrumentalists in a classroom and continued with 12 questions about personal quantitative perception of the noise level in the didactic/teaching environment. The interviewees expressed the possible presence of acoustic discomfort with a cross inside 5 boxes, from no discomfort to much discomfort. At the end, questions on the habits of using headphones and earphones to make phone calls, presence or absence of hearing problems and hearing aids, nature of noise nuisance in the workplace and study environment.

About vocal disorders of singers, it was used the "Voice Handicap Index" (VHI), a questionnaire to quantify the functional, physical and emotional impacts of a voice disorder on a patient's quality of life. It captures the patient's subjective rating a series of questions in order to quantify the disability and handicap levels. The original version of the VHI was introduced by Jacobson et al. [22] and it consists of 30 questions and a self-assessment of the severity of the voice problem as perceived by the subject.

The voice handicap index has been used in numerous studies as an indicator for finding evidence of voice disorders. Cross-cultural adaptations, translations and validations of VHI have been made for many languages; specific singers VHI were introduced to cater for the special needs of particular subgroups of singers: SVHI (singer voice handicap index), MSHI (modern singer handicap index), CSHI (classic singer handicap index).

For objective voice evaluation, a multiparametric software was used. It allows the analysis of specific parameters (*Jitt - Jitter Percent*; *Shim - Shimmer Percent*; *NHR – Noise – to - Harmonic Ratio* and *VTI - Voice Turbulence Index*) as well as broadband spectrography [23].

In addition, all participating singers underwent medical examination with Fibre Optic Laryngoscopy. For each singer, the sustained sung emission of the vowel \a\ was sampled in standing position, and standing or sitting while playing the instrument. Postural analysis, performed simultaneously, made it possible to visually evaluate the subjects in order to establish their real position in space with respect to an ideal position.

The study was carried out by collecting 20 samples for 17 subjects, including 7 men and 10 women. The difference between number of samples and number of subjects is because 3 of the singers analyzed are able to play both the instruments (piano and acoustic guitar). This allowed both to increase the number of samples and to compare the different variations of the parameters on the same subject in the following conditions: without instrument, with the piano and with the guitar.

Medical evaluations have been performed on 65 musicians including an otorine examination and an evaluation with subjective and objective audiometric tests. In subjective audiometry the subject's collaboration is always required and includes: the limestone tonal audiometric test which is based on the search of the auditory threshold for the pure tones emitted by the audiometer, sent by air and by bone; sopraliminar tonal audiometric test, that allows to reveal particular auditory phenomena, usually pathological, to stimulation intensity values that are found at supraliminal levels or even at intensity values close to the threshold of discomfort and pain; acufenometry which is a non-invasive diagnostic test consisting of the measurement of the frequency and intensity of a tinnitus; evaluation of frequency discrimination to determine if a patient is able to recognize two sequentially delivered stimuli as separate from each other. Objective audiometry does not require the participation of the patient because the stimulus responses are independent of his will and consists of: the impedenzometric examination which is an objective audiological diagnostic test that allows to evaluate the anatomical-functional state of the ear-tympanic system (tympanometry) and of the reflex cocleo-stapediale arc (reflexometry) measuring the resistance that meets the propagation of sound in passing through the middle ear (acoustic impedance); otoe-missions acoustic detection of acoustic signals generated by external hair cells, as an expression of the normal active mechanisms inside the cochlea; distortion products, evoked by two tones placed at the entrance of the cochlea, which at the output can produce components not present at the entrance or more simply modifications of the signals sent; ABR (audiometry brain response) able to detect the electrical potentials originating from the brainstem in response to an acoustic stimulation calibrated in intensity.

3 Results

On the 129 subjects analyzed, 81% are in low acoustic risk level ($L_{EX,8h} \leq 80$ dBA), 15% in medium risk level ($80 < L_{EX,8h} \leq 85$ dBA), and 4% in high risk level ($85 < L_{EX,8h} \leq 87$ dBA) (Table 3).

Table 3. Risk level for homogeneous noise exposure groups

Risk level	Homogeneous noise exposure groups
Low	Harp, Guitar, Harpsichord and historical keyboards, Double Bass, Lute, Electronic Music, Organ, Viola, Viola Da Gamba, Cello, Piano Accompaniment, Bowed string instrument ensemble, Theory, Offices
Mid	Oboe, Piano, Violin, Wind instruments ensemble
High	Voice, Percussion, Chamber music

The homogeneous noise exposure groups at mid risk level are Oboe, Piano, Violin, Wind instruments ensemble, while the homogeneous groups at high risk level are Voice, Percussion and Chamber music.

A survey method, used to measure sound insulation between classrooms, background noise and reverberation time, showed that values were not adequate for the purpose of the classrooms ($T_r = 1.2$ s, $R'_w = 49$ dB and $L_{eq} = 42$ dBA at windows closed and 52 dBA at windows closed).

The results of background noise levels measurement, performed in 16 classes, are shown in Fig. 1. They range from 35 dBA to 60 dBA, with a higher occurrence in the 50–55 dBA.

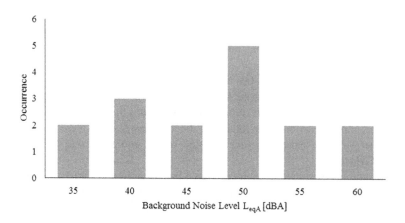

Fig. 1. Distribution of the background noise level

Moreover, from the questionnaires, a high discomfort was found both among teachers and students (respectively 60% and 33% of the interviewed people, Fig. 2).

The results show that among the group of students, 36% used wind instruments, 30% string instruments, 18% percussion instruments and the remaining 16% used the voice. The prominent age group ranged from 18 to 24 years. Among the group of teachers, 40% teach wind instruments, 20% percussion, 20% string instruments and another 20% singing. The age group is almost totally included between 45–54 years (80%), while only 20% belonged to the 65–70 range. The average level of acoustic discomfort perceived by the students stands on medium-high values. In particular, the students of percussion instruments (drums) are more susceptible to acoustic stress, according to the sound pressure level emitted by these instruments, followed by wind instrumental students (flute) and students of string instruments (classical guitar, electric bass, piano).

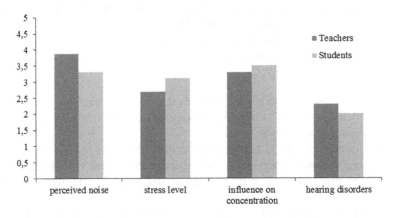

Fig. 2. Average values of answers on the noise survey

Students are unanimously agreed that a noisy environment affects concentration and performance, 66% considered the classroom too noisy, about 50% occasionally complains of tinnitus in the ears, 60% consider the noise from adjacent classrooms excessive, followed by noise from outside (25%) and the reverberation of the room itself (15%).

Among the teachers, over 80% considered the work environment too loudly. Consequently, teachers have to raise the middle tone of the voice, resulting in vocal fatigue (surmenàge). The teachers suffer even more from the inadequate acoustic isolation between the classrooms: the totality of the interviewees considered it insufficient. In the same way, the excessive reverberation of the room, present in 80% of the answers provided.

On 20 samples of singers examined, 5 are pianists and 15 guitarists. In the pianists group, neither significant vocal changes in the spectrogram and in the parameters examined, nor significant changes in posture that could affect the vocal emission, have been detected.

For guitarists, however, the postural changes correspond to the vocal changes; in fact, for 12 subjects in which the spectrogram was modified, the following postural alterations have been recognized: 8 rotations of the vertebral column to the left, 5 heads forward, 5 heads turned to the left and 5 shoulders raised, in the act of playing the instrument.

In addition, there was an insufficient significance level of the parameters examined except for VTI which significantly increased in 11 guitarists out of 15 with an average variation rate of about 23%.

Broadband spectrography for guitarists shows a general weakening of formants energy in 12 subjects in the act of playing guitar.

7 spectrograms show a lack in the formants above 4000 Hz and 2 spectrograms in the formants between 2000 Hz–3000 Hz, so it can be deduced that playing guitar during the singing performances leads to: greater slowness of the glottal closure (closing rate) found in the increase in the VTI value; decreasing of penetrance power related to energy decrease of the formants; vocal timbre modification especially when postural alteration consists in rotation of the column towards the left; generic correlation between posture required by the musical instrument and vocal alteration.

The results of the medical evaluation performed on 65 musicians with subjective and objective audiometric tests, show that 55% of them present hearing disorders without hearing loss, of which: 20% only distortion of loudness and pitch, 10% distortion; 10% only Tinnitus with Hyperacusis, 5% without Hyperacusis, 10% only Hyperacusis; combined symptoms: 45%. Moreover 40% of the musicians show a cochlear neurosensory hearing loss with recruitment, of which: 20% only pathological fatigue; 10% pitch and time distortion, 40% only Tinnitus and Hyperacusis, 30% combined symptoms. Finally, only 5% of the musicians show balance disorders with or without hearing loss.

4   Conclusion

The results show that specific acts have to be performed on the homogeneous noise exposure groups at mid risk level (Oboe, Piano, Violin, Wind instruments ensemble), and especially on group at high risk level (Voice, Percussion and Chamber music), to minimize occupational diseases. This goal can be achieved reducing exposure time, wearing ear protectors and improving room acoustics (reverberation time and sound insulation). This achievement could lead to a better perception of the acoustic quality of the environment, actually not adequate for the purpose of the classrooms for music education, as detected by the results of the questionnaires. Moreover this enhancement could achieve benefits in the vocal fatigue of the singers. Vocal effort dosimetry could be a effective technique to quantify vocal fatigue but at the moment, unfortunately, standardized reference values for assessment are not available. The study highlights that posture and voice are strictly coupled; the analysis of the singers' voices in relation to the posture allows to identify possible solutions to remedy occupational diseases.

Effective prevention can be achieved through adequate training and information to teaching staff and students, with the aim of raising awareness of potential hearing problems, extra-auditory effects and vocal tract diseases. Proper posture and a targeted acoustic restructuring of classrooms could improve comfort and performance of musician. Only an integrated approach would bring significant benefits to users, as well as optimize the costs of acoustic interventions.

References

1. Queensland Government. http://education.qld.gov.au/health/
2. Gutnick L. http://commons.emich.edu/cgi/viewcontent.cgi?article=1150&context=theses
3. Isaac MJ, McBroom DH, Nguyen SA, Halstead LA (2017) Prevalence of hearing loss in teachers of singing and voice students. J. Voice 31(3):379-e21
4. Olson AD, Gooding LF, Shikoh F, Graf J (2016) Hearing health in college instrumental musicians and prevention of hearing loss. Med Probl Perform 31(1):29–36
5. Cooper CL, Dewe P, O'Driscoll M (2001) Organizational stress: a review and critique of theory, research, and applications. Sage, Thousand Oaks
6. Cantley LF, Galusha D, Cullen MR, Dixon-Ernst C, Rabinowitz PM, Neitzel RL (2015) Association between ambient noise exposure, hearing acuity, and risk of acute occupational injury. Scand J Work Environ Health 41:75–83
7. Yoon J, Hong J, Roh J (2015) Dose - response relationship between noise exposure and the risk of occupational injury. Noise Health 17(74):43–47
8. Sound advice: Control of noise at work in music and entertainment, Heath and Safety Executive (2008)
9. Parra L, Torres M, Lloret J, Campos A, Bosh I (2018) Assisted protection headphone proposal to prevent chronic exposure to percussion instruments on musicians. J Healthc Eng 11
10. DPCM. Requisiti acustici passivi degli edifici, 5 December 1997
11. Legge 26 ottobre 1995, no 447, Legge quadro sull'inquinamento acustico
12. Decree of the President of the Province of Bolzano, 7 July 2008, no 26
13. UNI EN ISO 3382-1-2008. "Acoustics - Measurement of acoustic parameters of environments - Part 1: Showrooms"
14. UNI 11367 (2010) Building acoustics - Acoustic classification of real estate units - Assessment and verification procedure in progress
15. UNI 11532-1 (2018) Building acoustics - Internal acoustic characteristics of confined spaces
16. Ministerial circular no 3150 (22 May 1967) "Evaluation criteria and testing of acoustic requirements in school buildings"; Ministerial Decree 18 December 1975 "Technical standards relating to school buildings, including the minimum indexes of teaching functionality, building and urban planning- to be observed in the execution of school building works"
17. ANSI S12.60-2002 (2002) American National Standard Acoustical Performance Criteria, Design Requirements, and Guidelines for Schools
18. Building Bulletin 93 (28 May 2003) Acoustic design of schools - performance standards, UK, 2014; Arrêté du 25 avril 2003 relatif à la limitation du bruit dans les établissements d'nseignement JO
19. DIN 18041:2016-03 Hörsamkeit in Räumen - Anforderungen, Empfehlungen und Hinweise für die Planung (Acoustic quality in rooms- Specifications and instructions for the room acoustic design)
20. UNI EN ISO 9612 Acustica - Determinazione dell'esposizione al rumore negli ambienti di lavoro - Metodo tecnico progettuale
21. https://www.salford.ac.uk/research/sirc/research-groups/acoustics/psychoacoustics/sound-quality-making-products-sound-better
22. Jacobson B et al (1997) The voice handicap index (VHI): development and validation. J Speech-Lang Path 6:66–70
23. http://audiologistjobandnotes.blogspot.it/2012/01/mdvp-parameters-explaining-by.html

Searching for the Model of Common Ground in Human-Computer Dialogue

Clayton D. Rothwell[1,2(✉)], Valerie L. Shalin[1,3], and Griffin D. Romigh[4]

[1] Wright State University, Dayton, OH 45435, USA
[2] Infoscitex Corporation, Dayton, OH 45431, USA
crothwell@infoscitex.com
[3] Kno.e.sis, Dayton, OH 45431, USA
[4] Air Force Research Laboratory,
Wright-Patterson Air Force Base, Dayton, USA

Abstract. Natural language dialogue is a desirable method for human-robot interaction and human-computer interaction. Critical to the success of dialogue is the underlying model for common ground and the grounding process that establishes, adds to, and repairs shared understanding. The model of grounding for human-computer interaction should be informed by human-human dialogue. However, the processes involved in human-human grounding are under dispute within the research community. Three models have been proposed: *alignment*, a simple model that has been influential on dialogue system development, *interpersonal synergy*, an automatic coordination emerging from interaction, and *perspective taking*, a strategic interaction based on intentional coordination. Few studies have simultaneously evaluated these models. We tested the models' ability to account for human-human performance in a complex collaborative task that stressed the grounding process. The results supported the perspective taking model over the synergy model and the alignment model, indicating the need to reassess the alignment model as a foundation for human-computer interaction.

Keywords: Dialogue · Common ground · Human-computer interaction

1 Introduction

1.1 Background

The great promise of natural language interaction with computers is to handle complex commands with little to no device-specific training and this promise appears within reach. Industry analysts predicted that 33 million "voice-first" devices (e.g., Amazon Echo) would be in circulation by the end of 2017 [1]. A recent survey of 1,500+ technology experts on the Internet of Things predicts that speech interfaces will be one of the major advances between now and 2025 [2].

However, natural language interaction faces significant challenges, including the "conversational grounding" process of adding to, monitoring, and repairing shared understanding, known as *common ground*. Conversational grounding is a key feature

© Springer Nature Switzerland AG 2019
S. Bagnara et al. (Eds.): IEA 2018, AISC 827, pp. 33–42, 2019.
https://doi.org/10.1007/978-3-319-96059-3_4

of dialogue [3–5], and is a joint, rather than individual, process, and by implication requires methods that can analyze multiple dialogue participants. Moreover, the grounding process has a strong relationship to communication effectiveness and resulting performance metrics such as laboratory task completion time [4, 6–8].

Human-human grounding should serve as a model for how to design and envision human-computer grounding (e.g., [9]) although both the explanation and measurement of grounding is evolving. An influential approach for spoken dialogue systems (e.g., [10–14]) reflects the *alignment* model of grounding [5]. Alignment suggests that participants in a dialogue adapt to each other by increasing similarity of their phonetic, prosodic, lexical, or syntactic content [3]. Alignment occurs through priming, an automatic mechanism in which past experiences influence the likelihood of future contributions. Low-level alignment may propagate to the semantic level and a situation model (mental representation of the world), leading to common ground.

Syntactic alignment demonstrations [3] are suggestive of alignment at the situation model level, as syntax is at a high level close to the situation model level. Research on syntax has suggested that humans align to computers to a greater degree than they align to other humans [15]. Thus alignment provides an appealing, conceptually straightforward foundation for designing language-based human-computer interfaces.

However, the model of grounding is undecided [16]. Alternative models emphasize coordination and complementarity over alignment. Two coordination models have been proposed, distinguished by their emphasis on intentionality. *Perspective taking* is an intentional process that invokes Theory of Mind. Speakers know their audience does not share their perspective and can use this to design their contributions. Additionally, the audience can use the speaker's perspective to constrain interpretation [17]. Disrupting the ability to know your interlocutor's perspective disrupts collaboration [6, 18]. Attending to each other's perspective invokes an additional layer of collaborative management about the dialogue, called Track 2 dialogue [19]. Track 2 dialogue is talk about the dialogue itself including: acknowledgements of understanding, displays of non-understanding, and requests for clarification. Simulating Track 2 dialogue with computers is more difficult than using alignment.

Interpersonal synergy is a less examined theory that suggests coordination does not require intentionality [20]. Instead, coordination emerges from interaction as characterized by a complex dynamical system. This system is achieving stability in a specific context, and the coordinative interactions become cemented in interaction routines that are established by a particular system of interlocutors. The introduction of new interlocutors into established interaction routines disrupts communication [20].

The question remains, is alignment responsible for building up understanding and overcoming misunderstanding? If alignment does not influence the situation model, alignment should not predict task performance for tasks that require effective communication. Reitter and Moore [8] investigated the relationship between task performance and syntactic alignment for both short-term priming and long-term adaptation, finding no relationship for short-term priming which is most closely related to the paradigm in [3]. The current study extends these findings by analyzing additional linguistic levels and incorporating a different task to provide a more general conclusion regarding alignment.

1.2 Quantification of Recurrence

Grounding models had been investigated individually, until recent techniques allowed simultaneous comparison using recurrence quantification analysis (RQA) and cross recurrence quantification analysis (CRQA). RQA and CRQA are non-linear time series analysis from dynamic systems and provide metrics for characterizing the relationship between speakers' utterances, including the nature of the adaptations. Recurrence metrics quantify how much recurrence happens, the proportion recurrence appearing in longer sequences, the average length of recurrence sequences, and the variety in recurrence lengths. RQA measures recurrence within one time series (analogous to autocorrelation) and CRQA measures recurrence between two time series (analogous to cross-correlation). These methods were developed for continuous data, but have been adapted for categorical data in psychology [21], including for lexical analysis [7, 22] and syntactic analysis [23].

1.3 Grounding Models

Fusaroli and Tylén [7] modeled alignment and coordination using RQA and CRQA, and Rothwell and colleagues' [24] extension distinguished between perspective taking coordination and interpersonal synergy coordination. Fusaroli and Tylén [7] tested the alignment and coordination models by their relationship to task performance, with the rationale that the model that explains the most variance in task performance captures the grounding necessary for task completion. Figure 1 shows the same dialogue exchange with an illustration of each model. The alignment model examined recurrence between the two speakers, in which one speaker might adopt or repeat the wording of the other speaker (e.g., 'XYY' from speaker A to speaker B). Their coordination model examined recurrence in a speaker-independent manner, in which patterns of interaction between and across the two speakers might appear multiple times (e.g., 'YXZXY' occurs between A and B and later B and A.). Notably, their coordination model did not differentiate between perspective taking and interpersonal synergy. As a control, they also ran a baseline RQA of each speaker's self-consistency (i.e., lack of adaptation), then used the speaker that had highest rate of recurrence. Recurrence plots illustrating the alignment and coordination models appear in Fig. 2.

Rothwell and colleagues [24] extended this technique to differentiate types of coordination using statistical mediation analysis [25] for a task requiring extensive coordination. Their results converge with [7], who used a simple perception task. Both studies found that a coordination model accounted for performance better than an alignment model at multiple linguistic levels.

The current study expands on [24] through new analyses at the word level and the addition of the syntax level. Consistent with past findings, we hypothesized that the coordination model would have stronger relationship to performance than the alignment model. Furthermore, we used a statistical mediation analysis [25] to test if the perspective taking variant is the preferred coordination model. If so, perspective taking capability becomes a requirement for human-computer grounding and human-computer communication in general.

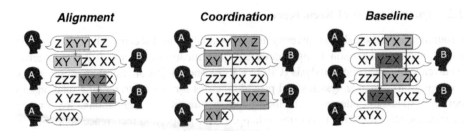

Fig. 1. Illustration of the recurrence models on the same dialogue (adapted from [20]). X, Y, and Z can stand for anything such as words, parts-of-speech, or voice intonation sequences. (Figure used with permission from John Wiley and Sons; Note: Originally [20] referred to coordination as interpersonal synergy).

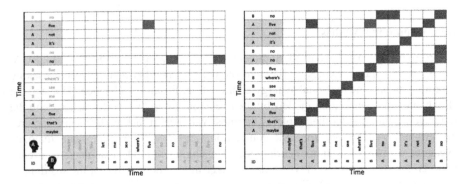

Fig. 2. Illustration of the word-level recurrence plots from the alignment analysis (left panel) and the coordination analysis (right panel). Green color shows the recurrence shared across the two plots. The right panel highlights the recurrence 'lines' that are parallel to the positive diagonal, which for coordination can occur across two speakers (i.e., 'five no').

2 Methods

2.1 Uncertainty Elicitation Task Corpus

We used a team dialogue corpus that has a referential communication component and a collaborative deduction component [26]. Participants (Ps) had many photographs from a labeled overhead perspective (i.e., satellite) as well as unlabeled street-level perspectives of buildings that they had to label together. Ps worked on computers in separate rooms and communicated over headsets. Street-level pictures and overhead pictures were obtained from Google Maps and all labels were removed. Referential communication was prompted because these 4 perspectives were split between Ps. Ps had to determine which building they discussed, as each P had unique views of each building. Collaborative deduction arose from the need to combine information from the different street-level perspectives to identify the location on the overhead perspective. These pictures led to conceptually complex and diverse dialogues. Ps discussed:

building features (e.g., siding, roof, windows, garage, 1 or 2-story), lot features (e.g., trees, yard, fence, driveway, garden, sidewalk,), street/neighborhood features (e.g., presence of stop-sign, power lines, parks), and car features (e.g., number of vehicles, vehicle type, color). A transcript excerpt is shown in Table 1.

Table 1. An transcript excerpt illustrating the referential communication (#14-18) and the collaborative deduction (#18-29) elements of the task.

#	Speaker	Transcript
14	B	um there's another one it's a small one-story house uh garage is separated
15	B	it's like really small and tiny
16	B	it's the yard is really big there's a tree in the front yard
17	A	I see it and there's like a stop sign
18	B	got it yeah I think it's at a corner
19	A	yeah I see the corner
20	B	okay
21	A	um
22	B	it's really small and very open
23	B	so I'm guessing it's either
24	A	it could be seven
25	B	seven
26	A	I don't know
27	B	I don't think it's seven it's either one
28	A	I- it can't be one because there's a pool in the back of one
29	B	true true true

The task was to label the street-level views with identification numbers from the overhead map by labeling through discussion. The trial was completed when both Ps had labeled all the buildings correctly. Performance was captured by task completion time. When Ps submitted their answers, they were provided with feedback on accuracy. The feedback was specific to each participant so one may be correct and the other incorrect, such as when Ps' grounding failed and they put down different numbers for the same building. There were 10 Ps organized into 5 teams of 2. Each team completed 8 trials with different stimuli for a total of 40 trials total in this corpus. Two outliers were identified as outside 1.5 times the interquartile range and removed from analysis.

2.2   Recurrence Analyses

Our analysis examined recurrence at the word level and the syntax level in search of the model that best predicts task performance. The word level was comprised of orthographic transcripts. The syntax level was created using the Stanford Log-Linear part-of-speech tagger (http://nlp.stanford.edu/software/tagger.shtml). Prior to analysis, RQA and CRQA require a number of parameters. We used values keeping with [7] and other categorical analyses [21, 22]. The radius value was set to 0, meaning only an exact match would be counted as a recurrence, which is appropriate for nominal data. The threshold for a line (parallel to the positive diagonal) was set at 2. Time delay was set to 1. Embed value was set to 1, so that single words and parts-of-speech were the units of analysis.

Three models were tested, as illustrated in Fig. 1 and described above. Individual recurrence metrics of recurrence rate (RR), determinism (DET), average line length (L), and entropy (ENTR) were calculated from recurrence plots created for each of the three models (i.e., alignment, coordination, baseline), for each trial in the corpus. To assess the relationship of each recurrence model to task performance, metrics from each recurrence model were used in separate regression models (not to be confused with adding/removing predictors from a single regression model). Linear models were evaluated through examining Adjusted R^2 values. These analysis procedures follow [7], assuming data from a between-subjects design. Subsequent tests address the repeated measures nature of this data. Analyses used R and the crqa package [27].

To test if coordination is the perspective taking variant, we completed a statistical mediation analysis [25]. This analysis used a model of Track 2 dialogue created with the Linguistic Inquiry and Word Count 2015 (LIWC) text analysis program [28]. LIWC uses word lists that measure different dimensions of text and our analysis relied on two word lists that may capture Track 2 issues of dialogue management [19]: Assent (e.g., agree, OK, yes) and Certainty (e.g., must, specific, clear). We focused on these two word lists because they may capture acknowledging understanding (e.g., backchannel communication) and signaling non-understanding (e.g., clarification requests). We followed the 'causal steps' procedure from [25]. The LIWC model of Track 2 dialogue was treated as the independent variable and the coordination recurrence models were treated as the mediator. In three steps, we tested (1) if the LIWC model accounted for performance, (2) if the LIWC model was related to the coordination model, and (3) if the coordination recurrence model mediated the LIWC model's relationship to performance.

In [24], the baseline model had shown high performance unexpectedly. One potential explanation is that the baseline model and coordination model are closely related. The current research investigated this by attempting mediation with the baseline model. If the baseline model does not mediate the relationship between LIWC and task performance when the coordination model does, this is evidence of a difference between the baseline and coordination models despite accounting for similar amounts of variance in task performance.

3   Results

The following results showed that the recurrence observed was not due to chance. The coordination model had stronger relationships to task performance than the alignment model for both word and syntax levels of analysis. Moreover, the coordination model accounted for variance in performance after the differences between the teams were statistically accounted for whereas the alignment and baseline models did not. Mediation analyses suggested that the coordination model was sensitive to Track 2 aspects of dialogue and the baseline model was not.

3.1   Control: Chance Analyses

A standard practice in recurrence quantification is testing if the structure of recurrence observed was greater than what would be expected by chance [7, 16]. Paired t-tests compared each recurrence metric to a 'shuffled' metric (i.e., calculated by randomly permuting each trial's dialogue) for each level of analysis. Due to space constraints, we simply report that recurrence structure was significantly different from shuffled controls for all models for all values, with the expected exception of recurrence rate (RR). All metrics were retained for the analyses.

3.2   Word-Level and Syntax-Level Analyses

The coordination model explained more of the variance in completion times ($AdjR^2 = 0.75$, $F(4, 33) = 28.0$, $p < .01$) than the alignment model ($AdjR^2 = 0.10$, $F(4, 33) = 2.02$, $p > .10$) and the baseline model ($AdjR^2 = 0.51$, $F(4, 33) = 10.83$, $p < .01$). The baseline model explained more variance than the alignment model, which was not significant. The syntax-level was similar to the word-level analysis. The coordination model explained more of the variance in completion times ($AdjR^2 = 0.68$, $F(4, 33) = 20.25$, $p < .01$) than the alignment model ($AdjR^2 = 0.03$, $F(4, 33) = 1.24$, $p > .10$) and the baseline model ($AdjR^2 = 0.56$, $F(4, 33) = 12.90$, $p < .01$).

3.3   Controlling for Team Differences

Controlling for team differences was needed as Team ID significantly predicted performance ($AdjR^2 = 0.43$, $F(4, 33) = 8.06$, $p < .001$). After regressing Team ID on performance, residuals were used to test if recurrence models remained significant (illustrated here for the word-level). The coordination model remained significant ($p < .05$), but the alignment and baseline models did not ($p = .41$ & $p = .12$, respectively).

3.4   Mediation Analysis: Track 2 Dialogue

Linear regression results for the LIWC Assent and Certainty lists showed both lists significantly predicted task completion times ($AdjR^2 = 0.49$, $F(2, 35) = 19.05$, $p < .001$). The coefficients for Assent and Certainty were both negative ($\beta = -0.83$, $p < .001$ & $\beta = -0.55$ $p < .001$, respectively), meaning that more instances of these words resulted in faster completion times. Mediation tests were performed at the word level using the coordination model. MANOVA showed the LIWC lists were related to the coordination model: Assent ($F(1, 4) = 5.23$, $p < .01$), and Certainty ($F(1, 4) = 9.67$, $p < .001$). The relationship between LIWC list and performance was mediated in the presence of the coordination model. Both lists ceased to be significant: Assent ($\beta = -0.27$, $p = .07$) and Certainty ($\beta = -0.07$, $p = .63$).

Additional mediation tests with the baseline model showed that the baseline model did not mediate LIWC factors. The Assent and Certainty word lists were not related to the baseline model ($F(1, 4) = 2.41$, $p = .07$ & $F(1, 4) = 2.31$, $p = .08$).

4   Discussion

The findings clearly supported the coordination model over the alignment model at the word level and the syntax level. Coordination accounted for 65% more of the variance in task completion times than alignment at both the word and syntax levels. Although the baseline model performed higher than [7], the pattern of findings for alignment and coordination was similar. Our higher-level, complex task with longer dialogues did not attenuate relationships between grounding processes and task performance. In fact, the relationships between coordination and performance found here were larger than those shown by [7]. In addition, the coordination model accounted for performance above team differences and the alignment model did not. The mediation tests supported the perspective taking variant of the coordination model. Mediation tests also showed that despite similarity in variance accounted for, the coordination and baseline models differ as the baseline model did not mediate Track 2 dialogue's relationship to task performance. The large relationship between perspective taking model and task performance/outcomes, as well as the relationship beyond team differences, strongly suggests that perspective taking should be the dominant model for human-human grounding processes and therefore for human-computer grounding.

This study has added to the results of Reitter and Moore [8] in a number of ways. The symmetric task here and the asymmetric task in [8] both found that syntactic alignment was not related to task performance. Furthermore, the current findings provided clear support for perspective taking as an alternative to alignment.

Other recent research agrees with these findings that communication processes are more complicated than priming-based alignment. Over time, research suggests that interlocutors diverge and become more complementary rather than become more similar [29]. A likely explanation is that interlocutors contribute new content rather than repeating content from previous contributions [30]. Table 1 contained an instance of new complementary content when Speaker A said "I see it and there's like a stop sign." The additional mention of the stop sign provided very strong evidence that he had identified the house that Speaker B was describing. Assertions like these have a Track 2 function because they invite Speaker B to object if necessary. Dialogue contributions often reflect different perspectives and interlocutors appear to maintain and keep track of multiple perspectives at the same time [31].

Here, by comparing the perspective taking model, the alignment model, and the interpersonal synergy model simultaneously, we have shown clear support for perspective taking. Findings also have implications for the design and development of natural language technologies. Establishing common ground is an inescapable problem for these systems and they should be based on a perspective taking model of the grounding process rather than alignment. In addition, while we recognize the challenges posed in designing these functions, the current findings have highlighted the importance of implementing conversation management that can perform Track 2 dialogue. Track 2 dialogue plays an essential role in language by discussing the communicative contributions themselves, signaling understanding of the intended meaning and providing opportunities to clarify the intended meaning when necessary.

5   Conclusion

For natural language interaction to succeed as a method of human-computer interaction, it must have the proper underlying model for the conversational grounding process and be informed by human-human interaction. We tested prominent models for this joint process, alignment, perspective taking and interpersonal synergy, in a complex collaborative grounding task. The key feature of this test was to discriminate between the models by their relationship to task performance. The results strongly supported the perspective taking model. These findings should encourage researchers to reassess the alignment model as the basis for their natural language technologies.

Acknowledgements. This research was partially supported by the Air Force Research Laboratory Contract#: FA8650-14-D-6501. We thank Sid Horton for valuable discussions.

References

1. Marchick A (2018) The 2017 Voice Report. https://alpine.ai/the-2017-voice-report-by-alpine/. Accessed 16 May 2018
2. Anderson J, Rainie L (2018) The Internet of Things Will Thrive by 2025. http://www.pewinternet.org/2014/05/14/internet-of-things/. Accessed 16 May 2018
3. Branigan H, Pickering M, Cleland A (2000) Syntactic coordination in dialogue. Cognition 75(2):B13–B25
4. Clark H, Wilkes-Gibbs D (1986) Referring as a collaborative process. Cognition 22:1–39
5. Pickering M, Garrod S (2004) Toward a mechanistic psychology of dialogue. Behav Brain Sci 27(2):169–190
6. Clark H, Krych M (2004) Speaking while monitoring addressees for understanding. J Memory Lang 50(1):62–81
7. Fusaroli R, Tylén K (2016) Investigating conversational dynamics: interactive alignment, interpersonal synergy, and collective task performance. Cogn Sci 40(1):145–171
8. Reitter D, Moore J (2014) Alignment and task success in spoken dialogue. J Memory Lang 76:29–46
9. Edlund J, Gustafson J, Heldner M, Hjalmarsson A (2008) Towards human-like spoken dialogue systems. Speech Commun 50(8):630–645
10. Brockmann, C., Isard, A., Oberlander, J. White, M.: Modelling alignment for affective dialogue. In: Proceedings of the workshop on adapting the interaction style to affective factors at the 10th international conference on user modeling (UM-05), pp 1–5 (2005)
11. Buschmeier H, Bergmann K, Kopp S (2009) An alignment-capable microplanner for natural language generation. In: Proceedings of the 12th european workshop on natural language generation, pp 82–89
12. Janarthanam S, Lemon O (2009) A wizard-of-Oz environment to study referring expression generation in a situated spoken dialogue task. In: Proceedings of the 12th european workshop on natural language generation, ENLG, pp 94–97
13. Tomko S (2006) Improving User Interaction with Spoken Dialog Systems via Shaping. Ph. D. dissertation. Carnegie Mellon University
14. Varges S (2006) Overgeneration and ranking for spoken dialogue systems. In: INLG 2006, pp 3–5

15. Branigan H, Pickering M, Pearson J, McLean J (2010) Linguistic alignment between people and computers. J Pragmat 42(9):2355–2368
16. Louwerse M, Dale R, Bard E, Jeuniax P (2012) Behavior matching in multimodal communication is synchronized. Cogn Sci 36(8):1404–1426
17. Keysar B, Barr D, Balin J, Brauner J (2000) Taking perspective in conversation: the role of mutual knowledge in comprehension. Psychol Sci 11(1):32–38
18. Gergle D, Kraut R, Fussell S (2004) Language efficiency and visual technology: minimizing collaborative effort with visual information. J Lang Soc Psych 23(4):491–517
19. Clark H (1996) Using Language. Cambridge University Press, Cambridge
20. Fusaroli R, Raczaszek-Leonardi J, Tylén K (2014) Dialog as interpersonal synergy. New Ideas Psychol 32(1):147–156
21. Dale R, Spivey M (2005) Categorical recurrence analysis of child language. In: Proceedings of the 27th annual meeting of the cognitive science society. Lawrence Erlbaum, Mahwah, NJ, pp 530–535
22. Orsucci F, Petrosino R, Paoloni G, Canestri L, Conte E, Reda M, Fulcheri M (2013) Prosody and synchronization in cognitive neuroscience. EPJ Nonlinear Biomed Phys 1(1):1–6
23. Dale R, Spivey M (2006) Unraveling the dyad: Using recurrence analysis to explore patterns of syntactic coordination between children and caregivers in conversation. Lang Learn 56 (3):391–430
24. Rothwell C, Shalin VRG (2017) Quantitative models of human-human conversational grounding processes. In: Gunzelmann G, Howes A, Tebrink T, Davelaar EJ (eds) Proceedings of the 39th annual conference of the cognitive science society. Cognitive Science Society, Austin, TX (2017)
25. Baron R, Kenny D (1986) The Moderator-mediator variable distinction in social psychological research: conceptual, strategic, and statistical considerations. J Pers Soc Psychol 51(6):1173–1182
26. Romigh G, Rothwell C, Greenwell B, Newman M (2016) Modeling uncertainty in spontaneous speech: lexical and acoustic features. J Aco Soc Am 140(4):3401–3401 (2016)
27. Moreno Coco M, Dale R (2014) Cross-recurrence quantification analysis of categorical and continuous time series: an R package. Front Psychol 5:1–14
28. Pennebaker J, Boyd R, Jordan K, Blackburn K (2015) The Development and Psychometric Properties of LIWC2015. Technical report. University of Texas at Austin, Austin, TX
29. Mills GJ (2014) Dialogue in joint activity: Complementarity, convergence and conventionalization. New Ideas Psychol 32(1):158–173
30. Tenbrink T, Andonova E, Coventry K (2008) Negotiating spatial relationships in dialogue: the role of the addressee. In: Proceedings of LONDIAL - the 12th SEMDIAL workshop, pp 193–200
31. Brennan S, Schuhmann K, Batres K (2013) Collaboratively Setting perspectives and referring to locations across multiple contexts. In: Proceedings of the Pre-CogSci 2013 workshop on the production of referring expressions, Berlin, vol. 1, pp 1–6

Designing Multi-modal Interaction – A Basic Operations Approach

I. C. MariAnne Karlsson$^{(\boxtimes)}$ ⓘ, Fredrick Ekman ⓘ,
and Mikael Johansson

Design & Human Factors, Department of Industrial and Materials Science,
Chalmers University of Technology, Gothenburg, Sweden
mak@chalmers.se

Abstract. A basic operations approach to designing multimodal user interfaces
has been explored in two experimental studies. By basic operations is here
meant the most fundamental operations that make up a(ny) interaction task.
Study A investigated if certain basic operations have a preferred modal coun-
terpart and execution and Study B what happens when several basic operations
are combined into tasks. The results suggest interdependencies between basic
operations and specific modalities but also indicate that users may prefer unimodal interaction or to choose modality as desired. Hence, further studies are
needed to assess the feasibility of a basic operations approach to multimodal
interaction design.

Keywords: Interface design · Multi-modal interaction · System input

1 Introduction

1.1 Background

Multimodal user interfaces are the result of combining multiple modalities in a
meaningful way to support efficient, safe, and satisficing interaction. It has for example
been suggested that multimodal interfaces will be easier to learn and use (Oviatt et al.
2004), more engaging, as well as allow a distribution of cognitive load (e.g., Oviatt
et al. 2003; Shi et al. 2007). This could for example involve hand gestures and/or
speech commands to control in-vehicle systems (Krahnstoever et al. 2002) when
undivided visual attention is needed on the primary task of driving (Carrino et al.
2012). Other research suggests that implementing a multimodal approach to human-
machine interaction (HMI) could facilitate a more 'natural' interaction, in particular
when users themselves can decide which modality to use (Oviatt 1999).

However, even though the pros and cons of multimodality have been investigated
in several studies many challenges remain in designing multimodal interaction, such as
choosing which modality for which interaction and deciding how (or not) to combine
modalities. Speech as a communicative mean could increase flexibility, but research
has also shown that it could create frustration, for example when a user does not know
which words to use in order to control a system (Oviatt et al. 2003). In addition, users
may be reluctant to change between modalities since it could add to cognitive load.

© Springer Nature Switzerland AG 2019
S. Bagnara et al. (Eds.): IEA 2018, AISC 827, pp. 43–53, 2019.
https://doi.org/10.1007/978-3-319-96059-3_5

The paper presents an approach to designing multimodal interaction based on the notion of 'basic operations'. The rationale behind is a search for a generic, structured user-centred approach instead of a design based on unique choices, task by task, and with the risk of facing non-coherent interaction and sub-optimisation. In order to investigate the feasibility of the concept from a user perspective, two complementary experimental studies, A and B, with a focus on system input, were carried out.

1.2   Tasks and Basic Operations

The concept of a task is fundamental to human-machine interaction (HMI). A task can be described as an activity necessary to achieve a goal using a device. In a Hierarchical task analysis (HTA) the task is broken down into sub-tasks and even further into lower-level sub-tasks. Basic operations here signify the most fundamental and generic sub-tasks that make up an(y) interaction task. Examples of such basic operations are:

- 'Activate' and 'deactivate' – where there are only two choices, both belonging to the same object/function.
- 'Increase' or 'decrease' – which refers to an operation where properties can be altered in one or two different directions.
- 'Choose' – where there is a choice between different alternatives (excluding activate/deactivate).
- 'Search' – an operation triggering a continuous operation performed by the system.

2   Study A

2.1   Aim

The aim of study A was to explore how users, given a predefined choice of modality, choose to execute predefined 'basic operations' with an imagined 'device' or function.

2.2   Study Design

Participants. The study involved altogether 20 participants (11 men and 9 women, aged between 20 and 65 years.

Procedure. The participants were asked to demonstrate their spontaneous choice, using one of three pre-defined modalities – speech, gesture, or haptics – in response to 12 different task scenarios, each including one basic operations. The scenarios were presented to the participants in different order. No specific context was introduced but the tasks were of a kind that the participants could perform in their everyday life, such as activating or deactivating a device, increasing/decreasing (e.g., volume/sound), selecting (e.g., an item from a list) etc. For each scenario, the participants were cued by a question, for example "*If you were to turn a device 'on'* (=operation), *how would you use speech* (=modality)?"

Equipment. In addition to video-camera and microphone for data collection, the set up included a physical button, dial knob and slider.

Data Collection. The participants' behaviours, speech commands as well as verbalised comments during the execution of the operations were video and audio recorded. After completing the tasks, the participants were interviewed on what modality was the most difficult/easiest for them to use considering the basic operations they had performed and the reasons why. They were also asked to rate how natural it felt to perform the respective operations on a scale from 1–5 (1 = very unnatural and 5 = very natural).

Analysis. The video and audio recordings were analysed and patterns were searched regarding participants' actual execution of the basic operations as well as their arguments and explanations of their choices. The responses to the questionnaire were compiled and mean value as well as standard deviation for each item were calculated.

2.3   Results

Observations. The observations revealed several similarities but also some differences across participants, operations and modalities. For example, when cued to 'activate'/ 'deactivate' something using touch/haptics, all participants chose to touch, turn or push a physical button. For speech, most proposed to say "Turn ON" or "ON" but for gestures a range of options were chosen including, for example tapping on or pointing at an imaged device. To perform 'choose', the most common choice for touch/haptics included pressing a button, for gesture pointing at an imagined link/object, and for speech "Select X". However, when cued to 'search', the participants had difficulties suggesting any option at all for touch/haptics. They proposed some gestures but it was obvious that they felt hesitant as to their efficiency. In this case, speech appeared as the most immediate choice and a number of synonym imperatives were suggested, such as "Search" and "Find". For "Navigate", chosen executions included for touch/haptics swiping fingers across an imagined touch screen, for gesture sweeping gestures in the air and for speech commands "Navigate to ..." but also "Scroll up/down ..." in an assumed list.

Questionnaire. The rating of how natural it felt to use a specific modality for a specific operation is presented in Fig. 1.

Interview. In the interviews, the participants were asked to explain their choices. One type of explanations to their choices was related to their experience/inexperience of using a specific type of modality in their everyday life. One participants explained: *"Touch is what we are used to the most, so that was easy to formulate. I think I came up with a solution instantly for touch ...//... since I'm used to it, it feels more natural."* and further that: *"The difficult one? Gestures because I'm not very used to it."*

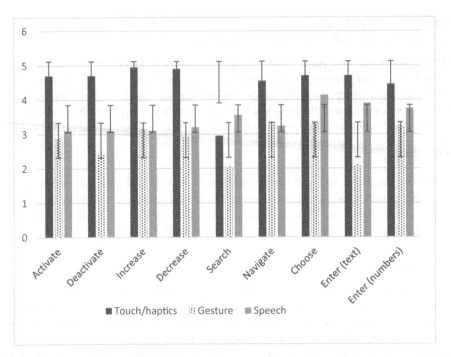

Fig. 1. Rating of how natural it felt to use a specific modality for a specific operation. (1 = very unnatural and 5 = very natural). Mean values and standard deviations (n = 20).

However, even though touch/haptics was commonly mentioned as the easiest way to complete an operation, some participants argued speech to be easier in some specific cases: *"Easiest? As in we don't have to think about? It would be touching because that is what we usually do. But speech doesn't require physical movements so that could be easier in some scenarios."* Hence, the perception of ease involved a cognitive as well as a physical dimension.

Gestures (performed "in the air") appeared particularly difficult as no obvious feedback was received during the specific trial: *"Always when I think of gesture, I somehow see the feedback of my movement in the screen. Just doing gestures and not getting feedback while interacting with something ... it's hard for me to imagine gestures on its own without any visual feedback."*

Another but related issue concerned the experienced need for precision in order to complete the operations: *"When I think about the gesture, I feel there are lot more options. It is a more unrestrained type of interaction. Touch was the easiest because it's the least open."*

In addition, several participants commented on the lack of context and how context would influence their response, some even had to imagine a context in order to perform the operation: *"Gesture was the most difficult for me because I kept guessing about the context."*

3 Study B

Based on observations and ratings of how natural different options felt, Study A indicated connections between certain basic operations and modalities; for example, for increase/decrease – touch/haptics, for search – speech, and for navigate – gesture. However, a task most often consists of several basic operations. A second study, Study B, was therefore completed to investigate the effect of combining several basic operations.

Different design principles were considered including temporally cascaded multi-modality and redundant multimodality. Temporally cascaded is sequential and means that one modality follows another in a defined order within a task and should by itself enable an understanding of which modality that follows (Muller and Weinberg 2011). Study B used a sequential multimodal approach, referred to as 'fixed multimodality', which shares similarities to temporally cascaded except that it does not give any hints to which modality that should follow in the interaction order. Finally, redundant multimodality is preference-based due to the user's possibility to freely choose a modality for every basic operation.

3.1 Aim

The aim of Study B was to investigate how users interact with a multimodal system as well as how 'natural' they perceived different multimodal interface designs, more specifically 'fixed multimodality' and 'redundant' compared to 'unimodal' (baseline). Questions posed were: Can combinations of basic operations and modalities be combined no matter how and whatever the task? and Do users experience a certain combination of basic operations and modalities as more natural than others?

3.2 Study Design

Participants. The study involved 20 participants, 12 men and 8 women, aged between 22 and 53 years.

Equipment and Set-Up. A graphical user interface was designed specifically for the experiment including four different applications: a weather-, a GPS-, a music- and a phone-app. These four applications created the basis for four different tasks: (task#1) check the weather forecast, (task#2) search for a geographical position, (task#3) play a song and (task#4) call a contact. The respective tasks consisted of one or a combination of different basic operations (e.g., task#1='choose'+'choose' and task#2='choose'+'search'+'choose'+'choose').

The set-up consisted of a touch screen, a (fake) sensor to 'register' gestures, a microphone for speech and a physical scroll wheel for haptic manipulations. Two of the modalities, gestures and speech, were not controlled by the participants but by a 'wizard' (sitting in an adjacent room seeing and hearing what the participants were doing and saying, by means of four cameras and one microphone).

Procedure. Four different sequences were specified, all which included four basic operations (see Table 1):

- Unimodal: A unimodal sequence that included one modality - touch – only (baseline).
- Fixed Conform: One fixed multimodal sequence that used four modalities according to the findings from study A.
- Fixed Contradict: A second fixed multimodal sequence that linked the four different modalities to basic operations in a way that contradicted the findings from study A.
- Redundant: A redundant multimodal sequence where the participants freely could choose which of the four modalities they wanted to use for each of the basic operations.

Each participant completed altogether three sequences, i.e. all completed the 'Unimodal' and the 'Fixed conform' sequences whereas half of the participants completed the 'Fixed contradict' and the other half the 'Redundant' sequence. The participants were randomly allocated to the respective groups.

Table 1. Basic operations and modality combinations for the different test procedures.

Basic operation	'Unimodal'	'Fixed Conform'	'Fixed Contradict'	'Redundant'
Choose	Touch	Touch	Speech	Free
Increase & Decrease	Touch	Haptic	Gesture	Free
Search	Touch	Speech	Touch	Free
Navigate	Touch	Gesture	Haptic	Free

Data Collection. The participants' behaviours and verbalised comments were documented by means of video and audio recordings. In addition, they were asked to fill in a usability questionnaire after each sequence. The questionnaire consisted of eight Likert statements and the participants rated their level of agreement with each of the statements on a five-point scale (1 = Totally Disagree and 5 = Totally Agree). Finally, personal interviews were completed in order to elicit more detailed information.

Analysis. The video recordings of the 'Redundant' tests were analysed, annotating which modalities the participants used to carry out specific basic operations as a way to reach further understanding of any preferred combination of modality and basic operation. The audio-recordings of the interviews were transcribed and analysed to identify patterns in the participants' responses. In addition, the participants' responses to the questionnaire were compiled and the mean value for each item were calculated.

3.3 Results

Observations. During the sequence 'Redundant' each participant's choice of modality was registered for each basic operation that made up the respective tasks. This revealed that even though some participant used primarily one and the same modality when completing a task, other participants used a more varied way of interacting with the system, and all participants used more than one modality. A certain pattern could be distinguished, in that the participants often used the same modality for the same basic operation. For example, touch was chosen 84% of the times that the participants 'chose' something; the scroll wheel was used 80% of the times that participants adjusted (i.e. for 'increase/decrease') the volume; and a majority of participants (65%) used speech to make a 'search'. For 'navigate' touch was used 48% of the times and gestures were used 41% of the times but in some cases the users chose speech to directly choose something they knew was in the menu.

Questionnaire. The result of ratings is presented in Table 2. Overall, the 'Unimodal' and the 'Redundant' sequences were assessed as the easier whereas 'Fixed Contradict' was rated the most difficult of the four.

Table 2. Usability questionnaire results. Mean values (1 = Totally Disagree and 5 = Totally Agree), n = 20.

Item	Unimodal	'Conform'	'Contradict'	Redundant
Learning how to use the different ways of manipulating the interface was easy	4.8	3.9	3.5	4.7
Using these ways of manipulating the interface was a frustrating experience for me[a]	1.8	2.2	2.7	1.5
These ways of manipulating the interface made me feel in control	4.2	3.7	2.8	4.3
I found it hard to understand in which way I should manipulate the interface[a]	1.3	2.4	3.0	1.6
These ways of manipulating the interface was very pleasant	3.8	3.7	3.3	4.3
I thought that these ways of manipulating the interface felt natural	4.3	3.7	3.0	4.3
I know how to use the different ways of manipulating the interface next time I try it	4.8	4.3	3.8	4.4
This way of interacting with an interface feels familiar	4.9	4.0	3.1	4.1

[a] Note the reverse phrasing.

Interview. Overall, the questionnaire results were confirmed in the interviews. Even though the participants explained that they would be able to use any modality in a multimodal system, many commented that they would prefer to be able to use whichever modality the wanted – whenever they wanted it: *"If I could choose which modality I would like to use, it would be favourable."* Furthermore, even if many of the participants thought that a multimodal system was more difficult to use and that it felt more unfamiliar, they were still positive towards the idea: "*Of course it makes more sense to only have uni-modality but it is not a sweet way of controlling it.*" For example, the participants generally felt that the 'Fixed Conform' interface would be quite easy to learn and to form a new habit. A majority of the participants felt though that what they are used to – or familiar with - is also what they experience as most 'natural': "*The unimodal interaction felt much more natural than the multimodal interactions because you are not used to it*" and some participants even expressed a dislike to modalities that they were unfamiliar with.

As in Study A, participants thought that without proper feedback, it was difficult to understand which modality was possible to use and when: "*It took a lot of focus because I tried to find feedback about how I should interact with the system.*" The feedback modality was furthermore noted as cue for the choice of modality that followed: "*If you are supposed to use speech commands maybe you should not get the feedback visually*". However, also the visual design of the specific system interface prompted the choice of modality; For example, when some of the participants saw a list they instinctively wanted to use touch.

The participants generally wanted consistency between the basic operations and modalities, i.e. they wanted to use the same modality for the same basic operation: "*Every time I navigate a menu I want to use the same modality*". What was perceived as the 'best' modality differed though depending on the task. Most participants thought it was difficult to use several modalities during the same sequence: "*I experienced that when I used several modalities that it disrupted my thought process. The more modalities, the more confused you get.*" Thus, transition between modalities was identified as a factor that seemed to affect how seamless the interaction felt but it was not only the transition itself that affected the user's experience but also which modalities and in which order. For one participant speech-to-touch felt more logical whereas another participant said "*To go from touch to speech feels more natural since I am used to activate the speech command by first pushing.*" Some participants found transitions more or less demanding from a cognitive perspective: *I thought it was easier when I did physical actions since they felt connected but when I also used voice it felt like I used another part of my brain*" but also physical transitions were mentioned: "*Speech and touch work well together but I do not think it feels good when I need to move my hand back and forth.*"

As in Study A, several participants chose to categorize the modalities as offering more or less precision and control. The different operations were also categorized according to how much control they felt was required: "*When I ask the system a question (searching), I am not looking for control.*" Even though the participants did not completely agree on which modalities that offered more or less control or how much the different basic operations required, some consensus was reached regarding "To choose", which was considered a very precise basic operation that demanded high level of control: "*It felt better when I searched for something and then was able choose with touch. If I choose with speech something can go wrong and then you may potentially activate a function that can be harmful*". "To search", on the other hand, felt like a non-precise basic operation and in this case, many participants preferred to use speech which was considered as a modality offering a low degree of control.

4   Discussion

The results indicate a possible interdependency between users' choice of modality and basic (input) operations; the participants tended to choose the same modality for a specific basic operation independent of task; when combining basic operations into tasks the participants preferred consistency (i.e. the same modality for one and the same basic operation); and in Study B the 'Fixed Conform' sequence was experienced as easier to carry out than 'Fixed Contradict'. However, the results also showed that participants preferred unimodal interaction or, if multimodality, redundancy rather than to be limited to predefined options. These findings would imply the proposed approach.

Also other studies have found that users may have specific preferences regarding their use of modalities related to **type of tasks** (Chen and Tremain 2005; Janowski et al. 2013). In addition to concluding that modality usage is determined by the task to be performed, Naumann et al. (2008) also argued that the choice of modality depends upon the (perceived) efficiency of the modality for achieving the task goal, something that was observed in both Study A and Study B where comments were made on the need (or not) for precision and control. Other studies have demonstrated that **task complexity** impact users' choice of and preferred input modalities; unimodal interaction for simpler tasks and multimodality for more complex tasks. The tasks in Study B were, in comparison, simple tasks and this may in part explain why 'Unimodal' interaction was preferred by the participants.

Another explanation could be related to effort. Sequential multimodality implies **transitions** between modalities and, according to the participants, transitions between modalities for completing a task was experienced as involving an effort (cognitive and/or physical) 'Unimodal' interaction required less cognitive transitions than the sequential multimodality options but at the same time the 'Fixed Conform' sequence was experienced as more 'natural' and easier to carry out than 'Fixed Contradict'. Perceived effort depended thus upon transitions between which modalities and in what order – something which must be considered in a basic operations approach to interaction design.

Familiarity was another factor found to affect the interplay between (preferred) modalities and basic operations. In the studies, participants used familiar interaction strategies even when interacting with non-specified devices (in Study A) and also rated these strategies as more 'natural'. Several studies have shown how experience influences users' choice of modality (e.g. Althoff et al. 2004), and expert users differing in their choices compared to novice. As the 'Unimodal' sequence was based on touch, which is today commonly used by a large number of people for interacting with smartphones, iPads etc., this may be another reason why 'Unimodal' received the highest usability scores.

Furthermore, sensory **feedback** when using a certain modality was in Study A and Study B considered a key issue, especially when interacting with gestures. The same observation was made by Hausen et al. (2013) who reported that when interacting with haptic, touch and gesture interfaces, participants lacked haptic feedback even when receiving visual feedback. However, feedback on an input influenced also participants' thoughts on which modality to use next in a sequence. That system output affects the users' choice of input has earlier been shown by, for example Tang and Lundgren (2016) as well as in studies by Bellik et al. (2009). In the present study, only system input was included but the results illustrate the necessity to consider the complete interaction loop, i.e. system input and system output, when designing multimodal interaction.

In summary, the results suggest an interdependency between preferred modality and basic (input) operations in multimodal interaction. However, the studies also indicate that from a user perspective, unimodal interaction or being able to choose modality as desired are preferred interaction designs. Hence, further studies are needed to assess the feasibility of a basic operations approach to multimodal interaction design.

References

Althoff F, McGlaun G, Lang M, Rigoll G (2004) Evaluating multimodal interaction patterns in various application scenarios. In: Camurri A, Volpe G (eds) Gesture-based communication in human-computer interaction. Lecture Notes in Computer Science, vol 2915. Springer, Heidelberg, pp 421–435

Carrino F et al (2012) In-vehicle natural interaction based on electromyography. In: Adjunct proceedings of the AutomotiveUI 2012. ACM, New York, pp 23–24

Chen X, Tremaine M (2005) Multimodal user input patterns in a non-visual context. In: Proceedings of the 7th international ACM SIGACCESS conference on computers and accessibility, pp 206–207

Hausen D, Richter H, Hemme A, Butz A (2013) Comparing input modalities for peripheral interaction: a case study on peripheral music control. In: Kotzé P, Marsden G, Lindgaard G, Wesson J, Winckler M (eds) Human-computer interaction – INTERACT 2013. Lecture Notes in Computer Science, vol 8119. Springer, Heidelberg, pp 162–179

Janowski K, Kistler F, André E (2013) Gestures or Speech? Comparing modality selection for different interaction tasks in a virtual environment. In: Proceedings of the Tilburg gesture research meeting

Krahnstoever N, Kettebekov S, Yeasin M, Sharma R (2002) A real-time framework for natural multimodal interaction with large screen displays. In: Proceedings of the 4th IEEE international conference on multimodal interfaces, Washington D.C. IEEE Computer Society Press, pp 349–354

Muller C, Weinberg G (2011) Multimodal input in the car, today and tomorrow. IEEE Multimed 18(1):98–103

Naumann AB, Wechsung I, Möller S (2008) Factors influencing modality choice in multimodal applications. In: André E, Dybkjær L, Minker W, Neumann H, Pieraccini R, Weber M (eds) Perception in multimodal dialogue systems. PIT 2008. Lecture Notes in Computer Science, vol 5078. Springer, Heidelberg, pp 37–43

Oviatt S (1999) Ten myths of multimodal interaction. Commun ACM 42(11):74–81

Oviatt S, Coulston R, Lunsford R (2004) When do we interact multimodally? Cognitive load and multimodal communication patterns. In: Proceedings of the 6th International Conference on Multimodal Interfaces. ACM, New York, pp 129–136

Oviatt S et al (2003) Toward a theory of organized multimodal integration patterns during human-computer interaction. In: Proceedings of the 5th international conference on multimodal interfaces. ACM, New York, pp 44–51

Shi Y, Taib R, Ruiz N, Choi E, Chen F (2007) Multimodal human-machine interface and user cognitive load measurement. IFAC Proc Vol 40(16):200–205

Tang L, Lundgren J (2016) Exploring multimodality and natural interaction in a driving context. Master Thesis, Chalmers University of Technology, Gothenburg

Scalable Auditory Alarms

Michael J. Waltrip[(⊠)] and Carryl L. Baldwin

George Mason University, Fairfax, VA, USA
mwaltrip@gmu.eduu

Abstract. In information dense, visually complex environments (e.g., automobiles, airplanes, and operating rooms) auditory alarms can direct attention to critical events. Effective alarms are designed to convey a hazard level appropriate for the situation they represent. Acoustic intensity plays an essential role in perceived urgency, with louder sounds generally being perceived as representing something more urgent. However, intensity may be dictated by factors outside the alarm designer's control (e.g., background noise, manufacturer's sound system specifications). Therefore it is essential to examine other acoustic parameters that can be used to convey scalable levels of perceived urgency. Previous work suggests that looming sounds, or sounds perceived to be coming towards a listener, may result in faster response times than other types of sounds. Further, people may respond faster to looming sounds if the rate of change (ROC) in the increase in intensity is more rapid, potentially making them a strong candidate for scalability. Sounds increasing in intensity quickly may be perceived as more urgent than other sounds. However, it is critical that ROC is not confounded by overall intensity, since the intensity-urgency relationship is well documented. The present study measured perceived urgency ratings of auditory looming stimuli with different intensity ROCs that were equated for overall intensity. Results revealed no differences in urgency ratings across the three ROCs. Current results indicate that overall intensity is more critical than ROC to achieving scalability in auditory alarms.

Keywords: Looming · Warning · Urgency

1 Introduction

Many operational environments (e.g., automobiles, airplanes and operating rooms) are visually and informationally dense. Effective auditory warnings are critical for both directing an operator's attention to time critical events (e.g. an imminent collision situation) and providing gentle reminders of less urgent events (e.g., low fuel). Effective design of auditory warnings for these complex environments is essential. Perceived urgency and hazard matching is one key component to effective alarms. Manipulating certain acoustic parameters such as intensity, frequency, and pulse rate can result in more urgent warnings (Hellier et al. 1993; Edworthy et al. 1991; Haas and Edworthy 1996; Baldwin and Lewis 2014). These acoustic characteristics can thus be used to achieve urgency scalability in order to represent a range of hazard levels.

Gray (2011) found that auditory looming warnings (sounds perceived to be coming towards a listener) were particularly effective as high urgency collision avoidance

© Springer Nature Switzerland AG 2019
S. Bagnara et al. (Eds.): IEA 2018, AISC 827, pp. 54–58, 2019.
https://doi.org/10.1007/978-3-319-96059-3_6

sounds. Further, his work suggested that looming warnings are more diverse because the rate at which the sound increases in intensity could be scaled to convey appropriate levels of urgency. Additionally, he argued that looming warnings are more urgent than abstract auditory warnings and more reliable because false alarms for the novel sounds are less likely. In his experiments using a driving simulator, Gray (2011) presented participants with a number of warnings including three auditory looming warnings that were varied on the perceived speed at which they approached the listener (by changing the rate of change in the intensity of the sounds). He found that auditory looming warnings, specifically two that were designed to convey accurate and early time to collision estimates, resulted in faster brake response times (BRTs) when compared to static warnings, pulsating warnings, or warnings linearly increasing in intensity.

However, as far as we know, the different looming sounds were not equated for overall intensity. Gray's results might be because some looming sounds were simply more intense than the other sounds. It is understood that more intense sounds (subjectively perceived as loudness) result in higher urgency ratings and decreased response times (Momtahan 1990; Haas and Edworthy1996). However, sound designers have little control over the comparative intensity of a warning due to factors such as noise inside and outside of the vehicle, and equipment manufacturer's sound system specifications. Because of this, the present study investigated whether various auditory looming warnings would differ on urgency ratings after being equated for overall intensity.

2 Methods

2.1 Participants

Fifty-eight participants completed the experiment ($M = 19.95$, $SD = 2.53$, 17 males). Participants were undergraduate and graduate students who participated for course credit or on a volunteer basis. All participants were 18 years or older and reported normal or corrected-to-normal hearing and vision.

2.2 Materials

Participants listened to auditory stimuli using a self-paced Qualtrics survey. Auditory stimuli were presented via two speakers placed on the left and the right of the laptop that the participants used. All relevant sounds were created using Adobe Audition CC 2015.2 and then equated for absolute intensity, measured by root mean square, using the Match Loudness function.

Participants were presented with 20 auditory sounds but this paper focuses only on three auditory looming sounds, and will only discuss those three sounds in the analyses. In this experiment we manipulated the rate of change in intensity for the looming sounds, see Gray (2011) for more details. The Loom Early sound had the fastest rate of change in intensity, followed by the Loom Veridical, and the Loom Late had the slowest rate of change in intensity.

2.3   Procedure

After giving informed consent, participants began by completing a sound check (selecting the correct word spoken in a test audio file) and answering demographics questions. They then began the first task of sorting each sound into one of three categories:

1. Highly urgent, time sensitive, warning
2. Medium urgency, non-critical, status notification
3. Non-urgent, non-critical, social notification

Next, participants were given the following definition of urgency, "Urgency is how time critical and important the signal is, with 100 being something extremely time critical requiring your immediate attention (e.g., imminent crash warning) and 1 being not at all important or time critical (e.g., notification of a junk email or text).". Participants then rated all sounds (repeated in a randomized order) for perceived urgency and then for annoyance using a sliding scale between 1–100. After completing these tasks participants were finished with the experiment.

3   Results

Average urgency ratings for each of the looming stimuli are reported in Table 1. To test for any differences in perceived urgency ratings across the three looming stimuli, we ran a one-way repeated measures ANOVA. There was no effect of stimuli on urgency ratings, $F(2, 114) = 0.633$, $p = .532$ (see Fig. 1).

Table 1. Average urgency ratings and standard deviations for each looming sound

Stimuli	Mean	Standard deviation
Loom True	37.19	24.66
Loom Early	35.47	22.51
Loom Late	38.21	26.32

4   Discussion

In this study we tested whether auditory looming warnings could function as a dynamic warning that could be scaled to convey differing levels of urgency by changing the rate of change in intensity after being matched for overall intensity. Using self-report measures of perceived urgency, participants listened to the looming sounds and rated them on an urgency scale from 1 to 100. We found no differences between ratings of urgency across the three sounds. This suggests that looming sounds, or sounds that appear to be coming toward you, may not be effective in conveying different levels of urgency without manipulating overall intensity. However, because sound designers often do not have control of the intensity at which a sound will be played when it is used as a warning, manipulations of overall intensity are not as reliable.

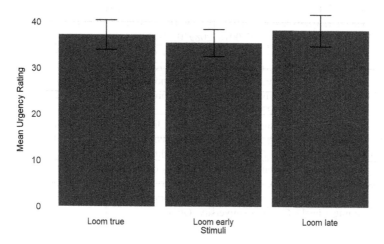

Fig. 1. Average urgency ratings for the three auditory looming sounds. Error bars represent standard error of the mean.

We do note that this investigation only manipulated one aspect of the looming sounds but that there are other manipulations that could be considered. For example, various looming sounds that start and end at the same intensities but differ on the length of the sound might be effective in conveying differing levels of urgency. Additionally, Gray (2011) used response times as his main dependent measure whereas we used urgency ratings. Though these two measures are correlated (Hass and Casali 1995; Baldwin et al. 2012), it may be that variations in the rate of change in intensity of looming warnings affect response times but not urgency ratings. More work is needed to fully understand the relationship between looming stimuli and perceptions of urgency.

In conclusion, it appears that the rate at which a looming sound increases in intensity is not effective in conveying differing levels of urgency if the sounds are matched for overall intensity. We advise that looming warnings matched for overall intensity not be used as forward collision warnings in vehicles until perception of looming stimuli is better understood.

Acknowledgements. The authors would like to thank Bridget Lewis for her help in sound creation. The authors would also like to thank Ian McCandliss and Fernando Barrientos for their data collection efforts.

References

Baldwin CL, Lewis BA (2014) Perceived urgency mapping across modalities within a driving context. Appl Ergon 45(5):1270–1277. https://doi.org/10.1016/j.apergo.2013.05.002
Baldwin CL, Eisert JL, Garcia A, Lewis B, Pratt SM, Gonzalez C (2012) Multimodal urgency coding: auditory, visual, and tactile parameters and their impact on perceived urgency. Work 41(Suppl 1):3586–3591. https://doi.org/10.3233/WOR-2012-0669-3586

Edworthy J, Loxley S, Dennis I (1991) Improving auditory warning design: relationship between warning sound parameters and perceived urgency. Hum Factors J Hum Factors Ergon Soc 33 (2):205–231. https://doi.org/10.1177/001872089103300206

Gray R (2011) Looming auditory collision warnings for driving. Hum Factors J Hum Factors Ergon Soc 53(1):63–74. https://doi.org/10.1177/0018720810397833

Haas EC, Casali JG (1995) Perceived urgency of and response time to multi-tone and frequency-modulated warning signals in broadband noise. Ergonomics 38(11):2313–2326. https://doi.org/10.1080/00140139508925270

Haas EC, Edworthy J (1996) Designing urgency into auditory warnings using pitch, speed and loudness. Comput Control Eng J 7(4):193–198. https://doi.org/10.1049/cce:19960407

Hellier EJ, Edworthy J, Dennis I (1993) Improving auditory warning design: quantifying and predicting the effects of different warning parameters on perceived urgency. Hum Factors J Hum Factors Ergon Soc 35(4):693–706. https://doi.org/10.1177/001872089303500408

Momtahan KL (1990) Mapping of psychoacoustic parameters to the perceived urgency of auditory warning signals. Carleton University, Ottawa

Visual Ergonomics

The Typographic Grid in the Editorial Project: An Essential Resource to the Graphic Consistency and Perception

Elisabete Rolo[(✉)]

CIAUD, Lisbon School of Architecture, Universidade de Lisboa,
Rua Sá Nogueira, Pólo Universitário do Alto da Ajuda,
1349-063 Lisbon, Portugal
erolo@fa.ulisboa.pt

Abstract. A fundamental aspect of ergonomics in communication design is related to the typography behaviour – namely legibility and readability.

In editorial design, one aspect that contributes significantly to the readability is the consistency throughout the publication, which confers coherence and uniformity to the content and generates security and familiarity from the reader to the graphic object. In this context, the use of **typographic grids** becomes a fundamental resource to the graphic project development.

Historically, there have been several manifestations in which Man, using mathematical thinking, conceived geometric systems to organize and/or define his spatial reality, for example the golden section, the Fibonnacci sequence, the Vitruvian canon or the *Modulor*. In editorial design, several studies, such as those of J. A. van de Graaf, Raul Rosarivo or Jan Tschichold, make it evident that the use of grids and systems of proportions dates back to medieval manuscripts.

In Modernism, this kind of systems was redesigned to suit the new vanguard reality, with grid systems reaching their peak during the Swiss Style. Nowadays, these are still an extremely useful resource, which continues to make perfect sense, despite the facilities provided by the computer media. However, their use must be decided in order to serve the content they are intended to disseminate.

The approach of the importance of the grid in editorial design presented in this article is made in a didactic and useful perspective to the teaching of the discipline.

Keywords: Typographic grids · History of the grid · Editorial design
Swiss style

1 Introductory Note

"Just as mathematics began with the measurement of objects and space, design began with the arrangement of objects in harmonious relationship to each other and to the space they occupied" (Hurlburt 1978, p. 9).

© Springer Nature Switzerland AG 2019
S. Bagnara et al. (Eds.): IEA 2018, AISC 827, pp. 61–72, 2019.
https://doi.org/10.1007/978-3-319-96059-3_7

In addition to aspects related to expressivity, the requirements of any editorial object are related, to a greater or lesser degree, by its legibility and readability – two concepts, often confused, but distinct from each other.

As David Jury states, "Legibility is the degree to which individual letters can be distinguished from each other" (Jury 2004, p. 58). Readability is related to the ability of reading. As the author also declares, "good typography encourages the desire to read and reduces the effort required to comprehend. Comprehension is the reason for all reading" (Jury 2004, p. 64). According to Ruari McLean, these are personal concepts, because when we say something is legible, it only means that we can read it under some specific conditions. Other people, under other conditions, might not experience the same (McLean 1980, p. 42).

The topics around legibility have been addressed and studied since the letterpress introduction by Gutenberg in the fifteenth century, therefore there are plenty of typographic rules intended to better achieve that legibility. However, as Unger (2007, p. 149) explains, legibility and readability do not depend only on the forms of typography. They depend considerably on the spaces between letters and words, spaces within letters, spaces between lines and space around the text as a whole.

According to Unger, it is at the macrotypography level (in the scope of editorial design and typography, and according to Hochuli (2009), we can consider that there are mainly two types of elements: macrotypography and microtypography. The microtypography, or typography of the detail, is related to the individual components – letters, space between letters, words, space between words, lines, leading and columns of text) that readers tend to better notice the spaces, since these are more evident than at the microtypography level (macrotypography – typographic layout – is related to the format, size and position of text columns and illustrations, with the organization of the hierarchy of headings, subheadings and captions (Hochuli 2009). "Readers have a tendency to look at the books at the macro level, so that this is where they consciously see space, whereas at the micro level they pay little or no attention to space and simply process it automatically" (Unger 2007, p. 151).

The importance of these aspects in readability and expressivity is the main subject of the article *White space in editorial design* (Rolo 2017), which focuses on white space and its various roles and potentialities in editorial design objects.

Legibility and readability, therefore, depend on how graphic material occupies the space in a publication, and we can consider that one of the aspects that most influences it is the coherence throughout the publication that, despite the unique character of each page, gives consistency and uniformity to the contents, or in other words, structure. And structure is one of the aspects that first influences readers. As Fawcett-Tang (2004: 58) states, "When potential readers first pick a book up, they usually begin by flicking through its pages to get some sense of the way in which it is organized".

This structural coherence generates a relationship of security between the reader and the graphic object, enabling him to know where each element is expected, bringing focus to the content rather than to the form, and also allowing a correct "navigation" throughout the publications. This "security" of the reader "within" the editorial object meets the designer Derek Birdsall's statement, who considers that "one of the paradoxes of book design is that it shouldn't be visible at all (...) It should look as if just happened. (...)" (Derek Birdsall, *apud* Fawcett-Tang 2004, pp. 90–91). This conception of editorial design as an invisible structure is also shared by Beatrice Warde in the text "The Crystal Goblet" (Warde 1995 [1955]) and by Bringhurst (2004, p. 17), who affirms, "typography with anything to say therefore aspires to a kind of statuesque transparency". It is through this "invisibility" that the reader abstracts from form and focuses on content, reaching the understanding of the message – the primary purpose of any editorial design object.

Since space and the disposition of graphic elements in it are so important in editorial design, we study in this brief investigation a fundamental instrument to achieve this organization and configuration: the typographic grid.

2   Definition, Elements and Types of Grids

There are numerous grid definitions. However, in a relatively consensual way, we can define this instrument as the geometric distribution of space into columns of text, space between them and distance from it to the limits (margins). The grid is, then, the structural element that defines the location of the elements on the page and which provides the white spaces also essential to good readability.

Generally, and in a present view, it is considered that the main components of a grid are margins, markers, columns, flowlines, spatial zones, and modules (Graver 2012; Tondreau 2009), as we can observe in Fig. 1.

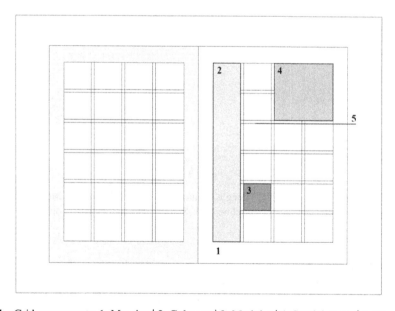

Fig. 1. Grid components: 1. Margins | 2. Columns | 3. Modules | 4. Spatial zones | 5. Flowlines. Source: The author based on: Graver 2012; Tondreau 2009.

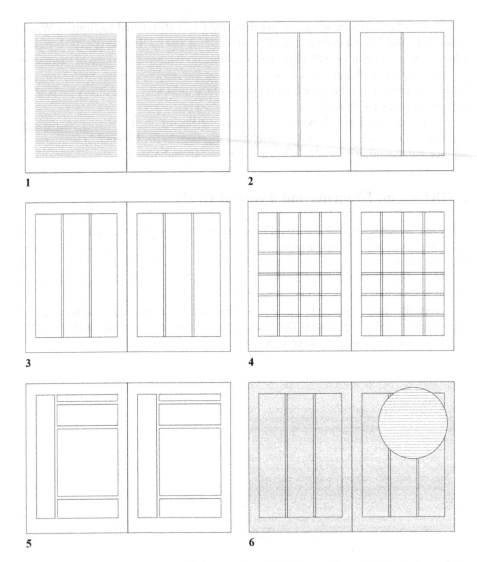

Fig. 2. Types of grid structures: 1. Single column | 2. Two-column | 3. Multicolumn | 4. Modular | 5. Hierarchical grid | 6. Baseline grid. Source: The author based on: Graver 2012; Tondreau 2009.

Several reference books in editorial design consider that there are essentially four types of grid structures: single column, multicolumn, modular and hierarchical. In addition to these, the baseline grid and compound grids are still considered. In Fig. 2 we can observe these variants.

In the single column grid, the main element is the text block. It is usually used for running text in works such as essays, reports or novels, etc.

Multi-column grids allow greater flexibility. They combine several columns of different widths and are very useful in magazines or websites.

Modular grids combine vertical and horizontal columns, forming modules and spatial zones. They are better controlling large amounts of complex information, such as in newspapers, and also for formatting graphical objects such as calendars, tables and graphs.

Hierarchical grids divide the pages into zones (usually horizontal, but not only) and are often used to organize websites into different content areas.

The baseline grid contributes to the alignment of the typographic elements by creating a series of horizontal lines based on the chosen font size and leading. It is especially important in multicolumn grids, to ensure that all rows adjust to the baseline.

Compound grids consist of the integration of two or more combined grid structures into a single system – for example, the combination of a two-column grid with a three-column grid. One of the prime examples of compound grids is the one created by Karl Gerstner in 1962 for *Capital* magazine (Fig. 3). This grid ("the matrix" as Gerstner used to call it) is based on a quadrangular area divided into 58 × 58 units. This quadrangular area can thus be divided into multiples of two, three, four, five and six units, creating areas with one, four, nine, 25 and 36 modules.

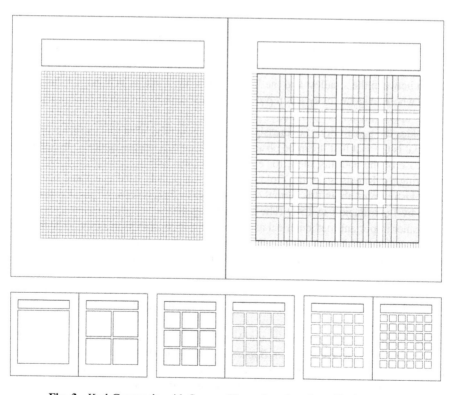

Fig. 3. Karl Gerstner's grid. Source: The author, based on: Haslam 2006.

3 Before the Grid

Historically, there have been various manifestations in which Man, using mathematical thinking, has devised grids to organize and/or to define his spatial reality. The pursuit for structure was notorious from the most remote civilizations, which were concerned with the construction and layout of their buildings and cities, being geometry one of the oldest disciplines in the mankind history. In this sense, the quest for ideal proportions dictated that the Golden Section or the Golden Rectangle and the Fibonacci sequence began to be used since classical antiquity to determine the forms of architecture, painting, and sculpture (Fig. 4).

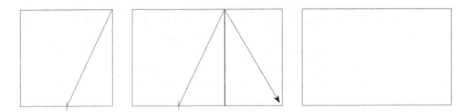

Fig. 4. The golden rectangle. Source: The author, based on: Haslam 2006.

The Golden Section is a real irrational algebraic constant symbolized by the Greek letter "Phi", in honour of Phideas, the sculptor who would have used it to design the Parthenon. It is based on the principle that when a form (or number) is divided, the relation between the smallest and the largest part is the same as between the larger part and the whole. Algebraically it is represented by the expression a: b = b: (a + b) and its value is approximately 1: 1,61803.

Besides the possibility of expressing it mathematically, what makes this proportion so special is the fact that it can be observed in human proportions and also in nature (as in the case of the nautilus shell, for example). (Elam 2001).

During the Renaissance, the Golden Section was widely used in art and one of the earliest testimonies about this formal order was set in the book *De Divine Proportione*, written by Luca Pacioli in 1509 (Hurlburt 1978, p.10).

Another important mathematical aspect in editorial design proportions relates to the dynamic or irrational rectangles – those that can only be subdivided into rectangles (and not squares because those are rational). The golden rectangle is irrational, such as the root-two rectangle – which forms the basis of the ISO series formats.

Also the square – the simplest of the rectangles – assumes a special importance in the way that led us to the modern grid, due to its stability and its relationship with the golden rectangle.

The square is also present in the Vitruvius Canon presented in the book *De Architectura Libri Decem* (*The Ten Books on Architecture*) and written by the architect Marco Vitruvio Polião around the first century BC. In this work, Vitruvius defends that architecture should be based on the proportions of the body human being. The height of a well-proportioned man is equal to the width of his opened arms. This height and width thus create a perfect square that encompasses the body, and the hands and feet touch a circle centred by the navel (Elam 2001, pp. 12–13).

The Vitruvius theories formed the basis to some Renaissance works. The treatise *De re aedificatoria* (*On the Art of Building*), written by Leon Battista Alberti (1404–1472), between 1443 and 1452, was the first theoretical work on the subject. Starting from the Vitruvius theories, he criticizes some of its aspects and postulate specific rules for building construction.

The Vitruvius canon was also used and represented in the fifteenth and sixteenth centuries by Albrecht Dürer and Leonardo da Vinci in "Human Figure in a Circle, Illustrating Proportions" (1485–1490)– one of his best known drawings and symbol of the Renaissance ideals (Fig. 5).

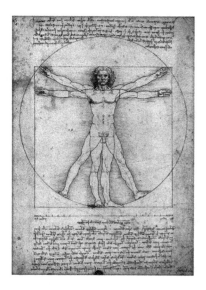

Fig. 5. Vitruvius Canon. Source: Zöllner 2004, p. 36.

Two thousand years after the Vitruvian Canon (in 1948), Le Corbusier, also an architect, presented the *Modulor* (Fig. 6) – a system of architectural proportions based on the golden section and human proportions. This system is based on three fundamental anatomical points: the top of the head, the solar plexus and the tip of the hand with the raised arm.

Initially designed only for architecture, the *Modulor* was later adapted to other areas, including the design of printed pages. However, according to Hurlburt (1978, p.17) his contribution to graphic design was unimpressive, with his greatest contribution being, perhaps, the inspiration it gave to the German and Swiss designers who later would create the modern modular and grid systems.

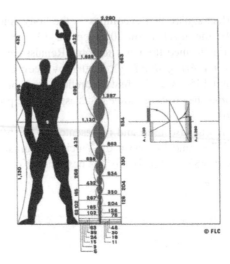

Fig. 6. *Modulor.* Source: Fondation Le Corbusier. Le Modulor (2018)

4 Historical Evolution of the Grid

"A page, like a building or a room, can be of any size and proportion, but some are distinctly more pleasing than others, and some have quite specific connotations" (Bringhurst 2004, p. 143).

The way to establish book formats and the positioning of text in a harmonious way has always been a concern, and there are standards of conformity in the medieval manuscripts that prove the existence of defined rules for doing so (Tschichold 1991 [1975], p. 60). However, those rules were workshop secrets and only through the analysis of a great number of books, it was possible to identify them.

Throughout history, several studies tried to decipher these secrets and, therefore, understand and describe the rules of proportion to create balanced pages. From those studies, we highlight J. A. van de Graaf's, Villard de Honnecourt's (1200–1250), Raúl M. Rosarivo's (1903–1966) and Jan Tschichold's (1902–1972), which arrived to very similar conclusions. In his work *The Form of the Book*, Tschichold introduced a chapter on "Consistent Correlation Between Book and Type Area", in which he describes his study and the way how it relates to those of its predecessors.

In Figs. 7, 8 and 9, we can observe the canons of Villard de Honnecourt, Raul Rosarivo and Jan Tschichold.

These systems and the studies about them have been important mainly to better understand the old books and the mathematical thought behind their proportions.

In a slight provocative perspective, Anthony Froshaug defends that the grid can be considered inherent to graphic design since the appearance of typography, once "Typography is a Grid" – as the typographic system itself is constituted of modular elements (characters) that interact with each other and are organized according to a structuring geometry. According to the author: "To mention both typographic, and, in

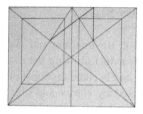

Fig. 7. Villard's Diagram. This is a canon of harmonious division named after its inventor, Villard de Honnecourt, an architect who lived and worked during the first half of the thirteenth century in the Picardy region of Northern France. His manuscript Bauhuttenbuch (workshop record book) is held at the National Library in Paris. Using Villard's canon, shown in bold, it is possible to divide a straight line into any number of equal parts without need of a measuring stick. Source: Tschichold 1991 [1975], p. 64.

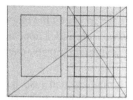

Fig. 8. Division of height and width of a page into nine parts, following **Rosarivo's** construction. This involves a 2:3 page proportion. This proved to be the canon used by Gutenberg and Peter Schöffer. Source: Tschichold 1991 [1975], p. 63.

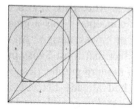

Fig. 9. The secret canon, upon which many late medieval manuscripts and incunabula are based. Determined by Jan Tschichold, 1953 – Page proportion 2:3. Text area and page show the same proportions. Height of text area equals page width. Margin proportions 2:3:4:6. Source: Tschichold 1991 [1975], p. 61.

the same breath/sentence, grids, is strictly tautologous. The word typography means to write/print using standard elements; to use standard elements implies some modular relationship between such elements; since such relationship is two-dimensional, it implies the determination of dimensions which are both horizontal and vertical" (Froshaug 1999, p. 177).

However, the grid as we know it nowadays was mainly a result of the early 20th century artistic and design thought, and of the particularities that led to the Bauhaus, the New Typography and, above all, the Swiss Style.

5   The Grid and the Swiss Style

The Swiss Style, Swiss School or International Typographic Style emerged in Switzerland in the post-World War II, following the avant-garde art movements of the Bauhaus and New Typography, and became the most outstanding design school of the 20th century. Mainly characterized by clarity and objectivity, it was of great importance during the 50 s and 60 s, throughout the entire world, and its influence continues to the present days. Its main features are the unity in the graphic elements layout, the clear and objective use of photography, the (visual and verbal) information disposition in a clear and factual way and the use of sans serif typography, with unjustified texts. All of this was achieved through the use of mathematically assembled grids, since the style proponents believed that this was the most readable and harmonious way to structure information in a progressive spirit era.

According to Hollis (2006, p. 177), the Swiss Style grid emerged as a result of various factors and influences. One of these influences was architecture. A facade of a modern building may resemble a page in which the elements are displayed according to an underlying structure, and in which the outer appearance is the result of that same structure. The fact that there were several architects practising graphic design reinforces this relationship even more.

Following this idea, modern urbanism can also have contact points to the typographic grid, as we can see, for example, in the Cerdà plan, for the city of Barcelona extension, in which a predetermined orthogonal organization dictates the layout of buildings and spaces between them.

Another factor influencing the Swiss grid is painting, especially that of Mondrian, which takes visual advantage of the idea of dividing space into grid units.

Also the work of the designers Max Bill, Lohse and Karl Gerstner – based on mathematical and programmatic concepts – was a great influence in establishing the grid as a fundamental element of Swiss design. Another aspect that favoured the introduction of the grid was the fact that most Swiss publications were trilingual.

As a practical working tool of modern graphic design, the grid was introduced quite gradually, but by the 1960s it had already become a standard procedure. Müller-Brockmann in his 1961 book – *Gestaltungprobleme des Grafikers* (*The graphic artist and his design problems*) – in which he presents his own working methods, presents the grid to a wide and international audience, according to the scope of the book. The grid arrived to determine the look and methods of graphic work of this style.

To fill a gap in practical information on this subject, the same author publishes in 1961 the book *Das Rastersystem ais Hilfsmittel bei der Gestaltung von Inseraten, Katalogen, Aussterlungen usw* (*The Grid System as an aid in the design of Advertisements, Catalogs, Exhibitions, etc.*), which, in his words, makes "an introduction to the spirit and application of the system, [collecting] 28 examples of grids, taken from his practical experience" (Müller-Brockman 1982, p.8).

According to Müller-Brockmann, this new tendency in graphic design was characterized by the arrangement of text and illustrations according to strict principles, trying to present the subjects in an objective way and trying to achieve uniformity in the composition of all the pages. In his work *Grid Systems*, Müller-Brockmann states that "Working with the grid system means submitting to laws of universal validity. The use of the grid system implies the will to systematize, to clarify/the will to penetrate to the essentials, to concentrate/the will to cultivate objectivity instead of subjectivity/the will to rationalize the creative and technical production processes/the will to integrate elements of colour, form and material/the will to achieve architectural dominion over surface and space/the will to adopt a positive, forward-looking attitude/the recognition of the importance of education and the effect of work devised in a constructive and creative spirit "(Müller-Brockman 1982, p.10).

However, according to him, the grid alone did not guarantee any success. "The grid system places in the hands of the designer no more no less than a serviceable instrument which makes it possible to create interesting, contrasting and dynamic arrangements of pictures and text, but which is in itself no guarantee of success. (...) Even in simpler solutions, the designer needs a good sense of composition and sensitivity to capture the rhythmic sequence of images and text" (Müller-Brockman 1982, p.75).

The dissemination of the Swiss Style through emigration and books translated in several languages, as well as its scientific, objective and functional nature allowed it to have a global reach. But, we can consider that more important than his visual legacy, was the attitude towards the profession that he introduced. Designers started to be seen not as "artists" but as information disseminators amidst various components of society. A more scientific attitude was adopted in solving problems, and the expressionism and eccentric solutions were rejected in a quest for clarity and order.

6   The Grid in the Present (or a Conclusion)

After its apogee with the Swiss Style, the use of the grid stopped being consensual. As Steven Heller puts it, "Yet once introduced as the panacea for graphic design clarity, rather than a simple organizational and compositional tool, the grid became a target of both love and hate. It was loved for bringing order to disorder and hated for purportedly locking designers into rigid confines (Heller 2012, p. 146).

In the 1970s, the rigidity that characterized the entire Swiss-style system was called into question, and its decline began to glimpse, giving way to the postmodern design, which refuted the grid and focused more on the graphic objects expressivity than on the organizational clarity. In this context we can highlight works from Wolfgang Weingart, Dan Friedman and April Greiman. This postmodern tendency was taken to extreme with the introduction of the computer in the graphic process, around 1985, bringing to light names like Catherine McCoy, Neville Brody and David Carson, among others (Meggs 2006).

Nowadays, the duality between using and not using a grid still remains. This instrument continues to be an extremely useful resource and its use, although due to the facilities provided by the computer is not so essential, continues to make perfect sense. In editorial design, the grid becomes even more necessary in periodical publications, because when designing such an editorial project, we are creating not only an object, but a system designed to serve variable, complex and diverse information, destined to be used by various designers. In other kinds of projects, expressivity assumes preponderance, and not following a grid may be the path that best serves the project. Like any other instrument in the graphic design process, the grid is not an absolute. It should be used flexibly, and when necessary, should be modified or even abandoned, in order to get a better solution. Functionality is the main feature of a grid system. And above all, a graphic work must honour its content and the message to communicate, regardless of how the information is organized.

References

Bringhurst R (2004) [1992] The elements of typographic style. Hartley & Marks Publishers, Point Roberts

Elam K (2001) Geometry of design: studies in proportion and composition. Princeton Architectural Press, New York

Fawcett-Tang R (2004) New book design. Laurence King Publishing, London

Froshaug A (1999) Typography is a grid. In: Looking closer 3: classic writings on graphic design. Allworth Press, New York

Graver A, Jura B (2012) Best practices for graphic designers, grids and page layouts: an essential guideline for understanding and applying page design principles (Best practices/Graphic designer). Rockport Publishers, Beverly

Heller S, Vienne V (2012) 100 ideas that changed graphic design. Laurence King Publishing Ltd., London

Hochuli J (2009) [2005] Detail in typography. Hyphen Press, London

Hollis R (2006) Swiss graphic design: the origins and growth of an international style 1920–1965. Laurence King Publishing, London

Hurlburt A (1978) The grid: a modular system for the design and production of newspapers, magazines, and books. Wiley, New York

Jury D (2004) About face: reviving the rules of typography. Rotovision, London

Fondation Le Corbusier. Le Modulor (1945). http://www.fondationlecorbusier.fr/corbuweb/morpheus.aspx?sysId=13&IrisObjectId=7837&sysLanguage=en-en&itemPos=11&itemSort=en-en_sort_string1&itemCount=35&sysParentName=Home&sysParentId=11. Accessed 29 May 2018

Mclean R (1980) The Thames and Hudson manual of typography. Thames and Hudson, London

Meggs P, Purvis A (2006) Meggs' history of graphic design. Wiley, Hobooken

Müller-Brockmann J (1982) [1961] Sistemas de grelhas. Editorial Gustavo Gili, Barcelona

Rolo E (2017) White space in editorial design. In: Advances in ergonomics in design, AHFE 2017, Advances in intelligent systems and computing. Springer, Cham

Samara T (2007) Making and breaking the grid. Rockport Publishers Inc, Gloucester

Tondreau B (2009) Layout essentials: 100 design principles for using grids. Rockport Publishers, Beverly

Tschichold J (1991) [1975] The form of the book: essays on the morality of good design. Lund Humphries, London

Unger G (2007) While you are reading. Mark Batty Publisher, New York

Warde B (1995) The crystal goblet or printing should be invisible. In: Typographers on type. W. W. Norton & Company, New York/London

Zöllner F (2004) Leonardo da Vinci 1492–1519. Taschen/Público, Köln

The Analysis Method of Visual Information Searching in the Human-Computer Interactive Process of Intelligent Control System

Xiaoli Wu[1,2]($\boxtimes$), Tom Gedeon[2], and Linlin Wang[1]

[1] College of Mechanical and Electrical Engineering, Hohai University,
Changzhou 213022, China
wuxlhhu@163.com
[2] Research School of Computer Science, The Australian National University,
Canberra ACT 2601, Australia

Abstract. As the back-end of system operation, visual information presentation of intelligent control system has become an important means and operation basis for people (operators) to obtain information, make inference and evaluate decision. The rational design of intelligent control system plays an important role in maintaining the safety and stability of production and operation in large enterprises. This paper presents the analysis method of visual information searching in intelligent control system. On the premise of visualization structure analysis of complex network information, taking reliability analysis of operator information searching as key point and main line, through studying characterization of information graphic elements relationship, information multidimensional attributes, information layouts and coding rules in information interactive interface, this paper establishes visual cognitive model of information structure which made information interaction more efficiently.

Keywords: Intelligent control system · Human computer interaction
Visual information searching · Visual cognition process

1 Introduce

With the rapid development of computer technology and information control theory, the control system becomes more complex and intelligent. In particular, major system areas such as large-scale production real-time scheduling, real-time monitoring of the entire process of manufacture, transportation hub monitoring, nuclear power control, environmental monitoring, aviation control which rely on computer technology totally are operated, supervised and decided by digital and intelligent industrial control system, as shown in Fig. 1. Especially in the background of large data, intelligent control system such as process monitoring of production line, intelligent traffic, police monitoring, satellite GPS, terrestrial geographic information system and other complex information display are characterized by a large quantity of information carrying capacity, complex information structure and task execution entering complex cognition. When the operator executes task such as producing or dispatching, the complicated human computer interactive process increases the cognitive load of the operator.

© Springer Nature Switzerland AG 2019
S. Bagnara et al. (Eds.): IEA 2018, AISC 827, pp. 73–84, 2019.
https://doi.org/10.1007/978-3-319-96059-3_8

Meanwhile, the execution task is difficult and the execution environment is complicated in the complex information system, which also brings the system unpredictability. The system is at high risk, so minor errors can lead to failure of the mission and cause major accidents easily. As the back-end of system operation, visual information presentation of intelligent control system has become an important means and operation basis for people (operators) to obtain information, make inference and evaluate decision.

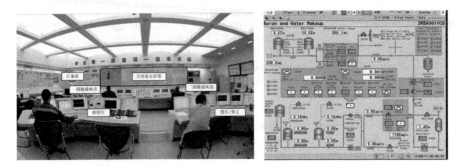

Fig. 1. Human computer interactive process of intelligent control system (Left) and displays of complex information (Right)

2   Background

2.1   Visual Information Interactive Research of Intelligent Control System

In recent years of visual information interactive research in intelligent control system, Burns [1] and Carvalho [2] established evaluation model and design method for the human-computer interactive interface controlled by nuclear power plant. Yim et al. [3] found the way to present interface design of complex information in limited screen space by information hierarchy visualization. Rydström [4] and Lee [5] studied in depth on the contrast method of digital interface, visualization representation method, experimental model and simulation environment of interface elements respectively for car navigation. Paul [6] studied the overload of information complexity in complex digital interface and established the pattern decision-making model from the perspective of time pressure to analyze the execution time required by different amounts of information. Zhang [7–11], Dai [10] and Li et al. [11] conducted a series study of analysis methods of reliability in digital industrial system, human cognitive behavior in large-scale digital control system and reliability of situational awareness of digital control room operator in nuclear power plant. That helped them put forward influence of human-computer interactive complexity, cognitive control mode of operator and identification method of human error. Besides, they studied evaluation method of digital human-computer interface by using mental load evaluation and eye-tracking technology. Li et al. [12] studied the human-computer interaction of the digital industrial system from the perspective of human-computer interaction, complexity and error and the diagnosis of task performance. Xue et al. [13–15] proposed scientific theory and design method for human-computer interactive interface design from the perspectives of cognitive load (CL) and situational awareness (SA) equilibrium, visual cognition brain mechanism and large data information visualization; Guo [16] studied the support technology of human-computer interactive interface design from the perspective of emotional needs, which provides a scientific method for interface design. Wu [17–22] put forward the corresponding relationship between error and cognition based on error-cognition correspondence model of human-computer interactive interface in complex system and carried out a series of experiments on visual searching. The above studies on digital control system and human-computer interactive interface provided scientific theoretical basis and technical means, but it ignored the study on association effect in cognitive behaviors, information characteristics and design factors.

2.2 Complex Information Structure and Information Presentation Research

In the research field of information structure in large data environment, there are some research achievements at home and abroad from aspects of information visualization, information complexity, information structure, information network analysis and information presentation. Reda et al. [23] studied how to realize extensible data visualization of the heterogeneous data set in a complex real environment, allowing users to collocate 2D and 3D data sets simultaneously and create information space from 2D to 3D. Cheshire et al [24]. illustrated the significance of large data and showed how to describe the data flow in different time scales based on London public transport system. Basole et al. [25] discussed the way to visualize large data with multiple associations through a variety of view methods based on node. Mueller et al [26] studied the visualization analysis of the data which using visual similarity matrix through graphic sorting algorithm. Kaushik et al. [27] studied the complexity of space from the perspective of information theory, and queried optimization through corresponding table. Wang et al. [28] showed the degree of regularity of images through the complexity of image space and located the calculation area of space complexity so as to extract information from complicated situation. Wang et al. [29] and Liu [30] explored network structure and node center of intelligent control system interface with the method of complex network. They also analyzed the distribution of network, the average path length and clustering coefficient. Kim et al. [31] put forward that it is better to simulate a system of connecting buyers and suppliers in a mesh way rather than linearity. They also established theory structure of social network analysis and key indicators of network structure construction. Ahn et al. [32] proposed a combination between interactive visualization and personalized search. Qi [33] designed the information management, information presentation and information dissemination in digital cultural heritage by adopting information visualization model and design principle. Anuar et al. [34] used the visual analysis tool kit to break through the traditional analysis methods of eye-movement data and proposed analysis method of space measurement of eye-movement data innovatively to summarize the searching strategy when users search for information in the map. Molnár et al. [35] compared the advantages and efficiencies of different scanning strategies by analyzing the horizontal and vertical eye-movement trajectories. Schreudera et al. [36] used eye-movement technique to test the degree of drivers' drowsiness during driving so as to improving the alarm system. Aloisea et al. [37] analyzed how to control the visual focus of the interface and how to adjust the interface element to cause the user's visual attention through eye-movement. Wang et al. [38] evaluated and analyzed the design plan experiment of interface layout by using eye-movement tracker. They chose reasonable optimized plan of space layout based on the evaluation of eye-movement data indicators. In the field of information structure and information presentation, scholars have accumulated research achievements on information visualization technology and information complexity analysis. In particular, mature complex network analysis method and information visualization model play important roles in establishing information interactive design method.

3 Analysis of the Human Computer Interactive Process of Visual Information

3.1 Human-Computer Interactive Process of Intelligent Control System

The complicated information in the system is analyzed and processed by the computer and finally presented to the operator to observe, analyze, judge the situation and make decision as visual information sources and bases. This information will be feedback from sensory channels through attention, perception, memory, and decision. Different from ordinary information system, it carries a huge amount of information. Its information structure is complex and it is presented to user (operator) in a dynamic and changeable way. The visual information interface of the intelligent control system includes multi-module information structure and multi-variate information unit. The interface expresses the complicated and changeable information in the form of character, text, image, icon, color and dimension through the navigation design and the structure of the information hierarchy; the operation of the system will show the dynamic of the interface and replace multi-hierarchy information content constantly; In the process of human-computer interactive monitoring mission, the operator needs to deal with large scale information and real-time monitoring at the same time.

All aspects of the control system information are gathered here. The internal operation of the system is presented in the form of information. So the system can interact with the outside world intuitively. Specifically, the information interactive interface takes responsibility for interactive tasks of whole system. It must present slight changes of the system. It is the only channel for users to communicate with the system, as shown in Fig. 2.

Therefore, as the back-end of system operation, visual information presentation of intelligent control system has become an important means and operation basis for human (operator) to obtain information, make inference and evaluate decision. In the large data environment, rational analysis and evaluation of visual information plays an important role in playing performance and precise implementation, also in maintaining the safety and stability of intelligent control system.

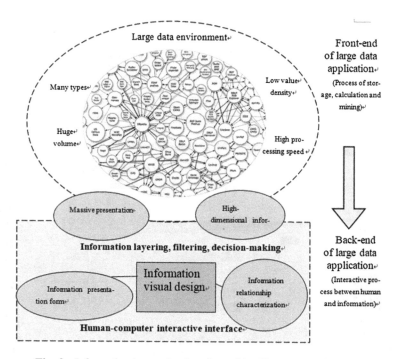

Fig. 2. Information interactive interface of intelligent control system

3.2   Cognitive Behavior Process of Visual Information

A system that hosts large data operations should use information for interaction and decision-making efficiently, which requires the information interactive interface to show "information exchange and communication, information circulating, updating and sharing among each other." For example, Trina Solar Technology Company established MES manufacture information management system for real-time monitoring of the entire process of production line, which is a typical intelligent control system. Figure 3 shows photovoltaic modules production line. The monitoring system involves a full set of technological process such as cell separation, welding, laminating, frame-making, curing, testing, and packaging. The monitoring system needs to record process parameters from the source of production process. The production presentation interface of each process can monitor the information in real time to response and adjust to abnormal condition.

The visual information presentation of the system can be visualized as monitoring, searching, response and execution based on information from the operator's records on visualized production process parameter, real-time process production status information and emergency decision, as shown in Fig. 4. The complicated information in the system is analyzed and processed by the computer and finally presented to the operator to observe, analyze, judge the situation and make decision as visual information sources and bases. This information will be feedback from sensory channels through attention, perception, memory, and decision. Therefore, it is necessary to classify,

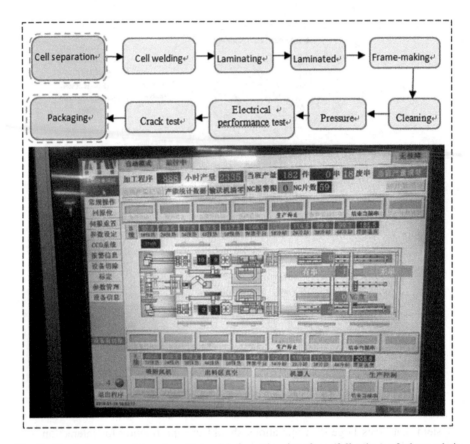

Fig. 3. Control process (above) and visual information interface (following) of photovoltaic modules production line

cluster and explore the reliability analysis method of visual information searching in intelligent control system. That should start with the cognitive behavior of visual information searching in the human-computer interactive process. Then we can establish a physiological evaluation model of visual information which is the premise of the interface design of information interaction.

4 The Analysis Method of Visual Information Searching

This paper presents the analysis method of visual information searching in intelligent control system. On the premise of visualization structure analysis of complex network information, starting with the cognitive behavior of visual information searching in the human computer interactive process, taking reliability analysis of operator information searching as key point and main line, classifying and clustering huge amount of information, this paper studies characterization of information graphic elements relationship, information multi-dimensional attributes, information layouts and coding

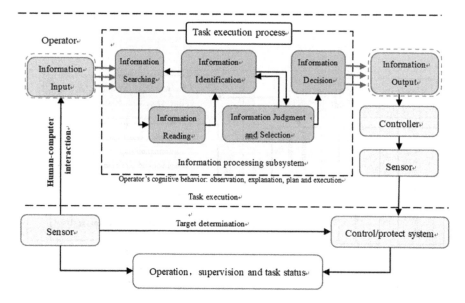

Fig. 4. Visual information cognitive behavior process of control system for real-time monitoring

rules in information interactive interface and establishes information structure model of visual cognition. Thus, operator can interact with control system efficiently.

4.1 Establishment of Graphic Elements Relationship of Visual Information

Liking a network, characterization of graphic elements relationship in information interface is nonlinear and complex. As optimized strategies of information visualization, solving network relationship characterization of information elements and establishing relationship efficiently among information nodes are helpful for reasonable information layouts and coding. When studying typical intelligent control system, we should analyze the structure of the massive and dynamic huge amount of information through visual cognition such as information searching, information recognition, information identification, information judgment and decision-making. Information visualization is a process to visualize the abstract information network. A huge amount of high-dimensional data information interface is characterized by irregularity, fuzziness and external "space complexity". When analyzing the complexity among graphic elements, we extract design factor of interface element, consider the problem of perfect matching, point set coverage and graph is not isomorphic and apply the theory of space complexity analysis, especially NP complete theory of complexity. The graphic element coding and layout in information interface should be designed according to the complexity of the graphic element which is analyzed by Pajek complex nonlinear network and the visual behavior relationship. That is benefit to establishing information adjacency matrix, breaking down the large data into several small internets and establishing characterization of visual graphic elements relationship.

The steps to establish an information space network are shown below (as shown in Fig. 5.):

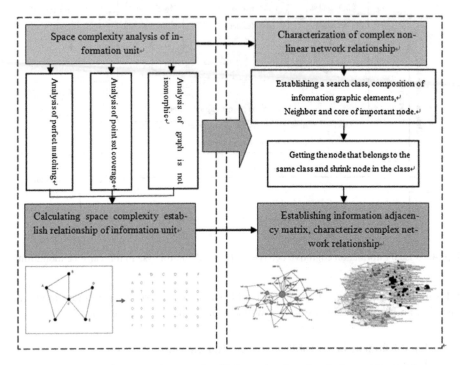

Fig. 5. Space complex network research of information interface in large data environment

The main research content is (as shown in Fig. 2.):

① Space complexity analysis of information unit in large data
② Establishing information node in the information graphic elements network
③ Characterization of complex non-linear network relationship based on information node

4.2 Establishment of Visual Information Structure of Human-Computer Interactive Process

It is an effective way to control cognitive load and ensure information perception advantage by adjusting information network structure and increasing information flow channel among information cluster. Through the method of increasing information dimension such as motion dimension, time dimension, image dimension, index dimension and constructing interactive system such as juxtaposition, covering, nesting, contrast and transition among information clusters vividly, this paper assembles static association information of complex network relationship into many dynamic organic wholes so as to regulate horizontal and vertical depth of information network. It needs us to study decompression form of high-dimensional information to express the high-dimensional data in low-dimensional space in different information clusters. In the visual interactive interface, the user's vision is just a small portion of the whole information presented while the user expects to "decompress" a lot of information from the small amount of data. Decompression form of high-dimensional information needs the study of implicit central manifold in high-dimensional or large data. Feature of high-dimensional information is characterized by low-dimensional variable. This paper mainly studies low-dimensional nonlinear approximation model of high-dimensional linear problem and changes high-dimensional linear problem into low-dimensional nonlinear problem. That is helpful for new research of Ultra-high dimensional data to reduce the dimension.

Information unit and characterization of graphic elements relationship based on multi-hierarchy structure need the visual structure model of information multi-dimensional attributes. Information is divided into three forms based on complex network theory, at the same time, the attributes of information are summarized as entity attributes, association attributes and time attributes. According to the attributes division

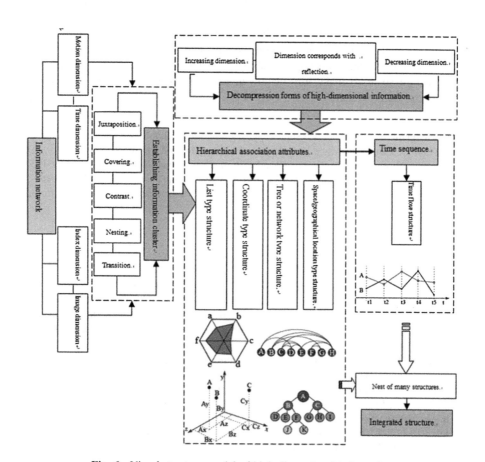

Fig. 6. Visual structure model of high-dimensional information

of information, this paper puts forward the mapping relationship between entity attributes and visual graphic, the mapping relationship between association attributes, time attributes and visual structure. We establish a visual structure mapping model through analysis of various association attributes, as shown in Fig. 6.

5 Conclusion

(1) This paper analyzes human-computer interactive process of intelligent control system and puts forward that visual information presentation of intelligent control system has become an important means and operation basis for people (operators) to obtain information, make inference and evaluate decision;

(2) This paper presents the analysis method of visual information searching in intelligent control system. On the premise of visualization structure analysis of complex network information, starting with the cognitive behavior of visual information searching in the human computer interactive process, taking reliability analysis of operator information searching as key point and main line, classifying and clustering a huge amount of information, through studying characterization of information graphic elements relationship, information multidimensional attributes, information layouts and coding rules in information interactive interface, this paper establishes visual cognitive model of information structure so as to make information interaction between operators and control system more efficiently.

Acknowledgement. This work was supported by Science and technology projects of Changzhou (CE20175032), Jiangsu Province Key Project of philosophy and the social sciences(2017ZDI XM023), the National Nature Science Foundation of China (Grant No. 71601068, 61603123), Overseas research project of Jiangsu Province (2017), Outstanding Young Scholars Joint Funds of Chinese Ergonomics Society- King Far (No. 2017-05), and Fundamental Research Funds for the Central Universities (Grant No. 2015B22714).

References

1. Burns CM, Skraaning G Jr, Jamieson GA et al (2008) Evaluation of ecological interface design for nuclear process control: situation awareness effects. Hum Factors 50(4):663–679
2. Carvalho PVR, dos Santos IL, Gomes JO et al (2008) Human factors approach for evaluation and redesign of human–system interfaces of a nuclear power plant simulator. Displays 29:273–284
3. Yim HB, Lee SM, Seong PH (2014) A development of a quantitative situation awareness measurement tool: Computational Representation of Situation Awareness with Graphical Expressions (CoRSAGE). Ann Nucl Energy 65:144–157
4. Rydström A, Broström R, Bengtsson P (2012) A comparison of two contemporary types of in-car multifunctional interfaces. Appl Ergon 43(3):507–514
5. Lee D-S (2009) The effect of visualizing the flow of multimedia content among and inside devices. Appl Ergon 40(3):440–447

6. Paul S, Nazareth D (2010) Input information complexity, perceived time pressure, and information processing in GSS-based work groups: an experimental investigation using a decision schema to alleviate information overload conditions. Decis Support Syst 49(1):31–40

7. Zhang L, Yang D, Wang Y (2010) The effect of information display on human reliability in a digital control room. China Saf Sci J 20(9):81–85

8. Zhou Y, Zhang L (2011) Analysis of nuclear power plant operators' cognitive control mode and error under stressful conditions. Mind Comput 5(1):1–14

9. Zou P, Zhang L, Jiang J (2013) Effects of complexity of human machine interaction on human error in digital control system. J Univ South China (Soc Sci Edn) 14(5):78–81

10. Dai L, Zhang L, Li P (2011) HRA in China: model and data. Saf Sci 49(3):468–472

11. Li P, Zhang L, Dai L, Hu H (2014) Human error identification of operator in digital main control room of NPPs based on simulator experiment. At Energy Sci Technol 48(11):2085–2093

12. Li Z (2011) Fault tree analysis of train crash accident and discussion on safety of complex systems. Ind Eng Manage 16(4):1–8 55

13. Niu Y, Xue C et al (2014) Icon memory research under different time pressures and icon quantities based on event-related potential. J. Southeast Univ (Engl Edn) 30(1):45–50

14. Zhou L, Xue C, et al (2013) Research of interface composition design optimization based on visual balance. In: The 2013 international conference on intelligent systems and knowledge engineering (ISKE2013), Shenzhen, pp 20–23

15. Li J, Xue C, Tang W, Wu X (2014) Color saliency research on visual perceptual layering method. In: Engineering psychology and cognitive ergonomics - 11th international conference, EPCE 2014, Held as part of HCI international proceedings. LNAI, Crete, vol 8510, pp 86–97

16. Guo F, Li M, Qu Q (2013) Design optimization of electronic commerce web page based on kansei engineering. Chin J Ergon 19(3):56–60

17. Wu X, Xue C, Wang H, Wu W, Niu Y (2014) E-C mapping model based on human computer interaction interface of complex system. Chin J. Mech Eng 50(12):206–212

18. Wu X, Xue C, Tang W (2014) Study on eye movement of information omission misjudgment in radar situation-interface. In: Engineering psychology and cognitive ergonomics - 11th international conference, EPCE 2014, Held as part of HCI international proceedings. LNAI, Crete, vol 8532, pp 407–418

19. Wu X (2015) Study on Error–Cognition Mechanism of Task Interface in Complex Information System. Doctoral thesis of Southeast University, Nanjing

20. Wu X (2017) Study on error-cognition mechanism of task interface in complex information system. Science Press, Beijing

21. Wu X, Chen Y, Li J (2017) Study on error-cognition mechanism of task interface in complex information system. In: Advances in safety management and human factors - proceedings of the AHFE 2017 international conference on safety management and human factors, Los Angeles, California, USA, 17–21 July, pp 497–506

22. Wu X, Xi T, Chen Y (2016) Study on design principle of touch screen with an example of Chinese-Pinyin 10 key input method in iPhone. In: Advances in ergonomics in design-proceedings of the AHFE 2016 international conference on ergonomics in design, Part VII, Walt Disney World®, Florida, USA, 27–31 July, pp 639–650

23. Reda K, Febretti A, Knoll A (2013) Visualizing large, heterogeneous data in hybrid-reality environments. IEEE Comput Graph Appl 33(4):38–48

24. Cheshire J, Batty M (2012) Visualization tools for understanding big data. Environ Plan B-Plan Des 39(3):413–415

25. Basole RC, Clear T, Hu M (2013) Understanding interfirm relationships in business ecosystems with interactive visualization. IEEE Trans Vis Comput Graph 19(12):2526–2535
26. Mueller C, Martin B, Lumsdaine A (2007) A comparison of vertex ordering algorithms for large graph visualization. In: Asia/Pacific symposium on visualization, Sydney, Australia, pp 141–148
27. Kaushik R, Naughton JF, Ramakrishnan R (2005) Synopses for query optimization: a space complexity perspective. ACM Trans Database Syst 30(4):1102–1127
28. Wang L, Yang F, Chang Y, Wang R (2008) An edge-detection algorithm based on spatial activity masking. J Image Graph 1:100–103
29. Wang J, Mo H, Wang F (2011) Exploring the network structure and nodal centrality of China's air transport network: a complex network approach. J Transp Geogr 19(4):712–721
30. Liu JL (2013) Research on synchronization of complex networks with random nodes. Acta Phys Sin 4:54–62
31. Kim Y, Choi TY, Yan T et al (2011) Structural investigation of supply networks: a social network analysis approach. J Oper Manage 29(3):194–211
32. Ahn J-W, Brusilovsky P (2013) Adaptive visualization for exploratory information retrieval. Inf Process Manage 49:1139–1164
33. Qi J (2006) Research on the design methods of digital cultural heritage based on information visualization. Tsinghua University (2006)
34. Anuar N, Kim J (2014) A direct methodology to establish design requirements for human–system interface (HSI) of automatic systems in nuclear power plants. Ann Nucl Energy 63:326–338
35. Molnár M, Tóth B, Boha R, Gaál ZA (2013) Aging effects on ERP correlates of emotional word discrimination. Clin Neurophysiol 124:1986–1994
36. Schreudera M, Ricciob A, Risetti M (2013) User-centered design in brain–computer interfaces—a case study. Artif Intell Med 59:71–80
37. Aloisea F, Aricò P, Schettini F (2013) Asynchronous gaze-independent event-related potential-based brain–computer interface. Artif Intell Med 59:61–69
38. Wang H, Bian T, Xue C (2011) Experiment evaluation of fighter's interface layout based on eye tracking. Electro-Mech Eng 27(6):50–53

The Ergonomics Experiment Research on Display Interface of Railway Transportation Dispatching Room

Zhongqi Liu[1,2], Xuemei Chen[1,2], Qianxiang Zhou[1,2(✉)],
Yuhong Chen[3], and Chenming Li[3]

[1] Key Laboratory for Biomechanics and Mechanobiology of the Ministry
of Education, School of Biological Science and Medical Engineering,
Beihang University, Beijing 100191, China
zqxg@buaa.edu.cn
[2] Beijing Advanced Innovation Centre for Biomedical Engineering,
Beihang University, Beijing 102402, China
[3] The Quartermaster Research Institute of Engineering and Technology,
Beijing 100010, China

Abstract. There are still some problems of the railway dispatching interface, such as function icons, the station chart and so on. To find the defects of the current dispatching interface and provide support for the improvement of interface, the eye movement experiment to the dispatching interface was conducted. Six railway dispatchers who have worked more than five years in this position participated in the experiment. The railway dispatch information was simulated with two displays and two patterns. One pattern was that the dispatching information was presented on two 32-inch displays placed side by side. This pattern was called the large display combination. Another pattern was that all information was presented on two 22-inch displays and was called the small display combination. The subjects were asked to do 12 kinds of typical railway dispatching tasks and the eye movement data was recorded. When the subjects completed the experiment, they were asked to make a subjective questionnaire that mainly related to the layout, color density and size of a variety of graphics, buttons, character and line and light flicker frequency on displays. The result of the task completion time showed that there is no significant difference with two patterns of display. There was no significant difference in the eye movement index of the fixation numbers, the average fixation time and the average scanning amplitude. Only the index of the average pupil size reached significant differences. The subjective data showed that the color of the interface element was the most seriously. Most of the subjects said that the color of the line of station chart is too bright and dazzling, the color of the graphic symbol is also some glare, and the color of the character of alarm information window is not easy to see. Based on above result, the conclusion can be made that within a certain range, the larger size display shows the better information; the color of the station chart need to be adjusted; the red color of the word is not easy to read, it should be changed to the color of obvious contrast to the background.

Keywords: Eye movement · Railway dispatching · Interface · Ergonomics
Evaluation

© Springer Nature Switzerland AG 2019
S. Bagnara et al. (Eds.): IEA 2018, AISC 827, pp. 85–96, 2019.
https://doi.org/10.1007/978-3-319-96059-3_9

1 Introduction

With the construction of "Silk Road" in our country and the economic development in all regions, the railway traffic undertakes the role of the mainstay. The station dispatching system is an important guarantee for the safety and efficiency of railway transport. The railway station interlocking interface is the technical means to achieve the signal, turnout and the path between the restrictive relationship and the operating sequence interlock control interface [1]. The interface not only has the interlocking control function of the relay interlocking device, but also utilizes the rapid information processing capability and storage capability of the computer to conveniently realize the function that the relay interlocking device is difficult to realize. Dispatching staff observe the situation of locomotives through the station interface to implement scheduling tasks. Choosing an operating interface that is more ergonomically relevant can significantly improve the operator's performance, reduce the error rate and accomplish the scheduling task safely and effectively [2–5]. It is perhaps because of the short history of the china railways and ergonomics. Ergonomics research on human-machine interface of railway dispatch room is still scarce, and most ergonomic studies were on the dispatching interface of the subway.

According to the problem of modern complex subway dispatching workstation interactive interface, Zhao Man et al. summarized the applicable design criteria of man-machine interactive interface of dispatcher workstation in which their research object is dual-screen [6]. They also analyzed ATS dispatching system by using abstract level method of ecological interface theory. Combined with the design thinking of the graphical user interface, they proposed a new method for the design of the human-computer interaction interface Based on the development platform of software System at ICS, Lin Xiaowei designed man-machine interface of the monitoring system and elaborated the architecture of integrated monitor HMI [7]. Based on these, they introduced emphatically the design principles, methods and functions of the HMI. By vividly displaying the dynamic configuration screen, the status of the device in the field can be truly reproduced while allowing Users issue control commands for field devices to achieve human-computer interaction. Based on TYJL-ADX computer interlocked operation display interface and the standard technical document "Technical Specifications for Display of Computer Interlocking Operation at Railway Station", Qi Zhihua et al. analyzed the difference between the existing operation display interface and the corresponding part of the specification content combined with the features and specifications of existing operation display interface of interlocking system [8]. They also expounded the optimized interlocking system operation display interface according to the specification standard. In order to meet the demand of railway dispatching and command, improve the efficiency of dispatching and dispatching, and work efficiency of handling emergencies, and more accurately and timely allocate of various resources, Zhou Feng made great suggestions on the system architecture, design features and difficulties of the large-screen system in the dispatch hall Analysis, introduced the design of large-screen dispatching system [9]. The reasonable layout of the ATS (Automatic Train Supervision, ATS) system function in the man-machine interface is closely related in traffic dispatchers' emergency response speed and disposal efficiency.

Liu Qinglei analyzed the status quo of standardization of ATS system dispatching interface in the industry and proposed the key content and construction principle of interface displaying standard, according to the actual working characteristics of dispatchers [10]. In the interface display standard scheme, the overall layout of each module of the system interface, the overall layout of the elements of the station yard map, the device coding rules, the device graphic marking and the definition of the status color are introduced.

It can be seen that although the interface design has made great progress in many environments and fields, it has not yet been popularized and applied in the design of railway traffic dispatch man-machine interface. The current form of railway dispatching man-machine interface, function icons, Station charts and so there are still some problems. With the railway train scheduling information management and intelligence, railway train safety is more and more important, therefore, ergonomics and theory applied to the station man-machine interface design is to enhance the safe operation of railway locomotives and scheduling efficiency Key factor. In order to make dispatchers quickly get the information of the running of vehicles on the railway from the dispatching interface and ensure the safe operation of the railway, in this study, the human-computer interface of railway dispatching was studied by eye movement. It is hoped that we can find the defects in the interface design through experiments and provide the basis for the improved design of railway dispatching man-machine interface.

2 Method

2.1 Participants

Six subjects participated in the experimental study. All of them were dispatchers who worked at the railway dispatching station for more than three years with an average working time of 9 years. From 28 years old to 40 years old, average 35 years old. Their visual acuity or corrected visual acuity was above 1.0, no astigmatism.

2.2 Apparatus

This experiment used the SMI head-mounted eye tracker manufactured by SMI Germany, as shown in Fig. 1. During the experiment, subjects wore a hat with an eye tracker, an infrared light source mounted on the hat, a half-reverse half-mirror and two cameras. Subjects looked the display through the front of the semi-reflective lens. Part of the light reflected by the semi-lens to the camera was recorded to determine the location of eyeballs and pupils to calculate the horizontal and vertical movement of the eyeball time, distance, velocity and pupil diameter. The other camera captured the object image of the subject watching. The two cameras' images were overlaid to determine the subject's gaze position. The camera tracked corneal reflexes on the iris and pupil to compensate for relative head movement. The eye tracker has a sampling frequency of 50 Hz, a resolution of $0.1°$, a stance position accuracy of $0.5°$, a horizontal tracking range of ± 30°, a vertical tracking range of ± 25° and a cap of a weight of 450 g with eye tracker mounted on it.

Fig. 1. Headmounted eye tracker of SMI

Eye tracker should be calibrated in the experiment. During the calibration process, the subjects needed to gaze orderly 5 points on a plane keeping head motionless. After the calibration, the experiment began. Eye tracker could automatically record the whole process of eye observation of external information, including the viewing target, viewing time, viewing order and path, pupil changes. These data could be pre-processed by the Begaze software included with the eye tracker.

In the experiment, four computer display were used that Include two 32-inch monitors and two 24-inch monitors.

The display interface for experiment was designed According to the specification of "Station Computer Interlocking Equipment Operation Display Specifications". Two 24-inch displays were combined to display the information of the railway station and two 32-inch displays were combined to display the same information of the railway station (Fig. 2). So the two 24-inch displays were called small display combination (SDC) and the two 32-inch displays were called larger display combination (LDC).

The purpose of the camera was to record the video information during the experiment. The video can be play backed to assist analysis the process of recording the video, recorded video playback can be assisted analysis.

2.3   Experiment Tasks

According to the scheduling tasks of railway station yard, 10 kinds of typical tasks were selected as experimental tasks, including basic approach operation, flexible approach operation, one-time approach operation, cancellation of approach operation and cancellation of guidance approach operation etc.

Fig. 2. The railway station information displayed on the combination of 2 dispalys

During the experiment, the content of the task was read out by the experimenter, and the subjects operated immediately after hearing the contents of the task. After completing one task, the experimenter read the next task, until the subjects completed 12 tasks. After the subjects completed 12 tasks and then completed once again from scratch and each subject fulfilled three times of 12 tasks. During the experiment, the eye tracker automatically recorded the subject's eye movement data and the video recorder recorded the whole process.

After the experiment was completed, each participant should fill in a subjective questionnaire. The questionnaire mainly involved in the investigation of the layout, color, density, size and lighting frequency of various figures, buttons, words, lines.

The experimental scene was shown in Fig. 3.

Fig. 3. The scenario of experiment

2.4   Experiment Procedure

When the subjects arrived at the experimental site, they do the experiment according to the following steps to experiment.

(1) The experimenter explained experimental task to the subjects;
(2) The subjects fill out the informed consent form;
(3) subjects practiced to adapt to the task;
(4) Subjects wearied the eye tracker to calibrate with 5 point calibration;
(5) Began the formal experiment and the first mission was carried out;
(6) The subject rested 5 min and then do the second mission;
(7) after another 5 min rest, did the third mission;
(8) subjective evaluation;
(9) The experiment was over.

3 Results

The experiment recorded the task completion time, the number of fixation points, the average fixation time, the average scanning amplitude, the average pupil size and subjective questionnaire data.

3.1 Task Completion Time

The subjects' task completion time data was shown in Table 1. Seen from the mean value of the task completion time, the time of SDC was longer than that of the LDC. The reason might be that the display elements on SDC were smaller than it was of the LDC and various display elements were more crowded, so the subjects were more difficult to find the target and they spent more time on search the target. but seen from the pairing t test results, there was no significant difference between SDC and LDC ($p > 0.05$), so there was no obvious difference of better or worse display judged from the task index of the task completion time.

Table 1. The completion time of subjects (ms)

The number of tasks	LDC	SDC
1	7059	14536
2	8210	21271
3	7345	29506
4	22842	32448
5	6675	22586
6	30422	25351
7	26565	24038
8	33278	26191
9	10930	7649
10	11990	8162
11	48714	34947
12	12386	7367
Average	18868	21171

3.2 Number of Fixation Points

The number of gaze points reflected the efficiency of the subjects searching for targets. The more fixation points indicated that the subjects found their targets after stopping observation at multiple locations. Under normal circumstances, the smaller the number of fixation points the better.

Table 2 was the subjects' gaze point data. From the overall data, we could see that the LDC has more fixation points than that of the large interface SDC. The reason might be that when the target is searched on the LDC, the subjects needed to stay several times in the search process to achieve the target because of the distance of the target. But paired t test results showed no significant difference(p > 0.05) kinds of the two kinds display.

Table 2. Number of fixation points

The number of tasks	LDC	SDC
1	14	9
2	8	12
3	12	19
4	23	27
5	9	17
6	28	20
7	18	15
8	23	20
9	10	5
10	11	6
11	41	33
12	18	12
Average	18	16

3.3 Average Fixation Time

The average fixation time data reflected the ease of interface information extraction. The larger the value, the harder it is to read the information. There are several reasons for this, which may be the reason of the light or the small target and the subjects are hard to see the target on interface. It also may be that the meaning of the target design is hard to understand for subjects. So the real reason for the longer average fixation time need to judge from the actual experimental mission scenarios.

The subjects' mean fixation time data are shown in Table 3. From the data of average fixation time, the time of SDC was larger than that of the LDC. The reason might be similar to the task compilation time. The elements on SDC were smaller and denser, so the information on SDC was hard to read and the subjects spent more time to read it. However, paired t tests showed no significant difference between the two kinds of displays (p > 0.05).

Table 3. Average fixation time (ms)

The number of tasks	LDC	SDC
1	293	329
2	302	210
3	286	363
4	272	310
5	300	258
6	285	288
7	325	266
8	249	298
9	397	330
10	279	312
11	246	312
12	321	487
Average	294	313

3.4   Scan Amplitude

The scan amplitude data of the eye reflects the rationality of the layout of the interface information. If the positions of the various information locations are not properly placed, the path traveled by the eyes in searching for the target information will be longer and the magnitude of the saccade will be larger. So the greater the scan amplitude, the layout of the interface information may be more unreasonable. On the other hand, if the distance between the targets of the interface increases, the magnitude of the scanning will also increase. Judging from the scan amplitude index, the smaller of the value, the better of the interface layout.

As shown in Table 4, the average scan amplitude was larger on LDC than that of SDC. Because the layout of LDC and SDC was the same, the reason of that result should be that the distance of the elements on LDC is larger resulted the subjects' longer scan path. However, the results of the paired t test showed no significant difference between the two types displays (p > 0.05).

3.5   Pupil Size

Pupil size in the interface evaluation can reflect the ease of information extraction. In general, if the information is more difficult to extract, the value will increase because the subjects strive to open their eyes to capture information to see the target details, which makes Pupil becomes larger. Pupil size is affected by many factors, if the environmental lighting is not good, the luminance of the display is weak, the target size is small, the target meaning is not easy to understand et al. will make the pupil become larger. Moreover, the people's motivation or interest, the suddenly irritation or scare also affect pupil size.

Table 4. Scan amplitude (°)

The number of tasks	LDC	SDC
1	27	17
2	19	23
3	21	24
4	28	31
5	23	20
6	37	13
7	34	23
8	44	63
9	26	8
10	16	10
11	54	40
12	10	8
Average	28	23

The subject's pupil size data is shown in Table 5. Seen from the average pupil size, it was larger of SDC than that of the SDC. It can be seen from the paired t test results that the significance level was reached ($p < 0.05$). The reason might be the size of the two kinds of displays. Compared to LDC, The elements size of SDC was smaller while the elements density was larger, making subjects to capture the information. It had to work hard to magnify the pupil to read the information. This change in the pupil is a negative factor for the operator, which will make the operator more likely to be tired when working for a long time and should be avoided as much as possible.

Table 5. Pupil size (px)

The number of tasks	LDC	SDC
1	71	77
2	77	83
3	72	74
4	68	80
5	62	80
6	64	82
7	63	74
8	70	90
9	70	77
10	68	77
11	65	73
12	69	75
Average	68	78

3.6 Subjective Questionnaire Data

Seen from the questionnaires, the subjects were satisfied with the interface design, including the alarm information window, station name, equipment status information window, early warning information window, common function button area and prompt information window on the operation interface, the layout of these elements of the spatial location; common function button area button density, size, location order; a variety of fonts and graphics on the size of the information; station map area line thickness and spacing etc.

Seen from the reflected problem by subjects, they all mentioned the irrational line color. Subjects reflected the site map area line bright and dazzling color; when they worked long time, they will feel uncomfortable and they suggest that gray color is more appropriate. They also reflected that some blue section of the station map area glare; the blue color of graphic symbols in station map was dazzling; the red color of font of the Alarm information window was not easy to see.

Therefore, based on the above problems, it is recommended d to change the color of the line pattern in the station map. At present, some colors may be too bright and may cause eye irritation. For a long time, it may cause eye fatigue. The line brightness may be controlled in a range of 1.7– 5.1 cd referenced the standard of GJB 455–1988 "Aircraft cockpit lighting basic technical requirements and test methods". The red color of various display information font color of station map was not easy to interpret and it should be changed to a more obvious contrast with the background color, such as black word with white background, white word with black background, yellow word with black background, green word with black background, black word with red background et al. Figure 4 is some examples of background and word color. n of some.

Fig. 4. Example of background and word color (Color figure online)

4 Discussion

The quality of the interface design will affect the users' performance. Station scheduling tasks require the operator to complete the task quickly and without error. By designing a station interlocking interface conformed to users' operation, the workload of the operator can be reduced and the task can be completed efficiently. The layout of the spatial location of the interface; function area button's density, size, order; the size of the fonts and graphics in; site map area line thickness and spacing and other factors reflect the quality of interface design. The eye movement data can reflect the degree of difficulty for subjects to finish the task. Eye movement data can objectively reflect the impact of different interfaces on the subjects.

The LDC and SDC have their own characteristics. The LDC's screen was large, so the word, icon and button in LDC were large and the line was thick. The information density on LDC is small, but the distance between the interests area was farther than that of SDC. The size of SDC screen was small, so the graphic on it was small smaller, the density of elements was larger, the distance between the area of interest was closer. These characteristics of LDC and SDC have different effects on task operations.

In addition to the pupil size indicators in the size of the interface was significantly different, the task of completion time, the number of fixation points, the average fixation time, these three indicators did not reach significant differences, However, the average of the three indicators consistently reflects the superiority of the LDC. Although the LDC increased the distance of eye sight switch, relative to other advantages, this is not an obvious problem. Therefore, based above results, it can be concluded that the LDC is more conducive to human access to information than that of the SDC, and more conducive to reduce people's workload, thereby reduce visual fatigue.

5   Conclusion

Through the analysis of eye movement test data, combined with the subjective questionnaire results, we can draw the following conclusions or suggestions.

(1) Within the scope of the screen size (for example, the common display size is now 24–32 inch), the bigger the screen size the better;
(2) It is suggested that the brightness of lines and lines in the station picture area should be changed. It is recommended that the brightness should be controlled in the range of 1.7–5.1 cd/m2 that refer to the Chinese standard of GJB 455–1988 "Basic Technical Requirements and Test Methods for Aircraft Cockpit Lighting".
(3) The font color of various display information is not easy to read in red color, and should be changed to other colors that have obvious contrast to the background, such as black word with white background, white word with black background, yellow word with black background, green word with black background, black word with red background et al.

Acknowledgement. This research was funded by National Key R&D Program of China (2016YFC0802807) and Electronic information equipment system research of Key laboratory of basic research projects of national defense technology (DXZT-JC-ZZ-2015-016).

References

1. Feng W (2005) Discussion on several technologies of computer interlocking system in railway station. The Dissertation of Nanjing University of Technology
2. Songtao L (2004) Interaction design essence. Electronic Industry Press

3. Qi Huang, Zhihua B (2012) Interaction design. Zhejiang University Press, Hangzhou
4. Yuanbo S, Min L, Lei S (2010) Engineering foundation and design. Beijing Institute of Technology Press, Beijing, pp 168-169
5. Qing L, Chengqi X, Hoehn F (2010) An interface usability assessment based on eye tracking technology. J SE Univ Nat Sci Ed 40(2):331–334
6. Man Z, Weining F, Jiancheng M (2014) Study on the design of man-machine interactive interface of subway dispatcher workstation. Comput Appl 23(6):49–55
7. Xiaowei L (2010) The design and realization of man-machine interface of subway integrated monitoring. Ind Control Comput 23(12):13–15
8. Zhihua Q, Yang P, Delong X (2017) The optimization design of computer interlocking system operation display interface. Railw Signal 53(7):1–5
9. Feng Z (2017) A brief analysis of the construction scheme of the large screen display system for railway passenger special scheduling. Railw Commun Signal 53(1):61–65
10. Qinglei L (2017) The standardized research of subway ATS system scheduling interface display in Tianjin shows. Mod Urban Rail Transp 9:5–9

Local Lighting Control in Open-Plan Offices: The Influence of Office Lay-Out

Christel de Bakker[1(✉)], Mariëlle Aarts[1], Helianthe Kort[1,2],
Alan Meier[3], and Alexander Rosemann[1]

[1] Eindhoven University of Technology,
Rondom 70, 5600 MB Eindhoven, The Netherlands
c.d.bakker@tue.nl
[2] Utrecht University of Applied Sciences,
Heidelberglaan 7, 3584 CJ Utrecht, The Netherlands
[3] Lawrence Berkeley National Laboratory,
1 Cyclotron Road, Berkeley, CA 94720, USA

Abstract. Highly granular lighting control involves switching on and off luminaires based on individual occupancy. The resulting high frequency of lighting changes can distract the office workers and negatively impact their work performance. In a cubicle office, this might be less of an issue than in an office without partitions, as users do not have an overview over the space here. We tested this control strategy in both office types and compared the results to determine the influence of office lay-out on the amount and acceptability of distractions that it poses. Our results indicated the opposite: occupants in the cubicle office were more often distracted and rated the distractions as less acceptable than in the bullpen office. As the job function types varied and the bullpen was consequently more dynamic, it seems that the type of work environment is of larger influence on users' satisfaction with local lighting control. However, more research is required to confirm this finding.

Keywords: Distraction · Occupancy-based lighting control · User satisfaction
Energy efficiency · Work performance

1 Introduction

The open-office environment has received much criticism over the years because it has been identified to negatively impact occupants' satisfaction with the work environment. Research identified noise and loss of privacy as the main sources causing this dissatisfaction [1, 2]. Satisfaction with lighting is typically not an issue in open-plan offices; its quality is more important to office workers in enclosed offices. However, the transition from enclosed to open offices has affected offices' energy consumption for lighting, as in these large offices it is more complicated to apply occupancy-based lighting control. In an enclosed, private office, one occupancy sensor suffices to determine the presence of the single worker and control the luminaires in the room accordingly. In open-plan offices, however, lighting control at this individual level requires a fine-grained sensor network. Although smart luminaires tends to be equipped

© Springer Nature Switzerland AG 2019
S. Bagnara et al. (Eds.): IEA 2018, AISC 827, pp. 97–106, 2019.
https://doi.org/10.1007/978-3-319-96059-3_10

with occupancy sensors, these just start to being implemented. In addition, luminaires and desks are typically not aligned in open-plan offices, hence some luminaires are also shared by multiple occupants, which complicates the design of highly granular lighting control. As a result, central lighting control tends still to be applied in open-plan offices.

Nevertheless, we find success stories of this type of lighting control in large office spaces, but they originate from studies in cubicle types; they showed energy savings of 41% and 60% [3, 4]. In these offices, luminaires and desks are aligned; as partitions limit the distribution of lighting from neighboring luminaires to surrounding desks, this alignment is required to provide occupants with sufficient illuminance. Hence, utilizing highly granular lighting control, or "local lighting control", entails less difficulties here, and is already being applied.

These studies evaluated local lighting control with occupants in a real office; the majority of them expressed satisfaction with the provided lighting conditions. However, they did not measure whether it distracted them from their work activities. With local lighting control, each time occupants arrive at or leave their desks leads to a change in the lighting situation. The respective luminaire is switched on or off, which occupants could perceive as distracting.

We studied this issue in an open-plan office without partitions (so-called bullpen office) [5]. The majority of occupants did not consider lighting changes due to vacancies as unacceptable; to a few users, however, they did. When comparing the acceptability ratings to those from other environmental sources, electric lighting scored the lowest, together with sound, and was rated on average as "just acceptable". Although the sample size of this study was limited, it suggests that this issue deserves further attention.

In cubicle offices, the lighting changes might be less noticeable to occupants as their view is blocked by partitions, and, as a result, less distracting. However, it is important to validate this hypothesis as distractions can negatively influence the work performance, as for example was found to be the case with noise [6]. In particularly in the open-plan office, it is highly undesirable that electric lighting poses another form of distraction as it already is perceived as a challenging environment to work in. In addition, employees form the major expense of companies; hence, their satisfaction cannot be endangered. Hence, user acceptance is highly important for the successful implementation of a strategy.

This motivated the replication of the bullpen study in a cubicle office, investigating the influence of office lay-out on users' perception of lighting changes. We chose to perform the cubicle study in Northern America, as they form the typical office lay-out here. The bullpen office was located in Northern Europe; hence, the results will also indicate whether cultural differences underlie lighting perception, which is also relevant for the transferability of other lighting control strategies across these continents.

2 Methodology

2.1 Study Design

The study in the cubicle office was conducted in January 2018 for three weeks; the European study had the same length in the same season (February 2016). We chose the winter season to minimize the influence of daylight. The cubicle testbed was located in Berkeley, California, the US, while the study in the bullpen office took place in the Netherlands. We first created a baseline through applying central manual lighting control for a week, followed by two variations of local lighting control, both for a week. We used a repeated measures design; participants experienced all three strategies.

2.2 Lighting Control Design

With the central manual control strategy, all luminaires were switched on by the occupant who arrived first in the office and switched off by the one who left the office the latest. We employed a time delay of five minutes during the first week local lighting control was tested, meaning that luminaires were turned off five minutes after occupants were detected to have left their cubicle. In the second week, this setting was changed to two minutes.

In the cubicle office, luminaires could already be controlled separately. Motion sensors above each desk allowed us to detect individual occupancy changes. In the open-plan office, off-the-shelve, plug-in switching nodes were placed at each luminaire, enabling them to be switched on and off separately. In addition to motion sensors, chair sensors were used in this office. More detailed information about this set-up is reported in Labeodan et al. [7].

In the open-plan office, local lighting control in the first week (local lighting control 1) involved that each luminaire was only attributed to one occupant, while in the second week all luminaires required to provide 500 lx horizontally on the desk of the occupant were switched on. In the cubicle office, we employed the same commissioning across the two weeks of testing local lighting control: all luminaires that contributed significantly to the horizontal illuminance on the desk of the occupant were controlled by the occupant's sensor. This resulted in the use of one or two luminaires per occupant. We used a Konica Minolta illuminance meter for these measurements.

2.3 Participants

The open-plan office space was shared by twelve occupants; nine agreed to participate (all male; median age category 40–49 years old). Their job function types were all technical and required much cooperation; work activities consisted of discussing, technical drawing, and planning. The cubicle office contained 17 workplaces; 11 occupants participated (7 male, 4 female; median age category 30–39 years old). They all held a research job function type and mainly performed individual computer work. In this office, two occupants opted out (halfway the second and third week, respectively). All participants signed an informed consent before participation; it explained the general set-up of the experiment, including that a new control strategy would be applied during the second and third week.

2.4   Experimental Space

Figures 1 and 2 show the lay-out of the open-plan and cubicle office, respectively. The open-plan offices was an enclosed space, while the cubicle offices were located in a larger space with additional cubicles on the right hand while surrounded by perimeter offices. It was equipped with pendant luminaires with T5HO 4-foot LED lamps (54 W). In the open-plan office, luminaires were ceiling-based (31x), each containing 2 * 36 W linear fluorescent lamps.

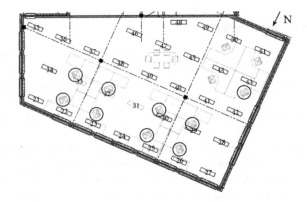

Fig. 1. Bullpen testbed with the luminaires numbered and the location of the participants circled (in the Netherlands)

2.5   Procedure

Before the start of the experiment, participants filled out a general survey assessing possible confounding variables. Surveys assessing distraction were distributed at Friday afternoon of each testing week. In the cubicle office, they were given the option to fill out this weekly questionnaire and a diary, or only a diary; four participants agreed to fulfil both. All participants in the open-plan office filled out the diary.

2.6   Measures

Distraction. The dependent variables (DV) included frequency of noticed change, frequency of distraction, and acceptability of distraction due to electric lighting. These three items were also assessed for six other environmental sources, namely 'Temperature', 'Odour', 'Ventilation', 'Occupancy', 'Sound', and 'Sunlight'. To assess them, we developed questions ourselves: "How often did you notice a change" and "How often did you got distracted" (answer options: never, sometimes, regularly, often, or always), and "If you got distracted, to which extent was this acceptable to you" (7-point

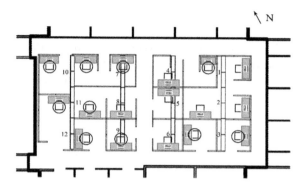

Fig. 2. Cubicle testbed with the luminaires numbered and the location of the participants circled (in the US)

Likert scale from Completely unacceptable to Completely acceptable), respectively. They were included in the weekly survey. This study does not report the three scores of the other environmental sources.

Possible Confounding Variables. In the pre-test survey, we assessed gender, age category, and vision (whether one wears glasses or contact lenses) as demographic characteristics. In addition, we assessed five other possible confounding variables: (1) productivity [8], (2) self-assessed general distraction, (3) privacy desire [9], (4) effect of environment on productivity [8], and (5) light sensitivity, using three items regarding light exposure sensitivity [10], with Cronbach's $\alpha = .536$. The weekly questionnaire started with a question regarding their concentration ability (four items: difficulties to concentrate, difficulties making choices, memory lapses, and difficulties to think clearly) on a 5-point Likert scale from Never to Always, Cronbach's $\alpha = .761$).

2.7 Analysis

In this paper, we compared the results from the weekly questionnaires regarding distraction between the occupants of the two office types. First, we compared the two cases on the potential confounding variables.

Preparatory Analyses. We detected significant correlations $>.30$ between the DV and three confounding variables: (1) age, correlating with frequency of distraction and acceptability of distraction, and (2) concentration ability, with frequency of distraction.

Distraction Analyses. To determine the differences between the two office cases, we used descriptive statistics and employed the ANCOVA procedure with frequency of noticed change, frequency of distraction, and acceptability of distraction due to electric lighting as DV, and the confounding variables correlating with the DV as covariates. Normality tests showed that the data was close enough to normality to use this parametric test.

3   Results

Figures 3, 4, and 5 report the Estimated Marginal Means (EMMs) of frequency of noticed change, frequency of distraction, and acceptability of distraction due to electric lighting, respectively, for both office cases and the three control strategies.

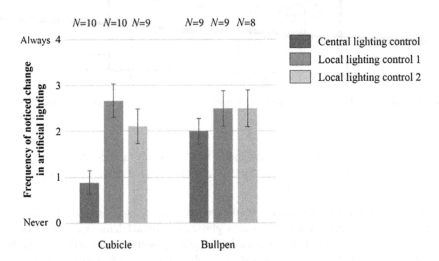

Fig. 3. Frequency of noticed change due to electric lighting with central lighting control, local lighting control 1, and local lighting control 2 in the cubicle and bullpen office. Acceptability scores are displayed as EMM's and error bars as SE's resulting from the ANCOVA post hoc analyses.

Figure 3 shows that with central lighting control, occupants also noticed some lighting changes, while all lighting remained switched on the entire day. Occupants of both offices noticed an increased number of times changes in electric lighting when local lighting control was applied; on average, they reported this to occur "often". This increase was relatively higher in the cubicle office. We found significant effects for both office type and control strategy (see Table 1 in the Appendix), but not for the inter-action effect between these two factors, which was of main interest to us.

Figure 4 teaches us that occupants in the cubicle office were more than "sometimes' distracted by the lighting changes that they noticed when local lighting control was applied. In the bullpen office, the frequency of distraction declined over the three weeks, suggesting an adaptation effect. Here neither the interaction effect nor any of the main was not significant (see Table 1 in the Appendix).

Figure 5 clearly indicates occupants in the cubicle office considered the distractions from local lighting control unacceptable, on average, while the participants from the bullpen office considered them acceptable. In both offices, the score increased during the second week of local lighting control, providing another indication for an adap-tation effect. The differences between the two office types were not significant (see Table 1 in the Appendix).

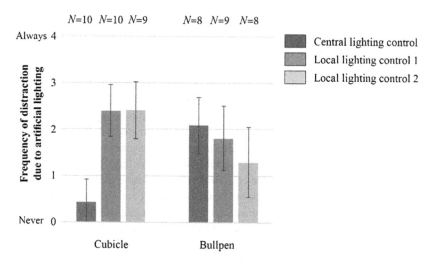

Fig. 4. Frequency of distraction due to electric lighting with central lighting control, local lighting control 1, and local lighting control 2 in the cubicle and bullpen office. Acceptability scores are displayed as EMM's and error bars as SE's resulting from the ANCOVA post hoc analyses.

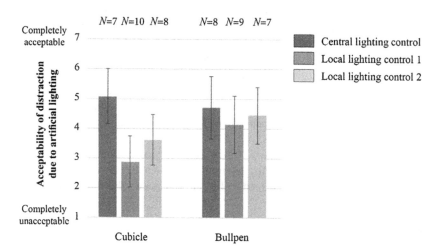

Fig. 5. Acceptability of distraction due to electric lighting with central lighting control, local lighting control 1, and local lighting control 2 in the cubicle and bullpen office. Acceptability scores are displayed as EMM's and error bars as SE's resulting from the ANCOVA post hoc analyses.

4 Discussion

Overall, our results suggest that the dynamic lighting conditions resulting from local lighting control were acceptable to the office workers of the bullpen office, but unacceptable to the participants in the cubicle office, while we expected the opposite.

First of all, when comparing the number of notice lighting changes between the two office cases, we see similar scores, while we expected the occupants of the cubicle office to less often notice them. In addition, occupants of the open-plan office were less often distracted by these changes, and evaluated them as more acceptable. This difference was not significant, but this is probably due to the large individual variability and small sample size. The two office cases were representative of the typical office lay-out, but had a different type of work environment, which can explain the unexpected results. In the open-plan office, interactions between occupants occurred very often, resulting in a highly dynamic environment. Occupants of the cubicle office had a more individualistic way of working; consequently, occurrences of any kind attract more attention. In addition, occupants of the open-plan office were used to changes in sunlight, as it had windows all along the façade. The cubicle offices, in contrast, were not exposed to direct sunlight. Our results suggest that the influence of office lay-out is less important on distractions posed by local lighting control than the type of work environment.

It has to be noted that the two office cases involved different type of occupancy sensors leading to different control behavior on false-offs. More false offs occurred in the cubicle office; that might have contributed to the lower acceptability scores here. It also caused the drop-out of the two occupants in the cubicle office. Thus, for local lighting control to succeed, it is highly important that suitable sensors are being used providing accurate information on occupancy.

In both offices, occupants got adapted to the lighting changes, resulting in higher acceptability scores during the second week. This also means that the shorter time delay applied during this week was not causing dissatisfaction, which is a highly positive finding for energy savings.

In addition to the weekly survey, we asked occupants to keep a diary about all moments they were distracted by any environmental source. However, only four participants of the cubicle test-bed agreed to fill-out this, so we did not use this data.

5 Conclusion

When applying highly granular lighting control in open-plan offices, the lighting levels change due to other co-workers leaving or arriving at their desk. Our study showed that these are being noticed by occupants, independently of office lay-out. They were sometimes considered distracting, but accepted by the occupants after a week of adaption time. Nevertheless, they were on average considered unacceptable by the occupants of the cubicle office, while the bullpen occupants evaluated them acceptable. This difference was insignificant, but before any conclusion can be drawn, a follow-up study is required where the type of work environment, meaning job function type and the amount of interactions are similar. This seemed to affect users' acceptance of distraction due to electric lighting changes instead of office type.

Acknowledgements. This work was in part funded by a Fulbright fellowship. It was also supported by the Assistant Secretary for Energy Efficiency and Renewable Energy, Building Technologies Office, of the U.S. Department of Energy under Contract No. DE-AC02-05CH11231.

Appendix

Table 1. Results from ANCOVA with frequency of noticed change, frequency of distraction, and acceptability of distraction as dependent variables (DV)

DV	Frequency of noticed change			Frequency of distraction			Acceptability of distraction		
Effect	df	F	p-value	df	F	p-value	df	F	p-value
Control strategy	**2**	**4.66**	**.017***	2	1.34	.285	2	.009	.991
Office type	**1**	**5.13**	**.04***	1	.001	.983	1	.331	.581
Age category	–	–	–	1	1.401	.264	1	.864	.380
Concentration ability	–	–	–	1	.375	.554	–	–	–
Control strategy* Office type	2	1.36	.273	2	3.43	.052	2	.361	.702
Control strategy* Age category	–	–	–	2	1.73	.202	2	.047	.954
Control strategy* Concentration ability	–	–	–	2	.95	.404	–	–	–

References

1. de Croon EM, Sluiter JK, Kuijer PPFM, Frings-Dresen MHW (2005) The effect of office concepts on worker health and performance: a systematic review of the literature. Ergonomics 48:119–134
2. Bodin Danielsson C, Bodin L (2009) Difference in satisfaction with office environment among employees in different office types. J Archit Plan Res J Archit Plan Res 26:241–2573
3. Rubinstein F, Enscoe A (2010) Saving energy with highly-controlled lighting in an open-plan office. LEUKOS 7:21–36. https://doi.org/10.1582/LEUKOS.2010.07.01002
4. Galasiu A, Newsham G, Suvagau C, Sander D (2007) Energy saving lighting control systems for open-plan offices: a field study. Leukos
5. de Bakker C, Aarts M, Kort H, Rosemann A (2017) Local lighting control in open-plan offices: acceptable or distracting? In: Healthy Buildings 2017 Europe, Lublin, pp 1–2

6. Banbury S, Berry D (2005) Office noise and employee concentration: identifying causes of disruption and potential improvements. Ergonomics 48:25–37. https://doi.org/10.1080/00140130412331311390
7. Labeodan T, De Bakker C, Rosemann A, Zeiler W (2016) On the application of wireless sensors and actuators network in existing buildings for occupancy detection and occupancy-driven lighting control. Energy Build 127. https://doi.org/10.1016/j.enbuild.2016.05.077
8. Oseland N, Hodsman P (2015) Planning for Psychoacoustics: a psychological approach to resolving office noise distraction
9. Kaya N, Weber MJ (2003) Cross-cultural differences in the perception of crowding and privacy regulation: American and Turkish students. J Environ Psychol 23:301–309. https://doi.org/10.1016/S0272-4944(02)00087-7
10. Smolders KCHJ, de Kort YAW, Cluitmans PJM (2012) A higher illuminance induces alertness even during office hours: findings on subjective measures, task performance and heart rate measures. Physiol Behav 107:7–16. https://doi.org/10.1016/j.physbeh.2012.04.028

The Application of Taguchi Method in Evaluating 3D Image Quality

Po-Hung Lin[1(✉)] and Hui-Hsuan Hsu[2]

[1] Department of Industrial Engineering and Management,
Ming Chi University of Technology, No. 84, Gungjuan Road, Taishan District,
New Taipei City 24301, Taiwan
frank.phlin@gmail.com
[2] Department of Industrial Engineering and Management Information,
Huafan University, No. 1, Huafan Road, Shiding District,
New Taipei City 22301, Taiwan

Abstract. With the promotion of the technology, the state-of-the-art TVs have already possessed the functions of 3D. This study collected relevant literature to identify the key factors affecting the image quality of 3D TVs. Then we apply Taguchi Method with large is better quality characteristic. Thirty subjects were recruited in this study and we ask them to fill out the overall image quality. Through the calculation of software, the S/N ratio were found, then we can get significant factors and contribution ratio and finally the optimal parameters were obtained. The result of this study can be a reference for future product design, in order to enhance customers' intention to buy 3DTVs.

Keywords: Taguchi Method · 3D · Image quality

1 Introduction

With the promotion of the technology, the state-of-the-art TVs have already possessed the functions of 3D. The previous studies indicated some advantages brought by a 3D display, like decreasing reaction time and increasing accuracy as well as depth perception [1]; its binocular cue provides information about 1.4 times more than the monocular one [2]; it heightens people's contrast sensitivity and cognitive ability [3]; it can simultaneously process multi-characteristic visual search [4]; and it performs better than 2D displays in searching performance [5].

Since 3D TVs have been the popular products in the market, theirs image quality assessment in related literatures are critical issues for users and manufacturers. Kooi and Toet [6] indicated that vertical disparity, crosstalk, and blur are important factors that have an effect on the visual fatigue of stereoscopic display. Wang et al. [7] also pointed out in their study that the interaction of luminance contrast between the shadow and the main background has an impact on depth. So and Chan [8] set five viewing angles to discuss the impact of text reading. Cai and Li [9] explored the impact of 15 display angles on readability. Based on above studies, disparity, crosstalk, contrast and viewing angle were the four critical factors to be investigated in this study.

© Springer Nature Switzerland AG 2019
S. Bagnara et al. (Eds.): IEA 2018, AISC 827, pp. 107–109, 2019.
https://doi.org/10.1007/978-3-319-96059-3_11

The Taguchi method is an engineering method created by Japanese quality expert Dr. Koichi Taguchi. The Taguchi method has gained popularity since the 1980s and has been widely used by industries. In today's industrial practice, employees are provided with simple and practical tools used in product design and process development, which along with technical and statistical methods can optimize product design and manufacturing process conditions, thereby quickly reducing costs and improving quality. The Taguchi method is intended to ensure the stability of product design quality with less volatility, insusceptibility to noises of all kinds during production processes, and the utilization of quality, costs, and benefits during product design processes in order to develop high-quality products under low-cost conditions [10].

2 Method

In this study, Taguchi Method was used and L9 orthogonal array (Table 1) with large is better quality characteristic (Formula 1) was applied in our experimental design. Thirty subjects were recruited in this study and we ask them to fill out the overall image quality questionnaire after the experiment. Through the calculation of MINITAB, the S/N ratio of overall image quality were found, then we can get significant factors and contribution ratios and then the optimal parameters were obtained.

$$SN = -10\log_{10}\left[\frac{1}{n}\sum_{i=1}^{n}\frac{1}{y_i^2}\right] \qquad (1)$$

Table 1. L9 orthogonal array used in this study

No.	A (Disparity)	B (Crosstalk)	C (Contrast)	D (Viewing angle)
1	1	1	1	1
2	1	2	2	2
3	1	3	3	3
4	2	1	2	3
5	2	2	3	1
6	2	3	1	2
7	3	1	3	2
8	3	2	1	3
9	3	3	2	1

3 Results and Discussion

After getting the response value of S/N ratio, the optimal parameter (A1, B2, C3, and D3) was obtained (see Table 2). Table 3 also shows the ANOVA results for S/N ratio. In overall image quality, the highest contribution ratio is crosstalk (61%). Although p value is larger than 0.05, 5.02% means that significant difference was almost found. It also means that crosstalk has the great impact in overall image quality.

Table 2. The optimal parameter for S/N ratio

Factor	Optimal parameter
Disparity	A1 (0%)
Crosstalk	B2 (2.02%)
Contrast	C3 (0.287)
Viewing angle	D3 (45°)

Table 3. ANOVA results for S/N ratio

Source	F value	P value	Contribution ratio
Disparity	1.884	0.346	6%
Crosstalk	18.103	0.052	61%
Contrast	8.469	0.106	29%
Viewing angle (Error)			4%

4 Conclusions

Experimental results of SN ratio indicate that A1, B2, C3, D3 is the optimal parameter for overall image quality. The result of this study can be a reference for future product design in order to enhance customers' intention for buying 3DTVs.

References

1. Barfield W, Rosenberg C (1995) Judgments of azimuth and elevation as a function of monoscopic and binocular depth cues using a perspective display. Hum Factors 37:173–181
2. Campbell FW, Green DG (1965) Monocular versus binocular visual acuity. Nature 208 (6):191–192
3. Yeh Y, Silverstein LD (1992) Spatial judgments with monoscopic and stereoscopic presentation of perspective displays. Hum Factors 34(10):583–600
4. Nakayam K, Silverman GH (1986) Serial and parallel processing of visual feature conjunction. Nature 320:264–265
5. Ntuen CA, Goings M, Reddin M, Holmesk K (2009) Comparison between 2-D & 3-D using an autostereoscopic display: the effects of viewing field and illumination on performance and visual fatigue. Int J Ind Ergon 39:388–395
6. Kooi FL, Toet A (2004) Visual comfort of binocular and 3D displays. Displays 25:99–108
7. Wang P-C, Hwang S-L, Huang H-Y, Chuang C-F (2011) System crosstalk issues on autostereoscopic displays. In: IS&T/SPIE Electronic Imaging – Science and Technology, San Francisco
8. So JCY, Chan AHS (2013) Effects of display method, text display rate and observation angle on comprehension performance and subjective preferences for reading Chinese on an LED display. Displays 34:371–379
9. Cai H, Li L (2014) The impact of display angles on the legibility of Sans-Serif 5×5 capitalized letters. Appl Ergon 45:865–877
10. Taguchi G, Chowdhury S, Wu Y (2004) Taguchi's quality engineering handbook, 1st edn. Wiley-Interscience, Hoboken

Personal Lighting Conditions to Obtain More Evidence in Light Effect Studies

J. van Duijnhoven[1(✉)], M. J. H. Burgmans[1], M. P. J. Aarts[1],
A. L. P. Rosemann[1], and H. S. M. Kort[2,3]

[1] Building Lighting Group, Department of the Built Environment,
Eindhoven University of Technology, Eindhoven, The Netherlands
j.v.duijnhovenl@tue.nl
[2] Research Centre for Innovations in Health Care,
University of Applied Sciences Utrecht, Utrecht, The Netherlands
[3] Building Performance Group, Department of the Built Environment,
Eindhoven University of Technology, Eindhoven, The Netherlands

Abstract. Research demonstrated a large variety regarding effects of light (e.g. health, performance, or comfort effects). Since human health is related to each individual separately, the lighting conditions around these individuals should be analysed individually as well. This paper provides, based on a literature study, an overview identifying the currently used methodologies for measuring lighting conditions in light effect studies. 22 eligible articles were analysed and this resulted in two overview tables regarding the light measurement methodologies. In 70% of the papers, no measurement details were reported. In addition, light measurements were often averaged over time (in 84% of the papers) or location level (in 32% of the papers) whereas it is recommended to use continuous personal lighting conditions when light effects are being investigated. Conclusions drawn in light effect studies based on personal lighting conditions may be more trusting and valuable to be used as input for an effect-driven lighting control system.

Keywords: Light effects · Measurements · Methodological issues
Individualized · Light exposure

1 Introduction

Research demonstrated a large diversity regarding effects of light (e.g. health, performance, or comfort effects). In the majority of the studies investigating the health effect, the health effect was often related to the photometric quantity illuminance or correlated colour temperature [1]; however, these aspects were mostly measured and included in the data analysis as average values (i.e., averaged over time or locations). Since human health is related to each individual separately, the environment around these individuals should be analyzed individually as well. Other environmental conditions (e.g., air pollution [2]) were already investigated at individual level. It is recommended to measure lighting conditions per individual as well since the impact of light is not identical for all people.

Health effects may be influenced, supported, or even controlled via a lighting control system which includes personal lighting conditions and personal health characteristics (either subjective or objective). In order to succeed, this lighting control system needs continuous information on the lighting and health conditions, both at individual level.

This paper provides an overview of the currently used methodologies for measuring lighting conditions in light effect studies as reported in literature. The methodological aspects that are being identified are the light aspects, and how, when, and where the light measurements were performed.

2 Method

The literature search was performed in September 2017 and did not have any restrictions on publication year to ensure all relevant articles were included in the search. The base of the literature search was the word combination: 'Alertness', lighting parameters (i.e., 'Lighting', 'Daylight', 'Light exposure', or 'Light'), and 'Office'. All three search aspects had to be present in potentially eligible articles. These search terms led to four possible combinations (i.e., Alertness – Lighting – Office, Alertness – Daylight – Office, Alertness – Light Exposure – Office, and Alertness – Light - Office). Inserting these four search combinations in four different scientific databases (ScienceDirect, Google Scholar, PubMed, and Web of Science) resulted into 141 hits of which 122 were eligible (based on abstract reading). After removing the duplicates, 22 papers were found to be relevant. See Fig. 1 for the search process.

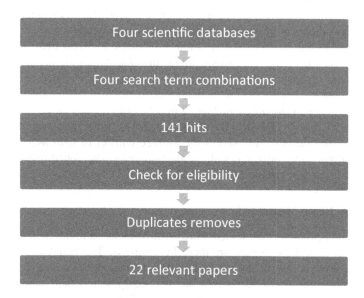

Fig. 1. Literature search process

3 Results

While mapping a certain luminous environment by measuring the lighting conditions, multiple aspects need to be taken into account. Which light aspects are being measured and how, when, and where are these measurements performed? Based on the performed literature study, the questions will be answered in the next paragraphs.

3.1 Which Light Aspects Are Being Measured?

As briefly mentioned in the introduction, mostly the illuminances or correlated colour temperatures were determined in order to investigate potential light effects. Besides these two light aspects, there are more light aspects which are important to consider while performing a light effect study. In the papers included in this literature study, these light aspects were measured and/or calculated: horizontal illuminance, vertical illuminance, brightness, irradiance, correlated colour temperature (CCT), spectrum, colour rendering index (CRI), reflectance, luminance, flicker, direction, daylight, glare, uniformity, and daylight factor (DF). Table 1 provides an overview of these light aspects per paper and also demonstrates how these light aspects were included (i.e., dark green means included and measured, light green means included but not measured, orange means not included, and red means not reported).

In the 22 eligible papers, horizontal illuminances was included and measured the most with 15 papers, followed by vertical illuminance in 14 papers. Correlated colour temperature is the light quantity which is often included in the papers but not measured. This is, for example, the case when an intervention study was executed comparing effect in two different light scenarios (e.g., CCTs of 4000 K and 6500 K [9]). The light aspect presence or absence of daylight is the only aspect which is reported to be not included. Light flicker is the aspect which is mostly not reported.

Figure 2 demonstrates the relative number of times a certain light quantity was included and measured (i.e. dark green colour in Table 1) per publication year. The relative number of times a certain photometric quantity was included and measured means the number of times it was included out of the total papers for that specific publication year (i.e., 1 in 2000, 1 in 2006, 2 in 2007, 1 in 2009, 1 in 2010, 1 in 2011, 3 in 2012, 4 in 2013, 1 in 2015, 1 in 2016, and 6 in 2017). 100% means in all the papers published in that year. Figure 3 shows the average number of light aspects included and measured per paper throughout all publication years. The dotted line is the linear trend line of these data points.

3.2 How Is Being Measured?

Once the light aspects are determined which will be included in the light effect study, the next question is how to measure these light aspects. The brand and type of the specific measurement instrument were often not reported (in 13 out of the 19 papers including light measurements, no measurement equipment details were provided). Light measurements can be executed using person-bound measurements (PBM, e.g. actiwatches [25], daysimeters [26], or Lightlogs [27]) or location-bound measurements (LBM) [28].

Table 1. Overview light aspects incorporated in light effect studies - based on the literature study. Dark green means included and measured, light green means included but not measured, orange means not included, and red means not reported.

| Literature | Light aspects | | | | | | | | | | | | | | |
| | Measured | | | | | | | | | | | Calculated | | |
	Illuminance (Hor)	Illuminance (Ver)	Brightness	Irradiance	CCT	Spectrum	CRI	Reflectance	Luminance	Flicker	Daylight	Glare	Uniformity	DF
[3]														
[4]														
[5]														
[6]														
[7]														
[8]														
[9]														
[10]														
[11]														
[12]														
[13]														
[14]														
[15]														
[16]														
[17]														
[18]														
[19]														
[20]														
[21]														
[22]														
[23]														
[24]														

The advantage of the PBM method is that the lighting conditions are continuously measured, at the position of the individual. Disadvantages of this method are the burden for the individual to continuously wear a measurement device [29] and the relatively high performance errors of the current wearables [30]. In order to measure lighting conditions with lower performance errors, highly accurate measurement instruments can be used for location-bound measurements (LBM). A disadvantage of this method is that the lighting conditions are being measured at certain locations only (i.e., not dynamically following the individual) and that these measurement instruments occupy locations (e.g. desks in an office) which cannot be used by building occupants.

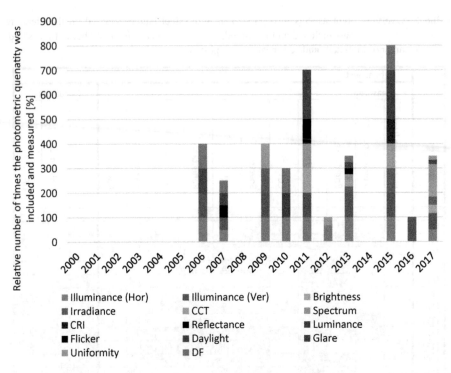

Fig. 2. Graph demonstrating the relative number of times a light quantity was included and measured in the 22 included papers of the literature study. For example, the value 100 for vertical illuminance in 2006 means that in 100% of the papers published in 2006 vertical illuminance was included and measured.

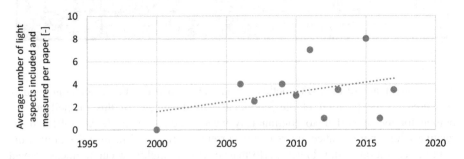

Fig. 3. Graph demonstrating the average number of light aspects included per paper throughout all publication years of the literature included in the literature search. The dotted line is the linear trend line of these data points. 14 light aspects were reported in literature (see Table 1) so that is the highest number possible in this graph.

The current literature search only revealed one study that applied person-bound measurements. In this study, a 'LuxBlick' was used to measure personal lighting conditions. This study did not report the accuracy of the measurement device and the corresponding advantages or disadvantages of this method [10].

3.3   When Is Being Measured?

The third aspect to be considered before performing the light measurements is the moment when the light measurements will be performed. Measurements can be performed at different measurement intervals: continuously (i.e., measurement interval ≤ 1 h), regularly (i.e., measurement interval >1 h), or only once (per light scenario). Only three papers mentioned that they measured lighting conditions with a measurement interval of less than or equal to 1 h [3, 13, 20].

3.4   Where Is Being Measured?

The fourth aspect regarding the light measurements is the location of the light measurements. A light effect study can be performed in a laboratory environment (lab study) or a realistic work/home environment (field study). 16 out of the total 22 papers included in the literature study described a light effect lab study.

Within the study environment (either lab or field), measurements need to be performed at certain locations. When these measurements are executed at a sample of locations inside the environment, often these measurements are averaged to determine an average lighting condition for the entire environment. In contrast, measurements can also be performed at the specific locations of the participants of the light effect study to be more accurate about the lighting condition for that specific participant. Measurements performed by person-bound measurement instruments or measurements performed at the specific workplace of that person (either vertically at eye height or horizontally at desk height) are assumed to measure personal lighting conditions. Thirteen papers in the literature study performed measurements at personal level of which 8 were measurements at eye height.

3.5   Overview

Table 2 provides an overview of the three methodological aspects (i.e., how, when, and where are lighting measurements performed) for the 22 included papers in the literature study.

Table 2. Overview methodological aspects (i.e., how, when, and where are light measurements performed) of included light effect studies - based on the literature study.

Literature	How — Measurement equipment details reported		How — Measurement method			When — Measurement interval				Where — Type of study		Where — Measurement location		
	Yes	No	Person-bound measurements (PBM)	Location-bound measurements (LBM)	Not reported	Continuously (measurement interval ≤ 1 hour)	Regularly (measurement interval > 1 hour)	Once (for each scenario)	Not reported	Lab study	Field study	Average	Personal	Not reported
[3]														
[4]														
[5]	No light measurements		No light measurements			No light measurements						No light measurements		
[6]														
[7]														
[8]														
[9]														
[10]														
[11]														
[12]														
[13]														
[14]														
[15]														
[16]														
[17]														
[18]														
[19]														
[20]														
[21]														
[22]														
[23]	No light measurements		No light measurements			No light measurements						No light measurements		
[24]	No light measurements		No light measurements			No light measurements						No light measurements		

4   Discussion

In the current literature study, the aspects which, how, when, and where do light measurements need to be performed were investigated. This paper gives an overview of these aspects and demonstrate multiple possibilities for each aspect. The decisions which need to be taken before performing light measurements may be based on standards, regulations or other literature. It is expected that, for example, illuminance is often included in light effect studies since this is the mostly used recommendation in current standards.

In 2002, a third photoreceptor was discovered explaining the mechanism of image-forming and non-image-forming effects through this intrinsically photosensitive retinal ganglion cell (ipRGc) [31]. In order to investigate light effects (excluding the biological effects such as tanning or the production of vitamin D due to the UV radiation in sunlight), it is essential to know the amount and type of light which enters the eye, i.e. lighting conditions vertically measured at eye level. In addition, Lucas et al. stated in

2014 that measuring and reporting light with photometric quantities will not be sufficient either [32]. Table 1 demonstrated the number and types of light aspects included in the 22 selected papers in the literature study. Although the CIE recommended to describe the total lit environment instead of individual elements within it [33], this literature study showed that none of the 22 papers included the broad range of light aspects. However, Fig. 3 showed that the average number of light aspects included and measured in the light effect studies increased over the years.

Regarding these different light aspects, it seems, especially for researchers with a non-technical expertise, that it is sometimes difficult to use the correct terminologies for certain light aspects. Van Hoof et al. and Aarts et al. provided tools to correctly measure and report all methodological aspects when performing a light effect study [34, 35]. In a previous review, these methodological issues of reporting light measurements were extensively highlighted [1]. In this literature study, the term brightness was reported to be measured in three papers and reported to be included but not measured in six papers. The question immediately arises how researchers defined this terminology and whether this term was related to, for example, illuminance measurements.

Since in 13 out of the 19 included papers (including light measurements), no measurement equipment details were provided, it can also be questioned whether the included light aspects were either objectively or subjectively measured. In two cases, both in the paper of Borisuit et al. [3], they mentioned that brightness and direction of light were subjectively measured (i.e., using an adapted version of the Office Lighting Survey [36]).

For both subjective as well as objective measurements, the choice when to perform light measurements may be influenced by the measurement equipment. For subjective measurements, the length of the questionnaire may influence the number of times the questionnaire is distributed, to limit the annoyance of filling in the questionnaires. For objective measurements, the choice when to measure may depend on the instrument properties. Wearables (PBM), for example, run on batteries and measuring at a shorter sample interval may shorten the battery life.

Besides the measurement equipment details, the moments for performing light measurements or the sample interval may depend on the lighting conditions in the environment as well. If the light effect study includes daylight availability, the ranges of lighting conditions vary more than compared with a situation without daylight. If daylight is available, the weather conditions (clear/overcast sky) also cause more variation in the lighting conditions during spring or summer. A wider range of lighting conditions may raise the necessity to measure the lighting conditions more often.

The decision to perform a light effect study in a laboratory or field environment depends mostly on the outcome measures, the aim or research question of the experiment and its hypotheses. While performing field studies, more measurements may need to be performed in order to identify potential confounders influencing the final results. The advantage of performing a lab study is that many parameters can be controlled to reduce the chances of having many confounders influencing the results. The large advantage of performing field studies is that the results were found and demonstrated in a realistic work/home environment. Then there is no need to doubt the possibility of extending the results to the realistic environment.

4.1 Practical Implications

This literature study demonstrated that in nearly all the studies light measurements were performed using location-bound measurement instruments and that these measurements were sometimes averaged over the entire environment. Since all individuals differ, each individual health differs, and this increases the importance to measure lighting conditions at individual level as well.

Many of the currently available wearables measure multiple light aspects (e.g. illuminance, or irradiances in different spectral bands); however, these measurement instruments suffer from higher performance errors compared to the location-bound measurement instruments. Van Duijnhoven et al. [28, 37] proposed a new non-obtrusive method to obtain personal lighting conditions using location-bound measurement instruments. This novel method may be a good alternative for less accurate person-bound measurement devices. This new method (location-bound estimations, i.e. LBE) consists of estimations based on location-bound measurements. Measurements at reference locations allow estimations of lighting conditions at other locations inside the building. The LBE was developed as a principle method and various methods of the LBE were already investigated. The accuracies of these LBE methods were determined based on two validation studies in offices. It is expected that, considering an effect-driven lighting system (an effect can be e.g. visual performance, health, or productivity), this LBE method will be a pragmatic approach of inserting personal lighting conditions into lighting control systems. The method may approach reality, is unobtrusive for the building occupants, and can easily be included in an Internet-of-Things-platform.

5 Conclusion

Light effect studies are a combination of measuring lighting conditions and health aspects. This multidisciplinary field of research requires knowledge of both fields. Therefore, it is highly recommended to perform light measurements and report these measurement methodologies as comprehensive as possible. Comprehensive descriptions of measurement methodologies enable researchers to understand, trust, and reproduce light effect studies.

This literature study showed that in ±70% (i.e., 13 out of the 19) of the papers including light measurements, no measurement details were provided. In addition, an average number of 3.4 (i.e., average of all values in Fig. 3) light aspects included and measured per light effect study suggests that researchers are not fully mapping the lit environment during their light effect studies as suggested by the CIE [33]. Furthermore, these light aspects were often averaged over time (i.e., only three studies applied light measurements with a measurement interval of less than an hour) or over location (i.e., six studies performed measurements at one location only).

Light effect studies are investigating potential effects of light, usually per individual. Each individual responds differently to light and lighting conditions should therefore be measured at individual level as well. Person-bound measurements (PBM), location-bound measurements (LBM), or location-bound estimations (LBE) can be applied measuring personal lighting conditions continuously for the entire study period. These obtained personal lighting conditions are essential information to draw conclusion within a light effect study. Conclusions drawn in light effect studies based on personal lighting conditions may be more trusting and valuable to be used as input for an effect-driven lighting control system.

References

1. van Duijnhoven J, Aarts MPJ, Aries MBC, Rosemann ALP, Kort HSM (2017) Systematic review on the interaction between office light conditions and occupational health: elucidating gaps and methodological issues. Indoor Built Environ. https://doi.org/10.1177/1420326x17735162, 1420326X1773516
2. Sbihi H, Allen RW, Becker A, Brook JR, Mandhane P, Scott JA, Sears MR, Subbarao P, Takaro TK, Turvey SE, Brauer M (2015) Perinatal exposure to traffic-related air pollution and atopy at 1 year of age in a multi-center canadian birth cohort study. Environ Health Perspect 123:902–908. https://doi.org/10.1289/ehp.1408700
3. Borisuit A, Linhart F, Scartezzini J-L, Munch M (2014) Effects of realistic office daylighting and electric lighting conditions on visual comfort, alertness and mood. Light Res Technol 47:192–209. https://doi.org/10.1177/1477153514531518
4. Boyce PR, Veitch JA, Newsham GR, Jones CC, Heerwagen J, Myer M, Hunter CM (2006) Lighting quality and office work: two field simulation experiments. Light Res Technol 38:191–223. https://doi.org/10.1191/13657828061rt161oa
5. Cajochen C, Zeitzer JM, Czeisler CA, Dijk DJ (2000) Dose-response relationship for light intensity and ocular and electroencephalographic correlates of human alertness. Behav Brain Res 115:75–83. http://www.ncbi.nlm.nih.gov/pubmed/10996410. Accessed 18 Jan 2017
6. Chellappa SL, Steiner R, Blattner P, Oelhafen P, Go T (2011) Non-visual effects of light on melatonin, alertness and cognitive performance : can blue-enriched light keep us alert ? PLoS One 6. https://doi.org/10.1371/journal.pone.0016429
7. de Kort Y, Smolders K (2010) Effects of dynamic lighting on office workers: first results of a field study with monthly alternating settings. Light Res Technol 42:345–360. https://doi.org/10.1177/1477153510378150
8. Eklund NH, Boyce PR, Simpson SN: Lighting and sustained Performance (n.d.)
9. Hoffmann G, Gufler V, Griesmacher A, Bartenbach C, Canazei M, Staggl S, Schobersberger W (2008) Effects of variable lighting intensities and colour temperatures on sulphatoxymelatonin and subjective mood in an experimental office workplace. Appl Ergon 39:719–728. https://doi.org/10.1016/j.apergo.2007.11.005
10. Hubalek S (2010) Office workers' daily exposure to light and its influence on sleep quality and mood. Light Res Technol 42:33–50. http://e-citations.ethbib.ethz.ch/view/pub:28947. Accessed 12 Apr 2016
11. Iskra-Golec IM, Wazna AMA, Smith L (2012) Effects of blue-enriched light on the daily course of mood, sleepiness and light perception: a field experiment. Light Res Technol 44:506–513
12. Kozaki T, Miura N, Takahashi M, Yasukouchi A (2012) Effect of reduced illumination on insomnia in office workers. J Occup Health 54:331–335. http://www.scopus.com/inward/record.url?eid=2-s2.0-84867347496&partnerID=tZOtx3y1

13. Maierova L, Borisuit A, Scartezzini J-L, Jaeggi SM, Schmidt C, Münch M, Moore RY, Eichler VB, Husse J, Eichele G, Oster H, Cajochen C, Khalsa S, Jewett M, Cajochen C, Czeisler C, Horne J, Östberg O, Roenneberg T, Wirz-Justice A, Merrow M, Kerkhof G, Korving H, Geest HW, Rietveld W, Katzenberg D, Reid KJ, Archer S, Hida A, Brown S, Duffy J, Rimmer D, Czeisler C, Emens J, Mongrain V, Lavoie S, Selmaoui B, Paquet J, Dumont M, Taillard J, Phillip P, Coste O, Sagaspe P, Bioulac B, Mongrain V, Carrier J, Dumont M, Horne J, Brass C, Petitt A, Schmidt C, Schmidt C, Goulet G, Mongrain V, Desrosiers C, Paquet J, Dumont M, Roenneberg T, Kumar CJ, Merrow M, Duffy J, Dijk D, Hall E, Czeisler C, Baehr E, Revelle W, Eastman C, Gunn P, Middleton B, Davies S, Revell V, Skene D, Santhi N, Cain S, Zeitzer J, Dijk D, Kronauer R, Brown E, Czeisler C, Kudielka B, Federenko I, Hellhammer D, Wüst S, Taillard J, Wittmann M, Dinrich J, Merrow M, Roenneberg T, Begeman S, Beld G, Tenner A, van der Meijden W, Phipps-Nelson J, Redman J, Dijk D-J, Cajochen C, Zeitzer JM, Czeisler CA, Dijk DJ, Danilenko K, Verevkin E, Antyufeev V, Wirz-Justice A, Cajochen C (2016) Diurnal variations of hormonal secretion, alertness and cognition in extreme chronotypes under different lighting conditions. Sci Rep 6:33591. https://doi.org/10.1038/srep33591

14. Mills PR, Tomkins SC, Schlangen LJM (2007) The effect of high correlated colour temperature office lighting on employee wellbeing and work performance. J Circadian Rhythms 5:2. https://doi.org/10.1186/1740-3391-5-2

15. Shamsul MTB, Nur Sajidah S, Ashok S (2013) Alertness, visual comfort, subjective preference and task performance assessment under three different light's colour temperature among office workers. Adv Eng Forum 10:77–82. https://doi.org/10.4028/www.scientific.net/AEF.10.77

16. Sivaji A, Shopian S, Nor ZM, Chuan N-K, Bahri S (2013) Lighting does matter: preliminary assessment on office workers. Procedia Soc Behav Sci 97:638–647. https://doi.org/10.1016/j.sbspro.2013.10.283

17. Smolders KCHJ, de Kort YAW, Cluitmans PJM (2012) A higher illuminance induces alertness even during office hours: findings on subjective measures, task performance and heart rate measures. Physiol Behav 107:7–16. https://doi.org/10.1016/j.physbeh.2012.04.028

18. Smolders KCHJ, de Kort YAW (2017) Investigating daytime effects of correlated colour temperature on experiences, performance, and arousal. J Environ Psychol 50:80–93. https://doi.org/10.1016/j.jenvp.2017.02.001

19. te Kulve M, Schlangen LJM, Schellen L, Frijns AJH, van Marken Lichtenbelt W (2017) The impact of morning light intensity and environmental temperature on body temperatures and alertness. Physiology 175:72–81. https://doi.org/10.1016/j.physbeh.2017.03.043

20. van Duijnhoven J, Aarts M, Rosemann A, Kort H (2017) Office light: Window distance and lighting conditions influencing occupational health

21. Viola AU, James LM, Schlangen LJM, Dijk D-J (2008) Blue-enriched white light in the workplace improves self-reported alertness, performance and sleep quality. Scand J Work Environ Health 34:297–306. http://www.ncbi.nlm.nih.gov/pubmed/18815716

22. Wahnschaffe A, Haedel S, Rodenbeck A, Stoll C, Rudolph H (2013) Out of the lab and into the bathroom : evening short-term exposure to conventional light suppresses melatonin and increases alertness perception: 2573–2589. https://doi.org/10.3390/ijms14022573

23. Yuda E, Ogasawara H, Yoshida Y, Hayano J (2017) Enhancement of autonomic and psychomotor arousal by exposures to blue wavelength light: importance of both absolute and relative contents of melanopic component. J Physiol Anthropol: 1–8. https://doi.org/10.1186/s40101-017-0126-x

24. Yuda E, Ogasawara H, Yoshida Y, Hayano J (2017) Exposure to blue light during lunch break : effects on autonomic arousal and behavioral alertness: 4–7. https://doi.org/10.1186/s40101-017-0148-4

25. Price L, Khazova M, O'Hagan J (2012) Performance assessment of commercial circadian personal exposure devices. Light Res Technol 44:17–26. https://doi.org/10.1177/14771535 11433171

26. Figueiro MG, Hamner R, Bierman A, Rea MS (2013) Comparisons of three practical field devices used to measure personal light exposures and activity levels. Light Res Technol 45:421–434. https://doi.org/10.1177/1477153512450453

27. Martin G (2015) LightLog – Brighten your day. http://lightlogproject.org/

28. van Duijnhoven J, Aarts MPJ, Kort HSM, Rosemann ALP (2018) External validations of a non-obtrusive practical method to measure personal lighting conditions in offices. Build Environ 134:74–86

29. van Duijnhoven J, Aarts MPJ, Aries MBC, Böhmer MN, Rosemann ALP (2017) Recommendations for measuring non-image-forming effects of light: a practical method to apply on cognitive impaired and unaffected participants. Technol Heal Care 25:171–186. https://doi.org/10.3233/THC-161258

30. Aarts MPJ, van Duijnhoven J, Aries MBC, Rosemann ALP (2017) Performance of personally worn dosimeters to study non-image forming effects of light: assessment methods. Build Environ 117:60–72. https://doi.org/10.1016/j.buildenv.2017.03.002

31. Berson DM, Dunn FA, Takao M (2002) Phototransduction by retinal ganglion cells that set the circadian clock. Science 295:1070–1073. https://doi.org/10.1126/science.1067262

32. Lucas RJ, Peirson SN, Berson DM, Brown TM, Cooper HM, Czeisler CA, Figueiro MG, Gamlin PD, Lockley SW, O'Hagan JB, Price LLA, Provencio I, Skene DJ, Brainard GC (2014) Measuring and using light in the melanopsin age. Trends Neurosci 37:1–9. https://doi.org/10.1016/j.tins.2013.10.004

33. CIE, CIE 218: Research Roadmap for Healthful Interior Lighting Applications - NSVV Nederlandse Stichting voor Verlichtingskunde, n.d. http://www.nsvv.nl/publicaties/cie-218-research-roadmap-for-healthful-interior-lighting-applications/. Accessed 10 Jan 2017

34. Hoof MBC, van Westerlaken J, Aarts AC, Wouters MPJ, Wouters EJM, Schoutens AMC, Sinoo MM, Aries MB (2012) Light therapy : methodological issues from an engineering perspective. Technol Heal Care 20:11–23. https://www.tue.nl/publicatie/ep/p/d/ep-uid/263359/

35. Aarts J, Aries MPJ, Diakoumis MBC, van Hoof A (2016) Shedding a Light on Phototherapy Studies with People having Dementia: a critical review of the methodology from a light perspective. Am J Alzheimers Dis Other Demen (2016). https://doi.org/10.1177/1533317515628046

36. Eklund NH, Boyce PR (1996) The development of a reliable, valid, and simple office lighting survey. J Illum Eng Soc 25:25–40. https://doi.org/10.1080/00994480.1996.10748145

37. van Duijnhoven J, Aarts M, Rosemann A, Kort H (2017) An unobtrusive practical method to estimate individual's lighting conditions in office environments. In: Proceedings of the 2017 IEEE 14th International Conference Networking, Sens. Control, Calabria, Italy, p 471–475

A 2018 Update on Computer Glasses for Use at Work in Norway

Magne Helland[1(✉)], Hanne-Mari Schiøtz Thorud[1],
and Hans Torvald Haugo[2]

[1] Department of Optometry, Radiography and Lighting Design,
Faculty of Health and Social Sciences, University College of Southeast Norway,
P. O. Box 251, 3601 Kongsberg, Norway
magne.helland@usn.no
[2] The Norwegian Association of Optometry, Øvre Slottsgate 18/20,
0157 Oslo, Norway

1 Introduction

On a more or less biannual basis, for thirteen years, The Norwegian Association of Optometry has performed consumer surveys in Norway. Typically, approximately 1300 randomly selected Norwegians, at the age of 15 years and above, have constituted the study participants. Professional telephone interviewers contacted them all. Among the questions addressed were also the questions listed below:

- *Do you have special prescribed computer glasses for use at work, fully or partly paid by the employer?*
- *Do you feel a need of a pair of glasses for use at the computer (computer glasses)?*
- *Have there been measurements at your work place for your special work glasses? (In particular, we have measurements of visual work distances, lighting and reflexes in mind.)*
- *Are the costs of an eye examination partly or fully covered by your employer?*

2 Purpose

The main purpose of this survey was to give an up to date number of the percentage of the Norwegian working population who use "special computer glasses for use at work" according to the Norwegian implementations [1–3] of the European Directive on computer work (EU 90/270/EEC) [4]. An alternative term for "computer glasses" is glasses for use at Visual Display Units (VDU).

3 Methods

In the first half of January 2018 Kantar TNS AS interviewed over 1300 persons by Computer Aided Telephone Interviewing (CATI). All participants were randomly selected, 15 years of age and above, and selected in such a way that they should

© Springer Nature Switzerland AG 2019
S. Bagnara et al. (Eds.): IEA 2018, AISC 827, pp. 122–124, 2019.
https://doi.org/10.1007/978-3-319-96059-3_13

represent the Norwegian population by their family's total income, geographically location and level of education. The person asked for by the interviewer was the person in the household who last had her/his birthday.

4 Results

For the 2018 survey a total of 1311 persons constituted the survey population. The percentages for those included in the questions related to computer glasses were based upon all participants who gave a positive answer on wearing spectacles, or spectacles in combinations with contact lenses (n = 958).

- *Do you have special prescribed computer glasses for use at work, fully or partly paid by the employer?* **Yes 17%**
- *Do you feel a need of a pair of glasses for use at the computer (computer glasses)?* **Yes 18%**
- *Have there been measurements at your work place for your special work glasses? (In particular, we have measurements of visual work distances, lighting and reflexes in mind.)* **Yes 34%**
- *Are the costs of an eye examination partly or fully covered by your employer?* **Yes 73% (fully 49%, partly 24%)**

In Fig. 1 the results for the question on whether people use "special" glasses at work or not, are presented for the seven surveys performed since 2005. For the surveys for 2014 and onwards the wording was altered to more precisely ask for special glasses prescribed for VDU work.

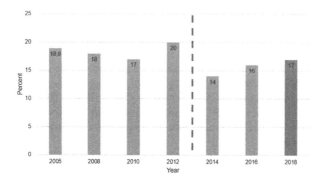

Fig. 1. Percent of people using glasses, or glasses in combination with contact lenses, having "special VDU-glasses for use at work, partly or fully paid by the employer". For 2005, 2008, 2010 and 2012 the numbers are higher and represent the percentage of people using optical corrections who also have "special glasses for use at work", which might also include special glasses for other work purposes than VDU work.

5 Discussion

In Europe work at VDUs has long been recognised as possible visually demanding. The European Directive on computer work (EU 90/270/EEC) was introduced in 1990 [4]. A few years later the Norwegian provision of the EU directive was established [1], and a separate set of guidelines developed by The Norwegian Labour Inspection Authority [2, 3]. Even though the Norwegian provision regulation has been altered since it was first introduced, the importance of optimized visual conditions at VDU work is still included at a part of the present regulation.

EU90/270/EEC [4] include recommendations for eye examinations of workers. If the results of such an examination show that it is necessary, and if normal corrective appliances cannot be used, workers must be provided with special corrective appliances appropriate for the work concerned.

6 Conclusion

According to the survey in 2018 approx. 17% of people using glasses, or glasses in combination with contact lenses, seems to use computer glasses after the regulations in the EU directive for computer work and the acting national provision.

References

1. The Norwegian Labour Inspection Authority. (Forskrift, best. Nr. 701) Forskrift om organisering, ledelse og medvirkning (2012). http://lovdata.no/dokument/SF/forskrift/2011-12-06-1355. Accessed 10 Apr 2018
2. The Norwegian Labour Inspection Authority. Arbeid ved dataskjerm (Work at Computer Screens). https://www.arbeidstilsynet.no/tema/ergonomi/arbeid-ved-dataskjerm/. Accessed 10 Apr 2018
3. The Norwegian Labour Inspection Authority. Synsundersøking og databriller (Optometric examination and Computer Glasses). https://www.arbeidstilsynet.no/tema/ergonomi/arbeid-ved-dataskjerm/synsundersoking-og-databriller/. Accessed 10 Apr 2018
4. EU 90/270/EEC Minimum safety and health requirements for work with display screen equipment (1990). http://eur-lex.europa.eu/legal-content/EN/TXT/?uri=CELEX:31990L0270. Accessed 10 Apr 2018

Visual Ergonomics in Control Rooms – An Example of Creativity in Practice

Jennifer Long[1,2(✉)], Russell Ockendon[3], and Fiona McDonald[4]

[1] Jennifer Long Visual Ergonomics, Katoomba, NSW, Australia
jlong@visualergonomics.com.au
[2] School of Optometry and Vision Science, UNSW Sydney,
Kensington, NSW, Australia
[3] Control Centres Australia, Newcastle, NSW, Australia
[4] Absolute Injury Solutions, Newcastle, NSW, Australia

Abstract. Control rooms are used in a variety of industries. Digital displays are often prominent as wallboards (overviews) and as multiple desktop displays. If the displays are not optimally configured for the work tasks and lines of sight, then individuals can develop visual and physical discomfort, and there may be adverse effects on work flow. This business case study reports the process used by an architect-ergonomist team to provide very early schematic design advice for 15 control rooms in which visual ergonomics was an integral component. End users were engaged in the design process by blending the requirements of ISO11064 for the conceptual design of control rooms with a modified participatory ergonomics approach. The principle observation is that the process engenders greater ownership of the design by the end users and pride in their new workplace when the control room is built. Engaging end users in the schematic design process also provides an opportunity for developing creative solutions to visual ergonomics design problems.

Keywords: Control rooms · Visual ergonomics · ISO 11064
Participatory ergonomics

1 Introduction

Control rooms are used in many industries including transport, mining, security, healthcare, utilities and entertainment. Digital displays are often prominent as wallboards/overviews (large displays located on walls that can be viewed by all operators within the control room) and as multiple desktop displays. The desktop displays may extend horizontally (displays either side of a central primary display) or vertically (monitors stacked in vertical tiers). Alternatively, an operator could have multiple larger size displays (such as 55 inch displays) on their console with information arranged in smaller windows horizontally and vertically on each display. Operators may also refer to other digital displays at their console, such as tablet devices, smartphones and touchscreens.

There is potential for operators to develop visual and physical discomfort if the displays are not optimally configured for the work tasks and lines of sight. For example, if an operator turns their head for prolonged periods of time to view a display located to their left- or right- side, then this could increase the muscle activity in the neck (cervical) and back (trapezius) muscles and contribute to musculoskeletal discomfort [1]. Similarly, it is more difficult for the eyes to focus (accommodate) when looking upwards [2]. Displays located above eye height will promote a change in eye gaze angle and head tilt [3], which could contribute to visual and musculoskeletal discomfort.

There is also potential for adverse effects on work flow when displays are not optimally configured. For example, console displays which obscure lines-of-sight to other operators may inhibit critical communication within the control room.

The International Standard for the ergonomic design of control centres ISO11064 [4] describes five phases for the control centre design process: (A) Clarify the purpose and constraints of the project (B) Conduct a task analysis of functional and performance requirements of the control centre (C) Conceptual design of the control centre (D) Detailed design of the control centre (E) Feedback post-commissioning to understand what works (and what does not work) with the design [4]. ISO11064 encourages a human-centred design approach whereby the physical and cognitive capabilities of users are considered in the control room design. The standard also recommends that designers should take into account organizational aspects such as how operators interact with each other, the business operations and management, as well as other factors such as operator "self-fulfillment, motivation and cultural considerations" [4]. To achieve these outcomes, ISO11064 describes various information-acquisition strategies, including the need for user participation during the design process and the formation of an interdisciplinary design team which includes user (operator) representatives. User participation could include questionnaires [5, 6] interviews [5, 7, 8], observations [5, 8] and participatory ergonomics [6].

Participatory ergonomics is an interactive process by which end users apply their expert knowledge of the work tasks to help solve workplace ergonomics problems [9]. The critical element of participatory ergonomics is the inclusion of key stakeholders, such as managers, operators and information technology workers (called the "working group") who receive training and information about ergonomics and design principles relevant to their workplace. Then through facilitated workshops with knowledge experts such as architects, ergonomists, industrial designers (called the "design team") the working group creates solutions to an ergonomics problem or design issue.

This business case study reports the process used by an architect-ergonomist team (called the "design team") for providing very early schematic design advice for 15 control rooms where visual ergonomics was an integral component. The process described in this paper corresponds to Phase C of the ISO11064 control centre design process and focusses on addressing visual ergonomics issues. This paper also presents generalised observations and lessons learnt from using the process for the very early schematic design of control rooms.

2   Control Room Projects

The control room projects selected for discussion in this paper are projects where there were visual ergonomics problems in the existing control room (such as glare from lighting, poor lines of sight within the facility) and these problems needed to be addressed in the new design, or projects in which there was a proposed complex array of visual displays or large wall-boards. Industries represented in these projects include energy production, energy supply, mining, healthcare (hospital) and transport.

The size of these projects ranged from a control room with two operators in the control room at any one time, to a control suite with more than 100 people working within the control room or interacting with operators from adjacent rooms or offsite.

3   Very Early Schematic Design Process

The design team comprised three core members: an architect, a generalist ergonomist with a background in occupational therapy, and a visual ergonomist with a background in optometry. Depending on the project scope, other members were co-opted to the design team or called upon for their specific expertise, for example, electrical and lighting designers, mechanical engineers, acoustic consultants, colour consultants and industrial designers.

The contracting organizations were encouraged to nominate representatives for the working group from a broad facet of work activities associated with the control room. These include operators representing different work functions within the control room, shift managers, overall managers, information technology workers, support workers who work outside the control room but who interact with the operators, base building design architects and project managers.

The very early schematic design process used by the design team blended the requirements of ISO11064 for the conceptual design of control rooms with a modified participatory ergonomics approach. The process included general ergonomics and visual ergonomics education, discussion of the features of other control room designs, facilitated workshops, evaluation of the preliminary design and consensus within the working group for the very early schematic design. The education component was integral to the design process because it provided an ergonomics framework for good design without placing constraints on how the control room should look. For example, the design team gave instruction to the working group for the optimal location of visual tasks. Then, through a workshop led by the design team, the working group decided how to arrange the visual displays at workstations and within the room so that they were visually comfortable and facilitated work efficiency.

Emphasis on individual elements within the very early schematic design process varied between projects according to the design requirements and the needs of the working group. Consequently, the process duration for projects ranged from 2 days to 10 days onsite. The need to modify the design process according to the project echoes observations by other authors that each control room is unique and that it is not possible to apply a generic solution to control room design [10, 11] nor is it wise to use a rigid process for generating designs [6].

4 Observations and Discussion

4.1 Ownership of the Design

The principle observation from these projects is that involving users in the very early schematic design engenders greater ownership of the design by the working group and pride in their new workplace when the control room is built.

Wilson reports a similar observation in the design a crane control room [6] and argues that although the design could have been devised by an independent consultant, it was the involvement of the workers which led to greater acceptance of the final control room design [6].

4.2 Generating Design Suggestions

The ergonomics education component of the process provided baseline knowledge for the working group. They could then draw on this knowledge when discussing design suggestions and understand the design constraints for good visual ergonomics [12]. For example, it is better to place computer displays below eye height because it is easier for the eyes to accommodate when reading from a display in a downwards gaze [2].

Most of the design suggestions in the projects were initiated by the working group. The workshop format provided a forum for the working group and the design group to discuss design options. It also helped to focus the working group's suggestions and resolve dilemmas between different design options. In their case study of a gas processing plant refurbishment Cordiner and Graves [8] stated that collaboration between the users and ergonomists generated trust because consultants did not simply put forward their own solutions, but took into account the user's views and generated solutions with the users.

There were occasions when the working group settled on non-optimal design solutions. For example, in one control room the users requested 12 displays on the workstation and this would require operators to make large head movements to view the displays. The design team could have refused to endorse this element of the design, but this would have undermined the consultation process. Therefore, the design team decided to acknowledge the rationale put forward by the working group for the large number of displays, incorporate it into the very early schematic design but with a caveat outlining the advantages and disadvantages for this design element. Wilson also reports a similar dilemma in his work assisting the design of a crane control room, and reports that the design team allowed a non-optimal solution because they did not want to override the working group and their ownership of the solution [6].

4.3 Communicating Design Options

There are myriad ways to communicate and test design options during the design process. Some examples include engaging users in building full-size prototypes [6], building small-size (Lego) scale models [13], constructing full-size 3-dimensional scale models and using tape on the floor to demarcate the location of walls [14] and developing 3-dimensional virtual representations to simulate human interaction within the control room [15].

In the control room projects described in this paper, designs were drawn with a CAD-program during the workshops. This allowed the working group and design group to visualize the designs, discuss "what if...?" and "what would you do...?" hypothetical scenarios to test the design (similar to that described by Seim and Broburg [14]) and then modify the design as the discussions progressed.

Strategies which were found valuable when testing the practicality of designs include mock-ups of the consoles and console displays to demonstrate how these affect line of sight within the control room, obtaining samples of furniture and equipment so that the working group could appraise the features and make more informed purchasing decisions (for example, to assess the impact of bezel width for viewing information displayed on a wallboard) and visiting the control room site to appreciate the size of the space and viewing distances. Three-dimensional rendered images were useful for illustrating the appearance of the control room after consensus for the design was achieved within the working group.

4.4   Demand Conflicts

While every endeavor was made to include design elements which met the operational demands of the control rooms AND facilitated good visual ergonomics, there were instances when compromises were required. For example, desktop displays stacked vertically in two tiers would reduce horizontal head rotations but would also obstruct an operator's clear view of the wallboard; consequently, operators would need to stand up or move away from their console to read the wallboard display. On one hand, such an arrangement is advantageous because it reduces sedentary behavior, and this may translate to positive health for the operators [16]. On the other hand, the obstructed lines-of-sight could be detrimental if operators adopt awkward postures to see around the console displays [17].

4.5   Technical Expertise

A tenet underpinning participatory ergonomics is that the user is an expert in the work tasks and they can provide valuable input to the design process [9]. There is also benefit including experts from a variety of professional backgrounds (such as ergonomists, architects, engineers and optometrists) who can provide evidence-based knowledge to substantiate design decisions. Teams of knowledge experts for control room design are also described by other authors [7, 18].

In some of these control room projects, business information technology (IT) professionals were also invited to the design discussions because they were able to provide advice about technology which could be integrated within the business computing systems. They were also able to make technology suggestions which addressed (or sometimes solved) design questions. For example, many of the control rooms used multiple computer applications which required dedicated displays, keyboard and mice for each application. Some of the business IT professionals were able to source multiple platform viewing software which increased the number of applications which could be accessed at any one time on a single display. This reduced the number of displays, keyboards and mice on the console and subsequently allowed a reduction in the size of the console.

4.6   Challenges

There were two principle challenges encountered during this process.

The first challenge relates to the composition of the working group. Contracting organizations were encouraged to nominate representatives for the working group and in some projects there were up to 15 people present. Other projects had only 2–3 people within the working group. Although there are not a set number of people required for working groups, more robust design discussions often occurred when there were more people present who represented a diversity of work roles associated with the control room.

While diversity among working group members is important, it is also essential to include key decision makers in the design process. Less-than-optimal designs occurred when key decision makers who were not actively involved in the very early schematic design process changed essential elements of the design (such as console orientation or lighting) without understanding the basis for the design considerations.

This type of challenge is also described by other authors for control room design [10] and for participatory ergonomics [9]. Strategies to minimize these problems include garnering active participation and support by senior and middle managers for the design process, and removing competing demands [9], for example, not expecting working group members to participate in the design process while maintaining their work output without staff relief.

A second challenge (and frustration) with some of the projects relates to the human-machine interface (HMI). There is ample literature describing optimal visual elements of an interface (such as [19]) and the visual ergonomics training helped the operators understand the visual ergonomics implications for interface design. However, visual ergonomics principles are not always incorporated within commercially available HMI products and this can affect the visual and physical comfort of operators. For example, in one control room the font size on the console displays was very small and the size could not be increased by the operators. It was possible to mount the displays on articulated arms so that operators could pull the displays closer when they wanted to read the text. In practice this took time, so operators instead leant over their console to read from their displays. This issue highlights a need for more visual ergonomics input during interface design to ensure that visual ergonomics principles are translated into the HMI design.

5   Conclusion

Control rooms are a complex working environment with high visual demands. Engaging end users in the schematic design process, as described in this paper, increases ownership of the design by the end users and provides an opportunity for developing creative solutions to visual ergonomics design problems.

The ergonomics education component of this process is integral to the design because it enables the working group to understand the scientific basis for the design. This ergonomics knowledge is not only applicable to the control room environment: it is directly transferable to other viewing environments such as working outside of the control room (for example, in the field) or using digital devices at home. Therefore the ergonomics knowledge obtained by the participants is knowledge which can be used throughout their life.

It is also important that the design elements agreed to by the working group are implemented during the build phase and that any design changes are reviewed by the working group before they are built. To that end, inviting participation by key decision makers, such as business managers, building design architects and project managers in the very early schematic design process helps these individuals understand the scientific basis for the design. This can translate to the construction of control rooms which are comfortable for the users and which facilitate good work flow, efficiency and reduce the risk of error.

References

1. Szeto G, Sham K (2008) The effects of angled positions of computer display screen on muscle activities of the neck-shoulder stabilizers. Int J Ind Ergon 38:9–17
2. Ripple P (1952) Variation in accommodation in vertical directions of gaze. Am J Ophthalmol 35:1631–1634
3. Burgess-Limerick R, Mon-Williams M, Coppard V (2000) Visual display height. Hum Factors 42(1):140–150
4. International Standards Organisation ISO 11064-1:2000 Ergonomic design of control centres- Part 1: Principles for the design of control centres. International Standards Organisation, Geneva
5. Falcao C, Soares M (2012) Ergonomic evaluation of the environment: a case study in a control room of the hydroelectric sector. Work 41:1449–1456
6. Wilson J (1995) Solution ownership in participatory work redesign: the case of a crane control room. Int J Ind Ergon 15:329–344
7. Shuman G, Walker-Peek S, Elbert G (2002) Centralized control room operates five process units. Hydrocarb Process 2002:43–46
8. Cordiner L, Graves R (1997) Ergonomic intervention during a gas processing plant refurbishment. Int J Ind Ergon 19:457–470
9. Burgess-Limerick R (2018) Participatory ergonomics: evidence and implementation lessons. Appl Ergon 68:289–293
10. Skrehot P, Marek J, Houser F (2016) Ergonomics aspects in control rooms. Theor Issues Ergon Sci 18:46–58. https://doi.org/10.1080/1463922X.2016.1159356
11. Graves V (2010) Ergonomic control room design improves operator comfort and safety. Power Eng 114:50–52
12. Long J, Helland M (2012) A multidisciplinary approach to solving computer related vision problems. Ophthalmic Physiol Opt 32:429–435
13. Bittencourt J, Duarte F, Beguin P (2017) From the past to the future: integrating work experience into the design process. Work 57:379–387
14. Seim R, Broberg O (2010) Participatory workspace design: a new approach for ergonomics? Int J Ind Ergon 40:25–33

15. Zamberlan M et al (2012) DHM simulation in virtual environments: a case-study on control room design. Work 41:2243–2247
16. Human Factors and Ergonomics Society of Australia (2015) Sedentary behaviour: HFESA position on prolonged unbroken sitting time. https://www.ergonomics.org.au/documents/item/184. Accessed 25 May 2018
17. Robbins R (2009) State-of-the-art control rooms. Control Eng 39–41
18. Hellen R (2014) Upgraded control room consoles improve ergonomics. Power 20–21
19. Hollifield B et al (2008) The High Performance HMI Handbook, 1st edn. Houston, PAS

The Relation of Visual-Digital Literacy in User Interaction with Mobile Devices

Patrícia Carrion$^{(\boxtimes)}$ ⓘ and Manuela Quaresma ⓘ

Pontifical Catholic University of Rio de Janeiro, Rio de Janeiro, RJ, Brazil
patriciatpc@gmail.com, mquaresma@puc-rio.br

Abstract. The rapid pace of technological innovation highlights the issues of users' relation with the digital sphere, and, in regards to graphical interfaces, shows the existence of a Visual-Digital Literacy. This paper proposes as an overall intention to investigate the impact of Visual Literacy in users' access to smartphones. To that effect, we hypothesize that a limited visual repertoire is a direct cause of users' deficiency in Digital Literacy skills. We defined two evaluation techniques: an Iconographic Comprehension Test and an experiment. The convergence of both sought to answer the following research question: how does Visual Literacy relate to users' digital proficiency? First, we applied the Comprehension Test to detect the proper understanding of symbols, i.e., to measure users' ability to interpret visual elements. The test proposed the classification of 48 subjects, ranging from 18 to 65 years old, in two extremes of Visual Literacy. Afterward, using extreme case sampling, we recruited 12 of said subjects to our experiment, seeking to assess the impact that Visual Literacy has on Digital Literacy, while users performed tasks on a smartphone. We used three methods of qualitative and quantitative data analysis to measure users' performance: Student's t-test and Pearson Correlation Coefficient, followed by Retrospective Think Aloud (RTA) protocol. The results showed that Visual Literacy does influence users' performance in the interaction with devices, proving that Digital Literacy relates to people's visual repertoires.

Keywords: Digital Literacy · Visual Literacy · Human-Computer Interaction

1 Introduction

Personal use technologies are everywhere, intertwined in the various aspects of everyday life. Mobile devices such as smartphones have been promoting massive transformations in society, starting from the accomplishment of daily tasks and the socialization between people to the acquisition of knowledge and the attenuation of geographical barriers. Although manifesting through a variety of senses, these technologies highlight the human tendency to visual stimuli, since device screens and their Graphical User Interfaces (GUIs) certify the presence of visual rhetoric [1].

In the digital sphere, visual information acts as active parts of the electronic text interaction process [2], which relies heavily on graphic symbols to inform the user of navigational paths. The user needs to understand the structure of the graphical interface, the meaning of an icon, feedback or other alerts and visual elements. This

© Springer Nature Switzerland AG 2019
S. Bagnara et al. (Eds.): IEA 2018, AISC 827, pp. 133–143, 2019.
https://doi.org/10.1007/978-3-319-96059-3_15

understanding of *visual reading* configures in a Visual Literacy. In this paper, we argue that this form of literacy, a sum of visual repertoires, is a skill inherent to digital users' fluency. Visual repertoires are a set of terms developed historically as parts of the common sense of culture [3]. Considering a mobile device's GUI, these terms, or elements, when interpreted by users, help them in the performance of tasks, and that is when we have the concept of *affordance*. We often use the notion of affordance to describe something that aids a user to do something. Thus, we can understand it as the relation established between an object and an organism that acts on the object.

For this study, we consider it relevant to argue about what constitutes the meaning of a visual element, stipulated through *cognitive affordance* [4]. For the affordance relation to occur, it is necessary that people, either in a designer (author) and user (reader) relationship, or by constructions and conventions within a social sphere, share the attribution of meaning to an object. If in the context of digital technologies the user demands unique visual abilities, therefore, we find it congruent to draw a parallel between Visual Literacy and the concept of Digital Literacy. Several studies on Human-Computer Interaction (HCI) fall within the scope of Digital Literacy, with researchers using cross-disciplinary approaches. In prior research, the conflict of generations in HCI, e.g., was a path deeply paved by the dichotomy of the Native and the Digital Immigrant [5]. Inequality in access to digital devices based on the age gap was also discussed in specific cases concerning the elderly [6] and children [7]. Moreover, other characteristics and social conditions were considered variables that intervene in the relation of users with devices. Amongst them, Verbal Literacy [8] and the different contexts, or cognitive and cultural backgrounds [9].

In short, there is a logical progression of the most fundamental properties of Digital Literacy, which starts from the basic access to devices to the comprehensibility of the paths of interaction through GUIs, for example. In this paper, we chose to relate the phenomenon of Digital Literacy, in the context of mobile devices – since mobile internet usage exceeds desktop worldwide [10] –, with that of Visual Literacy. We work with the hypothesis that a limited visual repertoire is a direct cause of users' deficiency in Digital Literacy skills. Here, we propose as an overall goal to research the impact of Visual Literacy, through the acquisition of repertoires, in the experience of users in accessing mobile devices such as smartphones.

2   Methodology

This descriptive research sought to establish the relationship between the variables Visual Literacy and Digital Literacy. In addition to being descriptive, this study assumes the characteristics of a *quasi*-experiment, considering the impracticability of total control of the sample, or even of the external influences on the dependent variables. This paper presents the analysis of a hypothetic-deductive model, and, in this section, we specify the techniques and procedures applied to the hypothesis testing. To understand how Digital Literacy relates to Visual Literacy, the convergence of two techniques was necessary. The first, called the Iconographic Comprehension Test, sought to evaluate how subjects interpret visual elements, more specifically icons, of graphical interfaces. With the results of the Comprehension Test, we evaluated the performance of our research subjects in an experiment.

2.1   Iconographic Comprehension Test

The Comprehension Test aims to detect the correct understanding of symbols, being an indispensable procedure in the development of images for public information [11]. Here, however, our goal was to measure the visual repertoire levels of subjects rather than the variants of a particular referent, as is usually the case. Thus, we chose not to spend the same rigidity of the tests of the genre, despite following the basic rules of application of this technique for the data collection and analysis. We developed this test plan to cover the following specific objectives: to measure the correctness and errors of the research subjects as to the meaning of commonly used icons; and to classify such individuals into two extremes of Visual Literacy. At one end, we expected to find the participants with high visual repertoire, and in another, those with a more limited repertoire. This segmentation, known as extreme or deviant case sampling, would be crucial to the recruitment of subjects for our experiment.

Participant Characteristics. To evaluate the levels of Visual Literacy, we stipulated some criteria for the selection of the participants. First, everyone should be smartphone users, since the subsequent experiment would occur by using that sort of device. Also, the subjects should either have or be pursuing a degree in higher education, so that very different degrees of schooling would not interfere in the control of the sample and, consequently, in the final results of the study. Self-selection bias is a well-documented research challenge, and, although we sought to balance the number of subjects in a total of approximately 40, distributed in a balanced way between male and female subjects and among six age groups, we do not claim that our sample accurately represents the whole population. We chose an estimated number of participants to guarantee a minimum population for a statistical treatment of the data.

Method (Test Design). The test consisted of presenting 40 images of digital icons for each test participant. Tests of this nature require around three to six variants for each referent [11], however, in the survey we chose to work only with icons commonly used in digital technologies, more specifically those present on operating systems for desktop computers of two popular platforms, Apple and Microsoft. For this reason, we selected only two variants for each referent, one representative of each platform. We chose Mac OS (from Apple) and Windows (from Microsoft) as our operating systems for icons selection since these are the most used platforms worldwide [12].

Icons Selection. For selecting the icons, we decided on a pre-selection of variants from 1984 to 2005 – due to the first operating systems of Apple, from 1984, and Microsoft, from 1985. The last variants observed were present in Windows XP (from 2001) and Mac OS X 10.4 Tiger (from 2005). The final year determined in this pre-selection is because the following versions of both systems were only released in 2007, when equivalent smartphone software began to circulate, influencing the visual repertoires of individuals. We had several icon referents fit for selection, but they were not all in balanced amounts between Apple and Microsoft's systems. Therefore, we chose only

20 of them, present on both platforms, as we matched each of the referents to two icons, tallying 40 icons. We based our choice of one icon variant per referent and platform according to the following Pre-Selection Method [11]: from each referent, we select the disparate graphic elements and, among them, those with better legibility.

Procedure. Before each icon, participants should conjecture as to the affordances of the element presented, i.e., indicate the icon's path of interaction when touched in a finger-operated UI or by a mouse cursor. As a result, we evaluated subjects' performance for visual comprehensibility; assigned them a score and then rated them about their inferred Visual Literacy. We sought to segregate the subjects' sample into two control groups: the participants with above average scores and those with below average scores. For this analysis, we scored the answers as *hits*, counting 2 points; *near hits*, 1 point; *errors*; and no answer, both without punctuation (no point, or zero). The maximum possible score achieved by each subject was 80 points. We computed the data in a spreadsheet in Microsoft Office Excel 2016 software, and we made five pilots for adjustments before conducting the test sessions.

Test Results – Selecting the Experiment Participants. We applied the Iconographic Comprehension Test remotely, via an online questionnaire, as a recruitment process to the experiment participants. We evaluated the visual repertoires of 48 subjects, being 24 male and 24 female, from six different age groups, ranging from 18 to 65 years old and above. Everyone claimed to be smartphone users and have or be pursuing a higher education degree in any field.

We selected the experiment participants based on the test scores. To do so, we performed a statistical analysis using a histogram and the Kolmogorov-Smirnov test (K–S), a test designed to assess normality, i.e., to verify the normal distribution of our histogram data. The result of this test showed normality distribution, allowing, therefore, parametric statistical inferences. After statistical analysis, we were able to highlight the extremes[1] of our sample. The subjects grouped at the ends of the sample were those that, from the mean, are at a standard deviation below or above. To facilitate the mention of these two groups in the course of this paper, we have chosen to name them as participants with *low* and *high visual comprehensibility*.

We identified 11 participants with low performance and 7 with high. For our experiment, we expected to evaluate the performance of 24 participants, among pilots, regulars, and backups. The decision to recruit backup participants started from the need to ensure that at the end of the study, approximately 14 of these subjects' performance were feasible for analysis. From 14, half should be ones with low visual comprehensibility and half with high. We defined this number since only seven subjects fit the high-comprehensibility group, and the final analysis would require two symmetric samples. To confirm the validity of the backups, we applied Student t-tests, which allowed determining which subjects outside our deviant sample we could consider, within a 95% confidence interval, as with values within the limits of one of the two

---

[1] By extremes, we mean the limits that are at or above standard deviation. They equate to the inflection point of a Normal [15].

extremities (low or high). Thus, we verified the existence of three potential backups for the low-comprehensibility group and nine for the high.

2.2   Experiment

We elaborated the plan of our experiment based on a format similar to that of a Usability Test [14], to collect empirical data on participants' behaviors and to continue the confirmation or refutation of our hypothesis. The purpose of this experiment was to explore how individuals use their respective visual repertoires in interacting with graphical user interfaces. To that effect, we targeted on evaluating only subjects' from our extreme case sampling (low or high-comprehensibility) while performing tasks on a smartphone. The confluence of the Iconographic Comprehension Test with the experiment resulted in the following questions:

- Do the different (deviant) visual repertoires interfere with tasks completion rates?
- Do the different (deviant) visual repertoires interfere with the number of touch-screen gestures during task performance?
- Do the different (deviant) visual repertoires interfere with how users navigate the interface and use shortcuts?

Method (Experiment Design). The experiment had as its essential goal to express the relations between two phenomena, in this case, Visual and Digital Literacy. Therefore, we work with some variables, being: the score of the participants in the Iconographic Comprehension Test, as an independent variable; and the task completion rates, the number of touchscreen gestures and users' overall interaction strategies – through qualitative analysis – as dependent variables. We decided to conduct the experiment sessions with the aid of an Eye-Tracking device since the use of this technology provides the monitoring of the subjects' visual attention. The experiment consisted of five tasks, presented to participants in a Nokia Lumia 820 smartphone with a Windows Phone operating system. The decision to work with this particular operating system was because it is a platform of low popularity in the market, in contrast to the Android and iOS devices [12]. For this reason, when performing the tasks using the Nokia Lumia 820 smartphone, the possibility of the users having prior knowledge of the system interface would be smaller, not influencing the experiment results. However, to ensure control of this variable, we asked the participants about any experience with Windows Phone devices and discarded participants whose response was positive.

Procedure. We gave the participants a summary of all crucial points of the survey – including purpose, type of data collected and procedures – and asked them to sign a term expressing Free, Prior and Informed Consent (FPIC). Due to the use of the Eye-Tracking device, we found necessary to have two moderators in the experiment room, one lead moderator, and one technical moderator. We defined as an introductory stage to explain the role of these moderators, as well as to point out and describe the equipment (see Fig. 1 – *left*); the participant's expected conduct; and the processes for performing the tasks, which we presented in the same order to all of the subjects. We compiled the experiment task list and presented the explanatory text of each task through the OneNote application. The five tasks were: (1) Add contact number;

(2) Disable mobile data; (3) Set a screen lock; (4) Save calendar appointment; and (5) Change system font size. As with the Iconographic Comprehension Test, we conducted pilots to determine the validity of our experiment's design. We recorded all sessions in both video and audio during the execution of the tasks, and audio only at the time of the RTA.

Regarding the smartphone device used, a Nokia Lumia 820, we made interface adaptations to the home screen (see Fig. 1 – *right*). The Windows Phone operating system has a UI based on Microsoft's "Metro" design project, whose home screen is made of live tiles. These dynamic tiles work as shortcuts, and can be added, rearranged or even removed by users. For this experiment, we chose to place all the shortcuts arranged on the initial screen to observe how the participants reacted to the visual organization of the tiles and its icons. The arrangement of the home screen shortcuts was the same for all subjects. Still, participants could have access to the features of the device in any way, at their discretion, and not only by selecting the icons of the home screen.

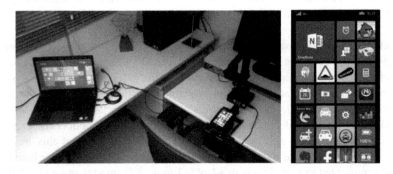

Fig. 1. *Left:* Experiment room equipped with a notebook, an external microphone, and the Tobii Pro X2–60 screen-based eye tracker. *Right:* Reordering of the Nokia Lumia 820 home screen shortcuts. We purposely enlarged the OneNote application icon to make it easier to access the tasks list.

Data Collection Methods and Analysis. To analyze our collected data, we considered necessary to manipulate the levels of the independent variable – the score of the participants in the Iconographic Comprehension Test – and to observe the result produced on the dependent variables. These variables were the users' performance in the experiment, as well as the interaction strategies in the course of the tasks. Performance data is based on user action metrics. We selected two metrics of a primarily quantitative nature and a qualitative one. Once we statistically validated the answers, we expected to crosscheck these data with the qualitative data obtained through the verbal protocol. Defining measurable issues was crucial for the data analysis planning since we needed to guarantee necessary conditions for inferring a causal relationship.

Task completion. As a criterion of analysis, we made use of the binary (or base-2) system, in which the participant received a score of zero, in case of failure in the completeness of the task, or one, if the subject completed it successfully. To determine completion, we instruct participants to announce during the session as soon as they believe they have reached the goal of the task. In a dropout situation, the subject would receive the score concerning the failure to completeness.

Touchscreen Gestures Count. Regarding touchscreen counting, we performed the convergence data analysis between the Iconographic Comprehension Test and the experiment following a mixed factorial model [15], which includes both between and within-subjects variables. To validate if there is the influence of the independent variable (Comprehension Test scoring) on the dependent (number of screen touches), we performed a Student's t parametric hypothesis test, with α of 5%[2]. With Student's t, we glimpsed to see if there was a statistically significant difference between the performances of the participants of the two groups analyzed within the experiment.

After verifying the existence of influence among the variables, we would apply another test to infer the intensity of this influence at a level of linear correlation. To do so, we used Pearson's Correlation Coefficient – or just Pearson's ρ – as a parameter, also between the Comprehension Test scores and the touchscreen gestures count. However, unlike what occurred in Student's t-test, this analysis was within-subjects, meaning we compared the results of each subject individually, not as a group. Pearson's Correlation Coefficient measures the degree of correlation (and the direction of this correlation, whether positive or negative) between two metric scale variables [16]. We based the test on the hypothesis that there should be at least a moderate to a strong correlation between the variables ($\pm < 0.5$).

Retrospective Think Aloud (RTA). Through the verbal protocol RTA, we sought to collect the qualitative material for analysis, complemented by the support of the eye-tracking device. We instructed the participants to perform the tasks without communicating with the moderator. Only at the end of the session, we requested the participants to explain the reasoning behind their actions. We used RTA with eye tracking so participants could provide context for their actions while observing video data collected by the eye tracker.

3   Results

We were able to experiment with 12 participants, six of low-comprehensibility and six of high. Among these 12, one from each group fits into the definition of a backup. We recruited them since some of the regular participants in our pre-selection were unable to attend the laboratory or did not respond to the invitation to schedule the session. We based our conclusions on the Visual-Digital Literacy relation on the empirical data collected in the experiment. In this, we observed the effect that the score on the Iconographic Comprehension Test produced on the performance of our subjects in the accomplishment of tasks.

---

[2] Alpha (α) is the percentage or margin of error accepted by the test and is equal to a 95% confidence interval.

Regarding task completion, we segregated the data by task, from one to five, and by subjects' sample, from low or high-comprehensibility. In comparing performance, low-comprehensibility subjects had less success in completing tasks (see Fig. 2).

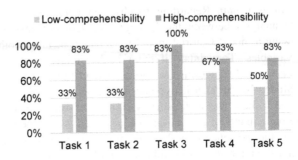

Fig. 2. Task completion rates between low and high visual comprehensibility subjects.

3.1   Student's t-Test and Pearson's Correlation Coefficient

Regarding the number of touchscreen gestures, the data crossing sought to validate if the first test scoring influences on the touchscreen gestures count. We verified that the test statistic exceeds the critical value, which rejects Student's t-test. By rejecting the null hypothesis, we can conclude that populations with low and high visual comprehensibility have statistically different averages; with the mean number of touchscreen gestures being higher for the low-comprehensibility population. This analysis, therefore, corroborates with the hypothesis of our paper, which points to the better overall performance of the subjects with a greater visual repertoire. Then, we calculated Pearson's Correlation Coefficient to measure the intensity of the independent variable's influence on the dependent variable. First, we had to deal with one outlier in our sampling, subject *P48*. In statistics, an outlier is an element in a data set that lies an abnormal distance from other values. Failure to detect these atypical values may lead to a compromised interpretation of test results. The analysis should reflect the majority of the data and not be influenced by points outside the curve [13].

The subject *P48* stood out among the low-comprehensibility subjects for not completing any of the five tasks proposed. Because he/she gave up after only a very few attempts, the participant also registered a low number of screen touches, which would mistakenly make him/her appear as a high-comprehensibility subject. Because it is a within-subject analysis, Pearson's ρ does not require a symmetric sample. Therefore, we chose to exclude *P48* from our sample since we considered it an outlier.

Regarding Pearson's Correlation, we found our sample coefficient (ρ) to indicate a moderate inverse correlation ($-0.7 > \rho < -0.5$). When the coefficient is negative, with the line declining downward, it means that one variable tends to increase as the other decreases. With this analysis, we attested that the higher the score of participants in the Comprehension Test, the smaller the number of touchscreen gestures in the experiment (see Fig. 3).

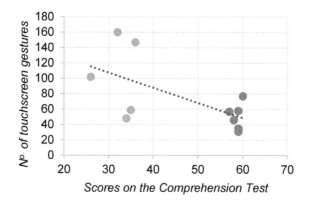

Fig. 3. Moderate-level inverse correlation: on the scatter plot we see that the number of touchscreen gestures in the experiment (Y) tends to decrease as the scores on the Comprehension Test (X) increase.

3.2   Retrospective Think Aloud (RTA)

During the RTA procedure, all 12 participants reported an initial shock with the Windows Phone System's GUI, with the six low-comprehensibility subjects having more resistance in establishing interaction strategies with the new interface. About *Task 1* ("Add contact number"), four of the five high-comprehensibility subjects who achieved task completion did so without major obstacles, except for some delay in finding navigation shortcuts on the home screen. Among the icons most searched by these participants were the ones that represented the notion of *address book*, *people* or *phone*. According to the subjects, the established search strategy was to make associations with the graphical representations presented on their Android or iOS devices.

In *Task 2* ("Disable mobile data"), four out of six low-comprehensibility participants failed completeness, with two stating to not knowing how to disable internet access on their own devices. Subject *P15*, e.g., explained that the internet is something that he/she uses "without realizing it". On that, he/she said, "If you ask me which signal is this (indicating a sign related to the mobile data) and why I'm using it, I do not know. I have no technical knowledge of the thing [sic]. I go on trying". Participants shared this sort of view throughout the sessions. Both low and high-comprehensibility people have justified many of their failures to the fact that they do not stop to reflect on how and why they perform certain actions on their own devices.

Of the 12 subjects of the experiment, only one, *P48*, was unable to complete *Task 3* ("Set a screen lock"). This participant was an outlier, as previously discussed. *Task 4* ("Save calendar appointment") had a good completion rate between both groups of participants. Only three of the 12 subjects failed task completion, two in the low-comprehensibility group and one in the high. At last, in *Task 5* ("Change system font size"), participants had to interact primarily with verbal texts in the settings menu. In the high-comprehensibility group, failure occurred only for one participant, who even went to *Settings*, but was distracted because "it had a lot of text". After exploring a few menu links, he/she gave up on finishing the task. Other participants voiced about the presence of much text in the interface by explaining: "When there is a lot of textual information, we have to stop and check everything up".

In the course of the sessions, it was tangible the greater ease of interaction of the users of high visual comprehensibility, a quality that resulted in better identification of shortcuts and icons; understanding the device features; in being straightforward while completing tasks. Low-comprehensibility subjects on many occasions seemed to explore the interface with little to no criteria, as one explained that if he/she had a whole day to accomplish the task, he/she would eventually do it: "I would end up accessing a link, one by one". On the other hand, the high-comprehensibility people were more likely to make associations and to evoke images from their repertoires.

4   Conclusion

In summary, by converging the two techniques applied in this study, we reached the goal of drawing an association between Visual and Digital Literacy. We also proved the influence of a high visual repertoire on the interaction of users with digital devices, particularly mobile devices. The experiment reinforced the validity of the deductions we raised, proving that there is a linear correlation between Visual and Digital Literacy. In an overview, high visual comprehensibility subjects obtained a higher degree of task completion and performed tasks with fewer touchscreen gestures, which indicates a more efficient performance. Although exposed to a different system than they were accustomed to, high-comprehensibility subjects were more successful in identifying pathways in navigation through shortcuts and icons, a behavior disparate to that of most low-comprehensibility people. In addition, we noticed that the low-comprehensibility subjects preferred to navigate through verbal texts, while the high-comprehensibility ones stated to be more inclined to graphical information.

With this paper, we advocate that research on Digital Literacy should be conducted with the purpose of promoting greater inclusion of people in the globalized world. For this, we find relevant to consider people with verbal communication challenges as well, as it happens with functional illiteracy and illiteracy in the strict sense. After all, when verifying the influence of Visual Literacy in the interaction of users with digital devices, it is valid to investigate if merely visual interfaces would be able to soften the barriers between illiterate people and technology. Finally, the techniques applied in this study also led to important learning. We chose to adapt the Comprehension Test, used in Informational Ergonomics, and the experiment, recurring in the fields of Ergonomics and Human-Computer Interaction. Here, instead of evaluating interface elements, we applied the techniques in the measurement of the performance of the research subjects. This applicability has proved to be successful in answering the questions of the research and could apply in future studies of the genre.

References

1. Burdick A, Drucker J, Lunenfeld P, Presner T, Schnapp J (2012) Digital humanities. MIT Press, Cambridge
2. Landow GP (2006) Hypertext 3.0: critical theory and new media in an era of globalization. JHU Press, Baltimore
3. Potter J (1996) Discourse analysis and constructionist approaches: theoretical background. In: Richardson JTE (ed) Handbook of qualitative research methods for psychology and the social sciences. British Psychological Society, Leicester, pp 125–140
4. Hartson R, Pyla PS (2012) The UX book: process and guidelines for ensuring a quality user experience. Elsevier, Morgan Kaufmann
5. Prensky M (2001) Digital natives. Digit Immigr Horiz 9(5):1–6
6. Claypoole VL, Schroeder BL, Mishler AD (2016) Keeping in touch: tactile interface design for older users. Ergon Des Quart Human Factors Appl 24(1):18–24
7. Beheshti J, Large JA (2006) Interface design, web portals, and children. Libr. Trends 54 (2):318–342
8. Chan AHS, Mahastama AW, Saptadi TS (2013) Designing usable icons for non e-literate user. In: Proceedings of the international multiconference of engineers and computer scientists (IMECS), vol. 2. Newswood Limited, Hong Kong
9. Dong Y, Lee K (2008) A Cross-cultural comparative study of users' perceptions of a webpage: with a focus on the cognitive styles of Chinese, Koreans and Americans. Int J Des 2(2):19–30
10. StatCounter GlobalStats. http://gs.statcounter.com/press/mobile-and-tablet-internet-usage-exceeds-desktop-for-first-time-worldwide. Accessed 20 Jan 2018
11. de Lemos Formiga E (2002) Comprehensibility assessment of graphical symbols through methods of informational ergonomics (Avaliação de Compreensibilidade de Símbolos Gráficos Através de Métodos da Ergonomia Informacional). In: Moraes A (org) Warnings and signaling project: informational ergodesign (Avisos, Advertências e Projeto de Sinalização: Ergodesign Informacional). iUsEr, Rio de Janeiro, pp 113–141
12. Net Market Share. https://netmarketshare.com/. Accessed 20 Jan 2018
13. Pituch KA, Whittaker TA, Stevens JP (2015) Intermediate statistics: a modern approach. Routledge, Oxfordshire
14. Rubin J, Chisnell D (2008) Handbook of usability testing: how to plan, design and conduct effective tests, 2nd edn. Wiley, Hoboken
15. Wickens CD, Gordon SE, Liu Y (1997) An introduction to human factors engineering. Addison Wesley Longman, Boston
16. Minitab. http://www.minitab.com/Accessed 19 Jan 2018

The Design Value of the Relationship Between Personal and Urban Data

Barbara Stabellini$^{(\boxtimes)}$, Paolo Tamborrini ,
and Andrea Di Salvo

DAD Department of Architecture and Design, Politecnico di Torino, Turin, Italy
{barbara.stabellini, paolo.tamborrini,
andrea.disalvo}@polito.it

Abstract. Everyday life is characterized by the interaction with an ever-increasing flow of digital data; the exponential diffusion of even more miniaturized and inexpensive sensors and the ease of connection to the Internet produce a vast amount of data, originating what is called "datization" of reality. Data belong to different typologies, but a great deal concerns the personal sphere where, in a more and more broad context of Quantified Self, people voluntarily records and tracks such data, archiving events and daily facts in a meticulous way.

However, when we talk about personal data, we have to consider the perception and the interaction between subject and three different but interconnected components: device, interface and data. In this context, design becomes a fundamental discipline, first of all trying to make the user active in the management of own data and helping him to understand them through information design tools. Secondly, data and information themselves become tools and materials for design, being a fundamental component of the project and not just its objective.

New design perspectives are opened up; starting from the tools of information design, it is possible to make immediately visible and understandable behavioural patterns of individuals, but also of a community, thinking on different scales that can range from small buildings to large cities. In this way, data can become a tool to preserve and improve individual well-being and of the society, acting with a bottom-up approach that starts with true citizen and inhabitant needs.

Keywords: Data visualization · Quantified Self · Sustainability

1 Introduction

"*Quantifico dunque sono*" [1], i.e. "I quantify therefore I am". Perhaps there is no more suitable slogan to tell about the world of Quantified Self, the motivations behind it and the thrusts that still lead many people to monitor their activities.

However, the visualization of this data can offer a reading of what really happens to your body. Without it, data collection would be an end in itself.

But it is also true, that without a proper relationship with the surrounding context, the reading of what was collected would be incomplete and sometimes insignificant.

© Springer Nature Switzerland AG 2019
S. Bagnara et al. (Eds.): IEA 2018, AISC 827, pp. 144–150, 2019.
https://doi.org/10.1007/978-3-319-96059-3_16

Exploring data through these tools offers the opportunity to read data in a more complete and meaningful way, while at the same time helping to interpret the relationships between the individual, the places where people live and individuals around them, thus favouring the implementation of small-scale improvements or development policies of the sustainable city. This process could generate a virtuous circle of exchange between user and city, with effect between personal/collective data that enables to affect mutually and to provoke behavioural changes in both views.

2 Quantified Self Scenario

The concept of self-tracking using digital technologies has recently begun, expanding his knowledge to people that were or could appear so distant from that phenomenon. Monitoring, measuring and recording elements of one's body and life as a form of self-reflection or self-improvement from the behavioural, psychological, medical point of view, are practices that have been discussed since ancient times. The introduction of digital technologies that facilitate these practices has led to renewed interest in what self-tracking can offer and to an expansion of the domains and purposes to which these practices are applied. Several reasons may influence a person to monitor himself; some of them simply do it to collect information about themselves as a way for remembering and recording aspects of their lives, or to satisfy their curiosity about the patterns in their behaviours. Others take a more specific approach, setting a goal and trying to give a meaning to the data acting to improve their health, physical fitness, emotional well-being, social relationships or work productivity. Finally, also the monitoring period can be different, ranging from very short periods, long periods, even up to years [2].

However, there weren't only new technologies, increasingly smaller and more efficient, to allow the spread of the Quantified Self movement [3], but we can identify other three factors: the integration of these technologies in the constantly connected computing devices, such as smartphones; the advent of social media which has spread the practice of sharing information, and the emergence of cloud that allows the access to a massive amount of data in order to map a collective intelligence [4].

2.1 Visualization of Personal Data

Despite nowadays' digital technologies allow data collection about many different aspects of people's life, making sense and reflecting on these large data sets is not an easy task. In fact, collecting and quantifying data is just one aspect of Quantified Self; the ultimate goal is to reflect upon one's data, extract meaningful insights, and make positive changes.

Data visualization is the way in which all the collected data are made available to the user; it can be a powerful tool for allowing users to identify patterns, establish comparisons and identify relations. It is possible to define information visualization as the use of computer-supported interactive visual representations of abstract data to amplify cognition; its purpose is not the production of the pictures themselves, but the insight [5]. The primary goals of this insight are discovery, decision making, and explanation.

By deepening the visualization of personal data, the methods of approach to it can be different. The use of wearable devices often brings with them already developed applications and platforms, often based on graphs and elemental forms, graphically attractive from the chromatic point of view. However, in any case, these apps are left in a less important level, as the majority of the attention of the project pass and on the product, and therefore on the wearable.

Those who are truly interested in the data, for a personal vision and reading, proceed with an analytical and manual approach, and then share their values in public, as witnessed by the countless interventions of MeetUp and conferences organized around the world by the Quantified Self movement. These visualizations tend to be mostly quantitative, using numbers, graphs and charts, without particular attention to the correlation if not purely mathematical between the data.

However, the recent dissemination and development of software development and coding skills have had a significant effect on information visualization practice [6]. In fact, with the ability to create and design visualizations in an increasingly accessible way, many artists and designers had begun to deepen disciplines that until a few years before had been the responsibility of engineers, mathematicians and statisticians [7].

The American designer Nicholas Felton is well known for the book he publishes each year, which details his self-tracked data in an attractive graphical representation. In the specific, the Annual Reports [8] document the measurements of a number of the author's activities over the course of the year; set out in maps and infographics, the reports reveal data gathered from everyday action (e.g. distance travelled on foot, the amount of time spent eating, travelling on public transport, the method of greeting different individuals, time spent with mom or other specific individuals, time devoted to reading or sleeping).

The project mentioned above is just one of many projects that shows a vision of the Quantified Self that tries to open outwards, but that continues to remain individualistic, creating connections only between the individual and his context, but not looking for feedback from the outside towards the inland.

On the other hand, Giorgia Lupi's approach is different, in fact at the TED2017 [9] she questions how it is possible to use personal data to create a connection between people. Starting from the experience of Dear Data [10] with Stefanie Posavec, this time the goal is to develop a system to generate a personal view of each person, a real personal portrait turned into wearable buttons, to spark conversations and connections for the five days of the event.

Finally, always taking up the concept of sharing, but only momentary of its data, an example is provided by the project developed by Clever°Franke [11] design for the electronic music festival Amsterdam Dance Event (ADE). The aim of it was creating a real-time visualization starting from the personal data of the subjects, each of them equipped with a particular sensor able to record movement, position, temperature, decibel and ID. The results are personal visualizations for each user and collective visualizations projected real-time during the event, creating a synergy between music and setting [12] (Fig. 1).

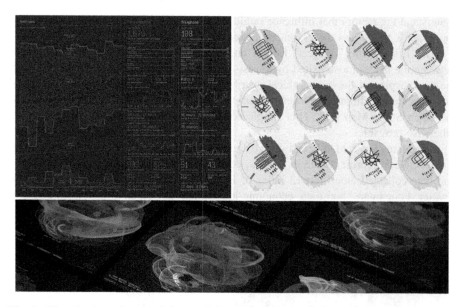

Fig. 1. Visualization of personal data: a. Feltron's Annual Report. b. Accurat, Data Portraits at TED2017 c. Clever°Franke, Red Bull Visualization.

3   City as a Datum

The case studies and the concepts described above keep the attention on a visualization that is purely an end in itself or organized for a specific purpose but of limited duration, excluding among all the usual graphics of the Quantified Self.

However, if aggregated personal data can offer an image of people in a space in a citizen model, a model that can also be fed by data from crowdsensing or open databases. The resulting city model can be used to provide personalized services to citizens and to increase people's awareness of their behaviours that can help promote common behavioural change [13].

3.1   Visualization of Urban and Personal Data

Through the tools of visual communication, such aggregation of data can be visualized and made available to understand the complex phenomena that take place in the city. This approach is what allows, and would allow more and more, a thorough reading of the territory, fostering the development of projects that respond precisely to the needs of citizens, as well as the development of new policies.

Some projects are already moving in this direction, trying to correlate and relate geography and people. An example is the project Cityways [14] developed by MIT Senseable City Lab; the goal is to explore the cities of San Francisco and Boston through the use of personal data collected by fitness tracking applications, widely used in the USA. By analysing these data and displaying them aggregated, it is possible to understand the factors that influence outdoor sports, such as time, urban morphology, topography, traffic and the presence of green areas.

The Eric Fischer' project, Locals and Tourists [15], highlights the difference between the places most visited by tourists and which instead by the residents through the map display of the geographical distribution of photographs. The author gathered metadata from millions of georeferenced online photographs and then traced them to the Open Street Map grid. This simple correlation allows a view of the city that is otherwise provable.

Indeed, the project On Broadway [16], developed for an artistic installation but with a new value; it represents life in the 21st-century city through a compilation of images and data collected along the 13 miles of Broadway that span Manhattan (Instagram, Twitter, Foursquare, taxi data, economic indicators). The result is a new type of city view, created from the activities of hundreds of thousands of people (Fig. 2).

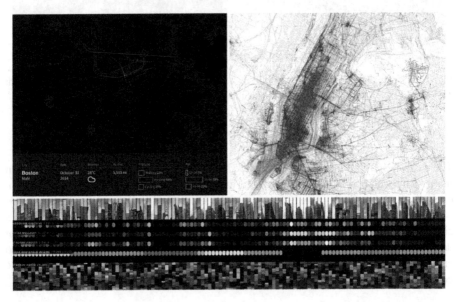

Fig. 2. Visualization of urban and personal data: a. MIT Senseable City Lab, Cityways b. Eric Fischer, Locals and Tourists. c. On Broadway.

4   Discussion

When we talk about design, and especially when we link the design to the world of data visualization, the focus tends to shift towards requirements that mainly affect aesthetics and beauty. However, we need to think about what the real components of design are and combine these aspects with the discipline of visualization, to understand all the needs. Indeed, information visualization seeks to achieve a balance between the requirements of utility, soundness and attractiveness. Utility corresponds to the basic notions of functionality, usability, usefulness and other quantitative performance measures; these aspects generally define the effectiveness and the efficiency of the visualization. Soundness is concerned with reliability and robustness. Attractiveness refers to the aesthetics aspect: the appeal or beauty of a given solution; aesthetics does not limit itself to the visual form but also includes aspects such as originality, innovation and user experience [17].

Visualization becomes part of the project. In fact, if on the one hand the attention can be turned to the visualization concerning usability and fruition, trying to develop a project that is as user-friendly as possible, on the other we need to consider what the implications might be which the project brings with it. It becomes fundamental to reason in a systemic way [18] paying attention to the organization, optimization and understanding of every single factor at play focusing at first on user requirements, then highlighting the best conditions and the most interesting facets to work on, while keeping an eye on their mutual relations. Data become as components that interact with each other in countless possible ways, and where the overall behaviour is not given by the simple sum of the behaviour of their constituent elements but depends strongly on their interactions.

The design of the collection, the analysis of the possible existing correlations and the development of a cascade system after the implementation of the project become fundamental phases, so that the project does not end itself, but well integrated with the existing system.

In this way the context represented by nowadays cities helps in changing the approach. If the project limits its boundaries only to personal data there will be changes, if they occur, in the personal sphere. Personal changes will of course have an impact on the whole city, but till today data visualization are not able the tell how and if that happens. If the project starts from issues that are both complex and related to a huge group of people, like citizens, personal data are supposed to influence all the system and data visualization should point out those relationships. For example, mobility system issues, a system that must be sustainable, are analysed in many ways, even using data visualization tools, but without linking and matching personal and urban data. Many visualizations are available, they represent: the updated state of the traffic, the use of public transport, the map of the services that promote sharing or pooling vehicles, even tools that show how the construction of a new subway line can change the system from the administration and social point of view [19]. All those visualizations are usually separated, barely readable and do not include a relationship with personal data that, on the contrary, could be essential not only to generate a personal positive, in this case sustainable, change but a new systemic citizen-based approach.

5 Conclusions

The visualization of personal data is of considerable interest, be it biological or physiological data, or data obtained from the use of social platforms or the geolocalizations of individual users. However, trying to extract the maximum potential from these data, aggregating them by relating them to a specific urban and territorial context, it can be seen how the projects developed today are configured only as beautiful representations of a state of the art, not always useful to activate subsequent decision-making processes, thus limiting the discipline of design to a mere decoration.

The growing availability of personal and urban data could be the key to the reading of complex phenomena; only through good design, to be carried out upstream and downstream, it is possible to bring out such stories from the data, thus extracting its value.

References

1. Saggio: Quantifico Dunque Sono: Il Personal Tracking E' La Religione Della Valley. http://www.mattscape.com/essay-quantifico-dunque-sono-il-personal-tracking-e-la-religione-della-valley.html. Accessed 21 May 2018
2. Lupton D (2015) The Quantified Self. Polity Press, Cambridge
3. Quantified Self. http://quantifiedself.com/. Accessed 21 May 2018
4. Lévi P (1996) L'intelligenza collettiva. Per un'antropologia del cyberspazio. Feltrinelli, Milano
5. Card SK, Mackinlay JD, Shneiderman, B. (eds) (1999) Readings in information visualization: using vision to think
6. Judelman G (2004) Aesthetics and inspiration for visualization design: bridging the gap between art and science. In: IEEE computer society, proceedings of the information visualisation, eighth international conference (IV 2004), Washington, DC, USA, pp 245–250
7. Viégas FB, Wattenberg M (2007) Artistic data visualization: beyond visual analytics. In: International conference on online communities and social computing, Berlin, pp 182–191
8. Feltron: feltron.com. Accessed 21 May 2018
9. Data Portraits at TED 2017. https://www.accurat.it/works/ted/. Accessed 21 May 2018
10. Lupi G, Posavec S (2016) Dear data: the story of a friendship in fifty-two postcards. Penguin, UK
11. Red Bull Visualization. https://www.cleverfranke.com/work/redbull-visualization. Accessed 21 May 2018
12. Red Bull at Night x ByBORRE - The Sixth Sense - Case movie. https://vimeo.com/221893073. Accessed 21 May 2018
13. Cena F, Matassa A (2015) Adopting a user modeling approach to quantify the city. In: Adjunct proceedings of the 2015 ACM international joint conference on pervasive and ubiquitous computing and proceedings of the 2015 ACM international symposium on wearable computers, Osaka, Japan, pp 1027–1032
14. Cityways. http://senseable.mit.edu/cityways/. Accessed 21 May 2018
15. Locals and Tourists. https://www.flickr.com/photos/walkingsf/sets/72157624209158632/with/4671589629/. Accessed 21 May 2018
16. On Broadway. http://www.on-broadway.nyc/. Accessed 21 May 2018
17. Moere AV, Purchase H (2011) On the role of design in information visualization. Inf Vis 10 (4):356–371
18. Bistagnino L (2011) Design sistemico: progettare la sostenibilità produttiva e ambientale. Slow Food, Bra
19. CityChrone. http://www.citychrone.org. Accessed 27 May 2018

Attention and Vigilance

A Large Scale Workplace Study

Adam C. Roberts[1,2(✉)] , George I. Christopoulos[2] ,
Hui-Shan Yap[1], Josip Car[3] , Kian-Woon Kwok[4],
and Chee-Kiong Soh[1]

[1] Civil and Environmental Engineering, Nanyang Technological University,
Singapore, Singapore
aroberts@ntu.edu.sg
[2] Nanyang Business School, Nanyang Technological University,
Singapore, Singapore
georchris7@gmail.com
[3] Health Services and Outcomes Research Programme, Lee Kong Chian School
of Medicine, Nanyang Technological University,
Singapore, Singapore
[4] Humanities and Social Sciences, Nanyang Technological University,
Singapore, Singapore

Abstract. Many previous studies have examined the effect of working environment on job performance. However, these are usually site-specific experiments examining office workers, concentrating on self-report measures and peer assessments. An area of particular interest is whether computerised tests could be used to identify deficits in performance and associate these with specific environmental problems.

We recruited over four hundred participants from several companies in Singapore, spanning a range of job types requiring different levels of visual attention, broadly grouped as technical workshop staff, office staff, and operational control room workers. Where possible, job types were matched across companies. Participants were given a series of psychological, environmental, and health-related questionnaires and computerised tests examining various aspects of visual attention (psychomotor vigilance task, go-nogo task and global-local change detection) as analogues of work performance.

Mixed effect models were used to examine the workers' performance, taking into account work-related, environmental, and health related factors. Results indicate variability across companies and job types, effects of shift work, and some effects of environment on vigilant and selective attention.

Keywords: Vigilance · Attention · Environment

1 Introduction

When designing a workplace, it is important to consider the job roles that will take place there. There is a tendency, however, to focus discussions about the work environment on the social, rather than the physical environment [1]. Indeed, collaborative

© Springer Nature Switzerland AG 2019
S. Bagnara et al. (Eds.): IEA 2018, AISC 827, pp. 151–158, 2019.
https://doi.org/10.1007/978-3-319-96059-3_17

work may benefit from open plan offices [2] whereas managerial work may be improved by individual offices that allow for private space [3]. But equally important are aspects of the physical environment that may have an impact on work. Sundstrom and Sundstrom [4] identified characteristics of the physical environment that could have outcomes for satisfaction and performance. These include ambient conditions such as temperature, air quality, lighting and noise. Similarly, Heerwagen et al. [1] identified poor ratings of these four factors as being associated with stress at work.

Previous studies mainly examine one or two of these ambient conditions in isolation, correlating these with subjective reports of productivity [5, 6]. There have been fewer studies examining the effects of these on cognitive performance, and these studies tend to be laboratory based and examining extreme conditions [7].

1.1 The Current Study

As part of a large-scale project to examine longitudinal health changes in workers in Singapore, many workers were recruited from several companies in Singapore and given a variety of cognitive tasks, questionnaires, health assessments, and environmental assessments. The overall aims of this program are to (1) examine the effect of the physical work environment on health and cognition, (2) identify resilient and sensitive populations of workers who are affected to a greater or lesser extent by the environment, and (3) offer possible interventional strategies in the form of guidelines, policies, and design recommendations. The results reported in this paper are a preliminary examination of a limited subset of these tests aimed at identifying the most salient features of the environment. Later work will examine a greater number of parameters, and include objective measurements of physical parameters of the workspaces.

2 Method

2.1 Participants

Four hundred and twenty-nine participants were recruited from three companies, across eight workplace locations within Singapore (company A: 2 locations, company B: 2 locations, company C: 4 locations). Workers from companies A and B comprised of office workers, control room staff, and workshop staff working in the transportation sector. Workers from company C were all control room staff working in the energy infrastructure sector. All workers were self-certified as fit and healthy, with normal or corrected-to-normal visual acuity. Participants were paid for their time.

2.2 Testing Environment and Procedure

All participants were tested on-site at their workplace to reduce effects of the localized testing environment. Where possible, questionnaires and cognitive tasks were administered on different days to avoid test fatigue. Cognitive testing took part in windowless meeting rooms to reduce visual distractions and noise, with small groups of between 2–4 workers at any one time.

2.3   Questionnaires

Participants were given two computer-based questionnaires to fill out. The first questionnaire examined health parameters including sleep quality (Pittsburg Sleep Quality Index, PSQI [8]), and physical and mental health (Short Form Health Survey, SF36 [9]). The second asked for subjective reports of the participant's direct work environment, and included ratings of air quality, background noise, lighting, surrounding greenery, and temperature on five-point likert scales. Demographic information was also recorded, including normal work space location, job category, and whether the participant was a shift worker.

2.4   Cognitive Tasks

Participants completed three cognitive tasks on laptops using the PEBL psychological test battery [8]: The Psychomotor Vigilance Task (PVT [9]), the Go/No-Go test [10], and the Global-Local Task [11].

PVT. The three minute PVT was administered to examine vigilant attention. This is a simple reaction time task, where a participant has to respond as quickly as possible to a target appearing on screen. Responses were categorized according to the speed of reaction time (RT < 150 ms [including negative response times] = "false alarm", 150 < RT < 500 ms = "correct", 500 < RT < 30000 ms = "lapse", RT > 30000 ms = "sleep").

Go/No-Go. The Go/No-Go task examines response inhibition and requires participants to either execute or inhibit a motor response. Participants had to watch a sequential presentation of letters appearing on the screen and respond to a target letter by pressing a button. In the first condition, participants responded to a high frequency target (P-Go) and had to inhibit responses to a low frequency target (R-NoGo). In the second condition, participants had to respond to the low frequency target (R-Go) and inhibit responses to the high frequency target (P-NoGo). Participants completed a short training block, followed by 160 trials in each condition. Accuracy and reaction time were recorded.

Global-Local. The Global-Local task uses Navon figures to examine global precedence. Many small letters are used to form a larger letter. In the first condition, participants are required to respond to the small letters (local stimuli) while inhibiting responses to the larger letters (global stimuli). In the second condition, participants are required to respond to the larger letters while inhibiting responses to the small letters. Participants completed several training blocks with isolated global and local stimuli, and combinations of global and local stimuli. The testing blocks comprised of 10 trials per condition. Accuracy and reaction time were recorded.

2.5   Analysis

All data analyses were performed in R [11]. For the questionnaire data, an initial data cleaning was performed, where participant responses were grouped into ordinal categories. Subjective reports of environmental variables were simplified into binary categories (satisfied vs. unsatisfied) using a median-split analysis. Self-reported sleep quality (PSQI) was simplified into a binary response by classifying a score of <5 to indicate good sleep quality and >5 to indicate poor sleep quality. The SF36 Physical and Mental components were again simplified into binary categories (good vs. poor physical and mental health). For the cognitive tasks, results were collapsed per person to give counts per response category (PVT), mean accuracy and reaction time (Go/No-Go) and global precedence (Global-Local) per participant. Finally, missing responses were removed from the analysis.

Mixed effect models with stepwise automatic model selection were computed to examine how cognitive performance was affected by subjective environmental and health factors. For each cognitive task outcome, the following model was applied:

$$Outcome\_variable \sim company + work\_category + IAQ + noise + lighting$$
$$+ greenery + temperature + PSQI + SF36Physical + SF36Mental$$
$$+ NightShiftWorker \tag{1}$$

A bidirectional stepwise AIC algorithm was used to remove terms from each model to give the best fit model for each outcome variable. Removed terms did not contribute to the interpretation of results.

3   Results

15 participants were removed from the analysis due to missing data in one or more input variables. Analysis was conducted on 414 participants. Ten best fit models were generated separately for the different subcomponents of the three cognitive tasks.

3.1   PVT

PVT model effects for false alarms, correct responses, and lapses are presented in Table 1. The stepwise AIC algorithm found an intercept-only effect for sleep responses, likely caused by a low number of responses in this category. In line with previous research [12], PVT false alarms and lapses were significantly affected by shift work, with shift workers having greater numbers of false alarms and lapses indicating possible sleep fatigue. However, this was not reflected in self-report sleep quality (PSQI). Several environmental factors were kept in the models, but the only environmental factor that met our $p < .05$ criterion was temperature, which impacted the number of correct responses. Participants who were more satisfied with the temperature in their personal work area made fewer correct responses. There was also an effect of work category, with workshop staff making fewer correct responses (post-hoc $t(406) = -2.92, p = .0037$).

3.2   Go/No-Go

Go/No-Go analysis was conducted on 401 participants due to missing reaction time data in 13 participants. Model effects for reaction times are presented in Table 2. There were some effects of environment on reaction times, but none reached the $p < .05$ criterion.

Table 1. PVT model effects.

	False alarm		Correct response		Lapse	
	F(406)	P-value	F(406)	P-value	F(406)	P-value
Intercept	122.86	<.0001	8808.57	<.0001	270.74	<.0001
Company	3.97	<.0001	4.57	.0109	6.37	.0019
Work category	2.75	.0652	5.13	.0063		
IAQ	3.767	.0530				
Noise					3.77	.0528
Lighting			0.19	.6628	1.20	.2746
Greenery					3.03	.0824
Temperature			4.55	.0335		
PSQI					2.04	.1541
SF36P	2.70	.1011	2.60	.1078		
Night shift worker	5.00	.0260			4.50	.0345

Table 2. Go/No-Go RT model effects.

	P-Go RT		R-Go RT	
	F(399)	P-value	F(397)	P-value
Intercept	25627.08	<.0001	45376.54	<.0001
Noise			3.62	.0579
Temperature	3.80	.0520		
SF36P			3.23	.0732
SF36 M			2.12	.1489

Model effects for Go/No-Go accuracy are presented in Table 3. There was an effect of work category, with workshop staff being less accurate in both P-Go (post-hoc t (396) = −4.34, $p < .0001$) and R-Go conditions (post-hoc t(394) = −2.25, $p = .0249$). The effects of environmental conditions were seen in R-Go accuracy, with workers being more accurate if they rated their lighting as satisfactory, but less accurate if they rated the amount of green space as satisfactory. Workers with higher mental health

Table 3. Go/No-Go Accuracy model effects.

	P-Go accuracy		P-NoGo accuracy		R-Go accuracy		R-NoGo accuracy	
	F(396)	P-value	F(398)	P-value	F(394)	P-value	F(397)	P-value
Intercept	55677.64	<.0001	455.77	<.0001	77387.52	<.0001	45356.60	<.0001
Company			4.83	.0084				
Work category	10.79	<.0001			3.26	.0395	2.63	.0734
Lighting					8.33	.0041		
Greenery					4.02	.0457		
Temperature	3.71	.0547			3.80	.0520		
PSQI					8.67	.0034		
SF36M	3.95	.0475	2.96	.0863			4.41	.0364

scores were more accurate in the P-Go and R-NoGo conditions. Finally, workers with a lower self-reported sleep quality (PSQI) were significantly less accurate in the R-Go condition, suggesting that sleep fatigue could be reducing their selective attention.

3.3   Global-Local

Global-Local analysis was conducted on 405 participants due to missing data in 9 participants. Global precedence was calculated as mean accuracy in responding to global targets with local distractors minus mean accuracy in responding to local targets with global distractors. Model effects for Global-Local accuracy are presented in Table 4. There was a significant effect of greenery only, where workers who were less satisfied with the greenery in their personal work area having a greater global precedence (least-square mean estimate: 0.03), whereas those more satisfied with the greenery in their personal work area had a negative global precedence (least-square mean estimate: −0.01), i.e. they had a small local precedence when confronted with conflicting local and global information.

Table 4. Global-Local Global precedence model effects.

	F(405)	P-value
Intercept	0.01	.9390
Greenery	5.34	.0213

4   Discussion

We gave several subjective health and environmental questionnaires and cognitive tasks to over four hundred workers as part of a larger work-based study on health and cognitive function at work. To deal with the large number of input parameters, we employed mixed effect models with stepwise automatic model selection to identify the models with the best fit.

Consistent with previous work [12], we found that shift workers performed poorly on the PVT, with a greater number of false alarms and lapses. This indicates that these workers are suffering from fatigue related to poor sleep. However, these effects did not show in subjective reports of sleep quality. This suggests that self-reported sleep quality should be supplemented with PVT scores in jobs requiring vigilant attention. However, PSQI score did have an effect on the responses to low-frequency stimuli in the Go/No-Go task, where workers with poor sleep quality were less accurate at responding in the R-Go condition.

Workshop staff differentiated themselves from the office and control room workers in both the PVT and Go/No-Go tasks, where they had poorer performance. This may be attributable to a lack of computer use in their day-to-day tasks or education level. However, this effect did not show in the Global-Local task. Alternatively, these participants may have a greater level of work fatigue caused by the manual nature of their jobs.

There were some effects of environmental variables on performance in these cognitive tasks. All five environmental variables studied were included in some way in the models generated, although the only variables that showed significant differences were temperature, lighting, and greenery in the personal work area. These showed surprising results, where satisfaction with temperature and greenery in the personal work environment decreased performance on cognitive tasks. It could be that a simple median split categorization of these parameters is too simplistic. Alternatively, a more objective approach using measured environmental variables may be more informative.

Overall, the results presented here show that cognitive performance is rarely studied in isolation. When testing workers in the field, many parameters can affect results, and these should be taken into account where possible.

Acknowledgements. This material is based on research/work supported by the Land and Liveability National Innovation Challenge under L2 NIC Award No. L2NICCFP1-2013-2. Any opinions, findings, and conclusions or recommendations expressed in this material are those of the authors and do not necessarily reflect the views of the L2 NIC.

References

1. Heerwagen JH, Heubach JG, Montgomery J, Weimer WC (1995) Environmental design, work, and well being: managing occupational stress through changes in the workplace environment. Aaohn J 43(9):458–468
2. Hua Y, Loftness V, Heerwagen JH, Powell KM (2011) Relationship between workplace spatial settings and occupant-perceived support for collaboration. Environ Behav 43(6):807–826
3. Brennan A, Chugh JS, Kline T (2002) Traditional versus open office design: a longitudinal field study. Environ Behav 34(3):279–299
4. Sundstrom E, Sundstrom MG (1986). Work places: The psychology of the physical environment in offices and factories. CUP Archive
5. Clements-Croome D, Baizhan L (2000) Productivity and indoor environment. Proc Healthy Build 1:629–634
6. Akimoto T, Tanabe SI, Yanai T, Sasaki M (2010) Thermal comfort and productivity-Evaluation of workplace environment in a task conditioned office. Build Environ 45(1):45–50
7. Qian S, Li M, Li G, Liu K, Li B, Jiang Q, Li L, Yang Z, Sun G (2015) Environmental heat stress enhances mental fatigue during sustained attention task performing: evidence from an ASL perfusion study. Behav Brain Res 280:6–15
8. Buysse DJ, Reynolds CF, Monk TH, Berman SR, Kupfer DJ (1989) The Pittsburgh Sleep Quality Index: a new instrument for psychiatric practice and research. Psychiatry Res 28 (2):193–213
9. Farivar SS, Cunningham WE, Hays RD (2007) Correlated physical and mental health summary scores for the SF-36 and SF-12 Health Survey, V. 1. Health Quality Outcomes 5 (1):54
10. Mueller ST, Piper BJ (2014) The psychology experiment building language (PEBL) and PEBL test battery. J Neurosci Methods 222:250–259

11. Dinges DF, Pack F, Williams K, Gillen KA, Powell JW, Ott GE, Aptowicz C, Pack AI (1997) Cumulative sleepiness, mood disturbance, and psychomotor vigilance performance decrements during a week of sleep restricted to 4–5 hours per night. Sleep 20(4):267–277
12. Bezdjian S, Baker LA, Lozano DI, Raine A (2009) Assessing inattention and impulsivity in children during the Go/NoGo task. Br J Dev Psychol 27(2):365–383
13. Navon D (1977) Forest before trees: The precedence of global features in visual perception. Cogn Psychol 9(3):353–383
14. Team RC (2013) R: A language and environment for statistical computing
15. Åkerstedt T, Wright KP (2009) Sleep loss and fatigue in shift work and shift work disorder. Sleep Med Clin 4(2):257–271

Effects of Differences in Vision upon Drivers' Spatial Cognition:

Focus on the Subjective and Objective Viewpoints

Katsuhiro Teranishi[1]([✉]), Tomonori Ohtsubo[2], Seishi Nakamura[2],
Yoshiaki Matsuba[2], and Miwa Nakanishi[1]

[1] Keio University, 3-14-1, Hiyoshi, Kohoku-ku, Yokohama,
Kanagawa 223-0061, Japan
`katsuhiro.teranishi@keio.jp`
[2] Mazda Motor Corporation, 3-1 Shinchi, Fuchu-cho,
Aki-gun, Hiroshima 730-8670, Japan

Abstract. To support spatial cognition by drivers, it is becoming common for cameras and monitors to be attached to automobiles to enable drivers to see perspectives (objective viewpoints) besides their own field of vision (subjective viewpoint). Previous studies have suggested that the difference between the subjective and objective viewpoints influences drivers' spatial cognition of their automobiles; however, the specific impacts on the human cognitive process of recognizing space, and on driving performance, have yet to be revealed.

Thus, this study was designed to experimentally assess the role of subjective and objective viewpoints in the cognitive process of driving and the level of driving performance.

The following results were obtained: (1) driving behavior with a subjective viewpoint tends to be more careful, as demonstrated by the rate of collision with dynamic objects. It was shown that a high cognitive load was applied in this case, but that subjective fatigue was small. It was thought that the subjective viewpoint makes a sense of ownership occur more readily than the objective viewpoint, so drivers tried to avoid collision by unconsciously recognizing their cars as part of themselves. (2) Driving with an objective viewpoint tended to be smoother, as evidenced by the frequency of collision with a wall. In addition, the cognitive load was also low.

Keywords: Subjective viewpoint · Objective viewpoint · Spatial recognition

1 Introduction

As a support for drivers, cameras and monitors have increasingly come to be attached to automobiles to give other perspectives (such as front viewpoint) besides their own vision (subjective viewpoint). Furthermore, there is a possibility that drivers' own view may be replaced by an objective view, during remote operation etc., in the future.

Previous studies suggest that subjective and objective viewpoints influence drivers' spatial cognition of their automobiles and the surrounding objects, including traffic lanes, oncoming traffic, and pedestrians in different ways [1]. However, neither the

© Springer Nature Switzerland AG 2019
S. Bagnara et al. (Eds.): IEA 2018, AISC 827, pp. 159–168, 2019.
https://doi.org/10.1007/978-3-319-96059-3_18

specifics of the impact, nor its effect on driving quality, are yet understood. Thus, this study was designed to experimentally assess the role that the subjective and objective viewpoints played in the cognitive process of driving and level of driving performance.

2 Method

We set a task of driving a car from the start of a displayed 3D virtual maze to the goal. This task was selected to have a certain cognitive load when looking for a road to the goal. During the experiment, we changed the vision presented to the participants and compared task-performance under each condition.

2.1 Experiment Environment

We used the two large displays (two 65-in. Displays, Panasonic Inc.) shown in Fig. 1 to completely cover the stable field of fixation by the participants. Participants performed tasks using a controller (Xbox controller, Microsoft) (Fig. 2).

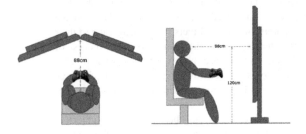

Fig. 1. Experimental environment

Fig. 2. Experimental photo

2.2 Experimental System

The maze was structured with 256 square units—16 columns and 16 rows. One side of each square unit (the road width) was 4 m. The wall of the maze was set at a height of 4 m and a thickness of 0.07 m. We prepared 7 patterns for the maze, each having the same difficulty based on the result of the preliminary experiment. The height of the car is 1.69 m, its length is 4.54 m, and its width is 1.84 m with reference to existing sport utility vehicles (SUVs).

In the maze, dynamic objects (objects imitating a bicycle or an elementary school student) appeared in the preset intersection. There are sound effects for when the car crashes against the dynamic objects and the wall.

2.3   Experiment Task

The task is to drive the car from the start to the goal as quickly as possible without collisions with the walls or dynamic objects. Participants are required to drive in the center of the road as much as possible. The task finished when the car reached its goal or when 8 min passed.

2.4   Experimental Conditions

We presented different views under experimental conditions, as shown in Fig. 3.

   i. Subjective view
      We set the perspective to the driver's viewpoint.
  ii. Objective view
      We set the perspective to a position where the whole of the car can be seen from behind.
 iii. Front view
      We set the viewpoint position to the front end of the car and presented a range of 150° to the left and right.
  iv. Subjective view with semitransparent walls
      We set the viewpoint to the same position as in i and the walls adjacent to the car are made semitransparent.
   v. Objective view with semitransparent walls
      We set the viewpoint to the same position as ii and the walls adjacent to the car are made semitransparent.

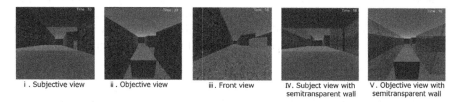

i . Subjective view   ii . Objective view   iii. Front view   IV. Subject view with   V. Objective view with
                                                                 semitransparent wall   semitransparent wall

Fig. 3. Experimental conditions

2.5   Experimental Procedure

Participants trained for more than 5 min under each condition prior to the experiment. We asked participants to rest with their eyes closed for 1 min before each task. After each task, participants answered questionnaires about the previous task. The task order was randomly set for each participant.

2.6   Recorded Items

We recorded operational logs, subjective-evaluation questionnaires, oxygenated-hemoglobin concentrations in the prefrontal cortex, and -electro-cardiograms (ECGs).

- **Operational logs.** Operational logs were recorded every 1/60 s, and we calculated the following driving-performance indices using these logs.

  a. Wall-collision frequency = (number of wall collisions)/(driving distance)
     Higher wall-collision frequencies are considered to represent the difficulty of spatial recognition of the positional relationship between the wall and the car.
  b. Deviation distance between the car and the center line of the road = (the average minimum distance between the center of the car and the center line)
     Participants are required to drive on the center line of the road as much as possible. Therefore, when the deviation distance between the car and the center line is small, participants are thought to correctly recognize the position of the car.
  c. Collision rate with dynamic objects = (number of collisions with dynamic objects)/(number of appearance times of dynamic objects)
     A high collision rate with dynamic objects indicates that such objects are not recognized properly.
  d. Stop frequency = (number of stops)/(driving distance)
     Stopping represents interruption or stagnation of movement. Therefore, higher stop frequencies indicate difficulties in decision making or cautious behavior.
  e. Braking duration during one braking operation = (average braking duration)
     Braking can be adjusted only in terms of duration, not in terms of strength. Therefore, drivers who frequently use short brakes are judged as careful while adjusting the speed, whereas those who use long brakes frequently are judged to be driving roughly with a lot of sudden braking.

- **Physiological indices.** The cerebral-blood volume and electrocardiogram were measured in order to examine the physical mental stress, cognitive load, and psychological change during the task. The first half of the task only was analyzed, since the influence of elements related to maze searching may differ during the latter half. Physiological-indicator data were standardized for the experimental participants.

  Participants were attached to electrocardiogram-measuring (ECG-100C, MP150, BIOPAC system Inc.) and functional near-infrared-spectroscopy (fNIRS) devices (OEG-16, Spectratech Inc.). The electrocardiogram was measured using a disposable electrode attached to three points of the left and right clavicle and the left rib. A bandpass filter of 0.5 to 35 Hz was applied and the sampling rate was 2 kHz. fNIRS recorded concentration changes of oxygenated-hemoglobin concentrations (Oxy-Hb) of 16 channels on the prefrontal cortex with a time resolution of 0.76 Hz and a sampling interval of 0.65 s. We analyzed only 12 participants from whom we were able to obtain appropriate data.

- **Subjective evaluation questionnaires.** Participants answered the following questions from −100 to 100 with Freescale.

  q1. Did you find it difficult to explore the maze? (−100: difficult, 100: easy)
  q2. Could you grasp the direction of goal? (−100: never, 100: always)
  q3. Was it easy to grasp the direction of the car on the lane of the road? (−100: difficult, 100: easy)

q4. Was it easy to grasp the position of the car on the lane of the road? (−100: difficult, 100: easy)

q5. Could you drive at the speed you wanted when going straight? (−100: never, 100: always)

q6. When turning right or left, could you drive as you wanted? (−100: never, 100: always)

q7. Were you fearful of any incidents arising from unexpected dynamic objects? (−100: never, 100: always)

q8. Was it easy to find dynamic objects? (−100: difficult, 100: easy)

q9. Did you drive faster than you imagined? (−100: slower, 100: faster)

q10. Did you drive smoothly during the whole task? (−100: jerky, 100: smooth)

2.7 Participants and Ethical Considerations

Participants were 18 students (21–24 years old; average: 22.7; standard deviation: 1.11). We instructed them not to take alcohol or caffeine starting the day before the experiment and confirmed that they had obeyed these instructions.

We explained the experiment orally and in writing to the participants and acquired signed consent forms. In addition, we analyzed the data after making personal identification impossible.

3 Comparison of Results for Subjective and Objective Views

Figure 4 shows the wall-collision frequency for the subjective and objective viewpoints. This frequency is found to be smaller using the objective viewpoint than the subjective one, showing that driving from the objective viewpoint makes it easier to grasp the distance from the car to the wall than it is using a subjective viewpoint.

Figure 5 shows the deviation distance between the car and the center line of the road for the subjective and objective viewpoints. From the results, the deviation distance between the car and the center line of the road is smaller using the objective viewpoint than the subjective one, showing that it is easier to conceptualize the position of the car in road from the objective viewpoint.

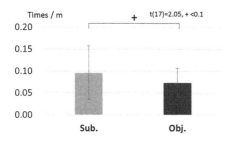

Fig. 4. Wall-collision frequency

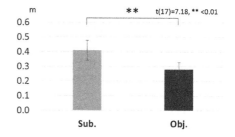

Fig. 5. Deviation distance between the center line of the road

The same result was obtained from items related to the subjective-evaluation questionnaire, namely "q4. Understanding the position of the car in the lane" (Fig. 6) and "q.6 the ease of control when turning left and right" (Fig. 7).

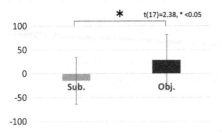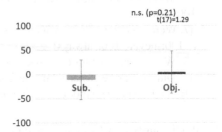

Fig. 6. q4. Was it easy to grasp the position of the car in the lane of the road? (−100: difficult, 100: easy)

Fig. 7. q6. When turning right or left, could you drive as you wanted? (−100: never, 100: always)

Thus, it can be said that the objective viewpoint is superior to the subjective one in terms of "static" spatial cognition. One factor contributing to this is that, from the subjective viewpoint, it is not possible to directly see the positional relationship between the walls and the car, but this can be seen from the objective viewpoint.

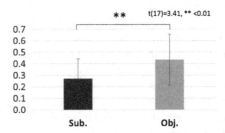

Fig. 8. Collision rate with dynamic objects

Figure 8 shows the collision rate with dynamic objects, as gauged from comparison between the subjective and objective viewpoints. From the results, the collision rate with dynamic objects is smaller from the subjective viewpoint than from the objective viewpoint, showing that the former is more appropriate than the latter in terms of awareness and action for avoid the collision with dynamic objects that appear unexpectedly.

The same result can be determined from the subjective-evaluation questions related to "q7. Frequency of potential incidents" (Fig. 9) and "q8. dynamic-object awareness" (Fig. 10).

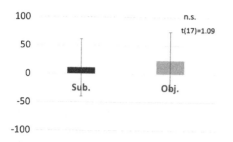

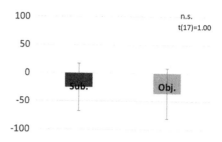

Fig. 9. q7. Were you fearful of any incidents arising from unexpected dynamic objects? (−100: never, 100: always)

Fig. 10. q8. Was it easy to find dynamic objects? (−100: difficult, 100: easy)

Thus, the subjective viewpoint can be said to be superior to the objective one in terms of "dynamic" spatial cognition.

Driving behaviors were analyzed to determine the factors contributing to the superiority of the subjective viewpoint in "dynamic" spatial cognition.

Figure 11 compares the stop frequency for the subjective and objective viewpoints. Stop frequency is shown to be higher in the subjective case than the objective one. Figure 12 compares the distribution of braking duration during one braking operation for the subjective and objective viewpoints; it can be seen that the subjective viewpoint has more short braking operations than the objective one.

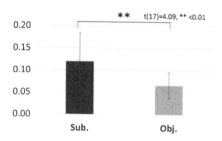

Fig. 11. Stop frequency

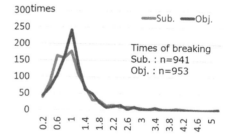

Fig. 12. Distribution of braking duration during one braking operation

Considering the above comprehensively, it is suggested that the subjective viewpoint resulted in a careful driving tendency represented by more frequent braking operations, stopping, and the like than the objective viewpoint. In addition, we believe that this careful driving tendency in the subjective viewpoint has led to the appropriateness of awareness and action for avoid the collision with dynamic objects that appear unexpectedly.

Furthermore, physiological indices were analyzed to consider the influence of each viewpoint on human cognition and psychology.

Figure 13 compares the activation amounts in the orbitofrontal cortex (OFC). Activation of the OFC is primarily involved to predict appropriate action with inferring and judging current or future situations based on experience [2]. As seen in Fig. 13, the cerebral-blood volume of the OFC from the subjective viewpoint was activated significantly compared to the objective viewpoint. This indicates that driving from the subjective viewpoint involves more active processing in terms of prediction for action selection than from the objective viewpoint. Figure 14 compares the activation amounts in the medial prefrontal cortex (MPFC), which is primarily involved in advanced cognition, judgment and decision-making [3]. As seen in Fig. 14, the cerebral-blood volume in the MPFC was significantly higher when using the subjective viewpoint than the objective one. This indicates that the subjective viewpoint involves more advanced cognition, judgment, and decision making than the objective one. Thus, it is suggested that a higher cognitive load was required for the subjective viewpoint than the objective one.

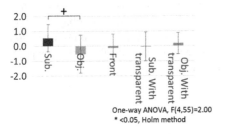

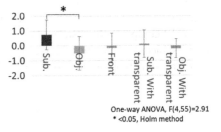

Fig. 13. Comparison of cerebral-blood-volume changes in the OFC

Fig. 14. Comparison of cerebral-blood-volume changes in the MPFC

Figure 15 compares the activation amounts in the left-ventrolateral prefrontal cortex (l-VLPFC). Activation of the l-VLPFC is primarily involved in linguistic working memory [4] and has a negative correlation with subjective fatigue [5]. There was found to be a higher cerebral-blood volume in the l-VLPFC for the subjective viewpoint than for the objective one. Thus, it seems that tasks are not related to linguistic working memory, so this result suggests that the subjective viewpoint involves less subjective fatigue than the objective one.

Seemingly contradictory results were obtained for the subjective viewpoint, with less conscious fatigue and a higher cognitive load than for the objective viewpoint. The experimental results showed that there were few collisions with dynamic obstacles, although there were many collisions with static objects when using the subjective viewpoint. This effect may result from careful and prudent driving behavior.

One possibility for why the subjective viewpoint encouraged careful driving behavior and whether or not subjective fatigue was suppressed despite this is that "a Sense of Ownership (SoO)" acted in the subjective viewpoint. SoO is the feeling that

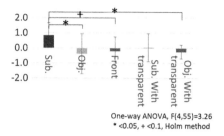

One-way ANOVA, F(4,55)=3.26
* <0.05, + <0.1, Holm method

Fig. 15. Comparison of cerebral-blood-volume changes in the l-VLPFC

one's body is one's own, and it is generated by sensory feedback; in other word, there is a need for no uncomfortable sensations to allow a SoO [6]. According to previous study, while it is difficult to have a SoO from an objective viewpoint, it has been suggested that it is easy to have one from a subjective viewpoint [7, 8]. Therefore, in this experiment, the subjective drivers are thought to have unknowingly recognized the car as a part of themselves. The other previous studies shows subjects are known to feel pain in reaction to an averse stimulus to an area recognized as part of their own body [9]. It is also known that the reaction time is faster than the reaction outside region [10]

Considering these things comprehensively, SoO is considered to allow in the subjective viewpoint, leading to cautious driving behavior and making efforts to avoid collisions between one's self and others. Also, since this is an unconscious action, it is possible that it might not result in subjective fatigue.

4 Conclusion

In this research, an experiment was conducted to clarify the influence of subjective and objective viewpoints upon drive behavior, cognition, and psychology at the time of driving.

The following features were found for each viewpoint. It was suggested that the subjective viewpoint easily enabled a sense of ownership; this was found to encourage careful and prudent driving behavior while leading to little subjective fatigue. On the other hand, from an objective viewpoint, seemingly smooth driving behavior was seen, especially with static objects, and the cognitive load was also low because the positional relationship between the car and the surroundings is explicit.

References

1. Shimizu S et al (2014) Development of wraparound view system for vehicles. Inst Image Inf Telev Eng 68(1):J24–J29
2. Ono T et al (2015) Mechanism of emotion and intelligent information processing. High Brain Funct Res 25(2):116–128
3. Murai T (2006) Intracerebral mechanism of emotional cognition and social behavior and its obstacles. Jpn J Cogn Neurosci 8(1):56–60

4. Paulesu E et al (1993) The neural correlates of the verbal component of working memory. Nature 362:342–345
5. Leby R et al (2006) Apathy and the functional anatomy of the prefrontal cortex–basal ganglia circuits. Cereb Cortex 16(7):916–928
6. Ueda S (2011) Mechanisms of extended body phenomena: the different between passive and active inputs, and relation to mirror system. J Grad Sch Humanit Sci 14:217–226
7. Preston C et al (2015) Owning the body in the mirror: the effect of visual perspective and mirror view on the full-body illusion. Sci Rep 5:18345
8. Maselli A et al (2013) The building blocks of the full body ownership illusion. Front Hum Neurosci 7(83):1–15
9. Durgin F et al (2007) Rubber hands feel the touch of light. Psychol Sci 18(2):152–157
10. Whiteley L et al (2007) Visual processing and the bodily self. Acta Psychlogia 127:129–136

Comparison of Rearview Options for Drivers Using a Virtual 3-D Simulation

Sora Kanzaki[1(✉)], Tomonori Ohtsubo[2], Seishi Nakamura[2],
Yoshiaki Matsuba[2], and Miwa Nakanishi[1]

[1] Keio University, 3-14-1, Hiyoshi, Kohoku-ku, Yokohama,
Kanagawa 223-0061, Japan
sora.0120@keio.jp
[2] Mazda Motor Corporation, 3-1 Shinchi, Fuchu-cho, Aki-gun,
Hiroshima 730-8670, Japan

Abstract. Mirrors are used for showing the backward view in conventional driving. However, it has become possible to present the rear space using cameras and monitors as international regulations have been changed in recent years. This increases the freedom of choosing the type of the backward view. The purpose of this study is to experimentally clarify how human spatial perception is influenced by the display method, especially the presentation position and the viewpoint. We prepared a setup in which a simulated vehicle moves while the participant views the front and backward virtual 3-D space. We analyzed the subjective evaluation, the performance, and the eye movement of participants during the task and searched for an optimum presentation position and viewpoint. As a result, we found that it is easier for participants to understand the space with an integrated monitor rather than separated monitors. They could better unify the viewpoints of the rearview and the side view, objectively or subjectively, when these views are independent. In contrast, we suspect that either viewpoint was suitable when presenting the entire backward view with one camera. Also, we suggest that their ease of understanding the space varies considerably depending on the visible field in the bird's-eye view.

Keywords: Camera monitor system · Spatial cognition · Viewpoint

1 Introduction

In 2015, the United Nations Economic Commission for Europe changed the international vehicle regulations to allow rearview mirrors to be substituted with cameras and monitors [1]. As a result, it became possible to design and manufacture a vehicle without rearview mirrors, which was obligatory until then.

For drivers, backward visibility plays an extremely important role in allowing them to understand the space on the rear side correctly and safely; therefore, this requirement attracted attention and expectation about methods of presenting a new rearward view [2].

A larger degree of freedom in choosing both the monitor position and the viewpoint of the image (the subjective and objective viewpoints) is offered to drivers when

© Springer Nature Switzerland AG 2019
S. Bagnara et al. (Eds.): IEA 2018, AISC 827, pp. 169–179, 2019.
https://doi.org/10.1007/978-3-319-96059-3_19

presenting them the rear vision using cameras and monitors compared with that using conventional rearview and side mirrors.

Therefore, the purpose of this study is to reveal how human spatial perception is influenced by the display method, especially the presentation position and the viewpoint, via a simulation.

2 Method

2.1 Experimental Task

The participants simulated driving a vehicle without colliding with other dynamically moving vehicles presented in front and rear views in a virtual three-dimensional (3-D) space. Their aim was to avoid contact with other vehicles coming from any direction while passing 10 checkpoints in the order specified in a circular space with a virtual radius of 600 m.

2.2 Experiment Environment

The experiment used two 65-inch displays (TH-65LFE8 J, Panasonic Inc.) placed 88 cm away from participants to fill their guidance field of fixation completely [3]. Participants sat down and performed tasks using an Xbox controller (4N6-00003, Microsoft). The simulator was constructed with Unity 2017.2.0f3 (64-bit). Figure 1 shows the experimental setup. Figure 2 shows the photograph of a participant taking part in an experiment.

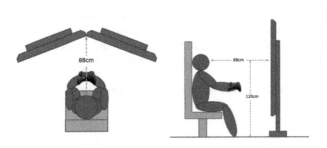

Fig. 1. Experimental environment

Fig. 2. Photograph of the experiment

2.3 Experiment Conditions

We used 10 different methods to present the backward view shown in the display. Figure 3 shows the simulated view through the windshield as presented on the display. Figure 4 shows the rearview monitors and also the overall image presented to the driver, comprising the view through the windshield with the monitors superimposed in different positions.

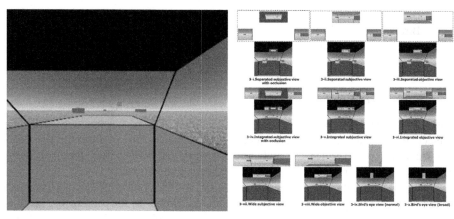

Fig. 3. The simulated view through the windshield

Fig. 4. The upper panels show the monitors and the lower panels show the overall image presented to the driver, comprising the view through the windshield with the monitors superimposed

The 10 experimental conditions shown in Fig. 4 are

i. Separated subjective view with occlusio
Rearview as if the driver looked back with occlusion in front and two normally positioned side-and-rear monitors

ii. Separated subjective view
Rearview as if the driver looked back in front and two normally positioned side-and-rear monitors

iii. Separated objective view
Rearview as if looking from the outside in front and two normally positioned side-and-rear monitors

iv. Integrated subjective view with occlusion
Rearview as if the driver looked back with occlusion in front and two next to-positioned side-and-rear monitors

v. Integrated subjective view
Rearview as if the driver looked back in front and two next to-positioned side-and-rear monitors

vi. Integrated objective view
Rearview as if looking from the outside in front and two next to-positioned side-and-rear monitors

vii. Wide-angle subjective view
Rearview as if the driver looked back (wide angle) which combine side-and-rear monitors in front

viii. Wide-angle objective view
Rearview as if looking from the outside (wide angle) which combine side-and-rear monitors in front

ix. Bird's-eye normal view
   Bird's-eye view in front
 x. Bird's-eye broad view
   Broad bird's-eye view in front.

2.4 Experiment Procedure

Tasks were completed in about 6 min per condition. The participants answered the subjective evaluation questionnaire (Sect. 2.5) immediately after performing each task. They closed their eyes for 2 min before the next task, which was for a different condition. This was repeated for all 10 experimental conditions.

Naturally, for retaining experimental generality, the order of the experimental conditions for each participant should be variable. However, all participants did the final task with a bird's-eye (broad) view because this condition displays completely the movement of other vehicles from above and it was thought that seeing this view earlier would influence their performance for the other conditions.

2.5 Record Items

Operation logs, subjective questionnaires, and eye movement data were recorded. Operation logs were recorded every 1/60 s. Subjective questions were scored from −100 to 100 with a free scale. Table 1 shows the 10 questions. Eye movements were recorded every 1/30 s with an eye mark recorder (EMR-9, Nac Inc.).

Table 1. Subjective questionnaire

No.	Question	Score
Q.1	Were you able to control your vehicle as you wanted throughout the task?	−100: difficult, 100: easy
Q.2	Were you able to successfully avoid other vehicles throughout the task?	−100: difficult, 100: easy
Q.3	Did you pay too much attention to the front or back throughout the task?	−100: front, 100: back
Q.4	Have you experienced in being afraid of any potential incident by the other vehicles in the front?	−100: always, 100: never
Q.5	Were you able to successfully avoid other vehicles in the front?	−100: difficult, 100: easy
Q.6	Was it easy to grasp the direction of other vehicles in the backward?	−100: difficult, 100: easy
Q.7	Was it easy to grasp the speed of other vehicles in the backward?	−100: difficult, 100: easy
Q.8	Was it easy to grasp the distance to other vehicles in the backward?	−100: difficult, 100: easy
Q.9	Have you experienced in being afraid of any potential incident by the other vehicles in the backward?	−100: always, 100: never
Q.10	Were you able to successfully avoid other vehicles in the backward?	−100: difficult, 100: easy

2.6 Participants and Ethical Considerations

Participants were 14 students (21–26 years old; average: 22.5; standard deviation: 1.24). We explained the experiment orally and in writing to the participants and acquired signed consent forms. In addition, we analyzed the data after making personal identification impossible.

3 Results

3.1 Performance

Figures 5, 6 and 7 shows participant's reaction rate and collision rate against other vehicles in separated type. Based on occlusion condition, Fig. 4 shows that attention to the backward has increased since the reaction rate decreases and the collision rate did not change when it becomes transparent. However, when the objectivity is reached, the collision rate does not change but the reaction rate has further decreased. It shows that the attention to the rear decreased or it was able to judge whether it was necessary to avoid other vehicles instantaneously.

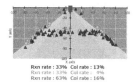
Rxn rate : 33% Col rate : 13%
Rxn rate : 33% Col rate : 4%
Rxn rate : 63% Col rate : 16%

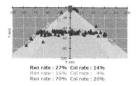
Rxn rate : 27% Col rate : 14%
Rxn rate : 16% Col rate : 4%
Rxn rate : 70% Col rate : 20%

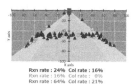
Rxn rate : 24% Col rate : 16%
Rxn rate : 16% Col rate : 0%
Rxn rate : 64% Col rate : 21%

Fig. 5. Separated subjective view with occlusion

Fig. 6. Separated subjective view

Fig. 7. Separated objective view

Figures 8, 9, and 10 show participant's reaction rate and collision rate for an integrated monitor. Similar to the results for the separated monitor, the participants' attention lessened or they were quickly able to judge how to avoid other vehicles. However, we think that the change occurred not because the participants' attention lessened but because it was easier for them to understand what was happening behind them, because the collision rate decreased under both subjective and objective modes compared with separated monitors.

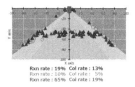
Rxn rate : 19% Col rate : 13%
Rxn rate : 10% Col rate : 5%
Rxn rate : 65% Col rate : 19%

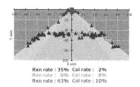
Rxn rate : 35% Col rate : 2%
Rxn rate : 0% Col rate : 8%
Rxn rate : 63% Col rate : 10%

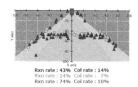
Rxn rate : 43% Col rate : 14%
Rxn rate : 14% Col rate : 7%
Rxn rate : 74% Col rate : 10%

Fig. 8. Integrated subjective view with occlusion

Fig. 9. Integrated subjective view

Fig. 10. Integrated objective view

174   S. Kanzaki et al.

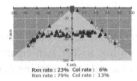

Rxn rate : 23% Col rate : 6%
Rxn rate : 79% Col rate : 13%

Fig. 11. Wide-angle subjective view

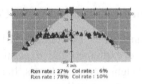

Rxn rate : 27% Col rate : 6%
Rxn rate : 78% Col rate : 10%

Fig. 12. wide-angle objective view

Figures 11 and 12 show participants' reaction rate and collision rate for a wide-angle view. We thought that attention to the rear view increased because the collision rate is considerably lower than that for either the narrow-angle separated monitors or the narrow-angle integrated monitor. There was no difference between the subjective mode and objective mode in the wide-angle view.

Figures 13 and 14 show participants' reaction rate and collision rate for a bird's-eye view. This type was most unique. Figure 13 indicates that the normal bird's-eye view was disadvantageous for judging the movement behind the simulated vehicle because the reaction rate is low and the collision rate is high. In contrast, Fig. 14 shows that the reaction rate was the highest and the collision rate was the lowest when the bird's-eye view was broader. This condition was thought the most advantageous view of all 10 considered.

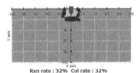

Rxn rate : 32% Col rate : 32%

Fig. 13. Bird's-eye normal view

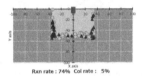

Rxn rate : 74% Col rate : 5%

Fig. 14. Bird's-eye broad view

3.2 Subjective Questionnaires

Figures 15 and 16 show the evaluations from the subjective questionnaires for separated monitors and an integrated monitor. The left chart shows the result for the entire task and for vehicles in front, whereas the right chart shows the result for vehicles behind.

Fig. 15. Separated view

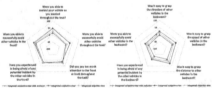

Fig. 16. Integrated view

Figure 15 reveals that there were no differences for the entire task and vehicles in front for separated monitors. By contrast, for the backward view, the objective mode was highly evaluated for avoiding other vehicles and the subjective mode was highly evaluated for understanding the position of other vehicles.

For an integrated monitor, Fig. 16 shows that the evaluation was in order from highest to lowest for transparent, objective, and occlusion views for vehicles behind.

Figures 17 and 18 show the evaluation from the subjective questionnaires for the wide-angle and bird's-eye views.

Fig. 17. Wide-angle view   **Fig. 18.** Bird's-eye view

According to Fig. 17, all the scores for both the wide-angle subjective view and the wide-angle objective view were higher than those for the previous conditions. Comparing the subjective mode and the objective mode for a wide-angle view, there is no difference except for the question about the distance from other vehicles.

Figure 18 depicts that the evaluation for a normal bird's-eye view was low overall while the scores for a broad bird's-eye view were highest for all questions compared with all other conditions.

3.3   Eye Mark

Separated Monitors. Figures 19, 20, and 21 show the distribution of the gazing point of one participant for the three conditions with separated monitors. Figures 22, 23, and 24 show histograms of gaze time for five people for the three conditions with separated monitors.

Figures 19 and 20 show that the gaze points for both the occluded and transparent conditions with separated monitors were on the trajectory connecting the center of the

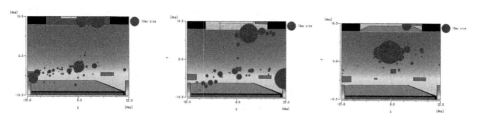

Fig. 19. Separated subjective view with occlusion

Fig. 20. Separated subjective view

Fig. 21. Separated   objective view

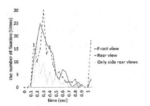

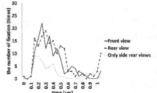

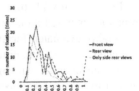

Fig. 22. Separated subjective view with occlusion

Fig. 23. Separated subjective view

Fig. 24. Separated objective view

display and each monitor. Figure 21 depicts that in objective mode there were many gaze points toward the center monitor and the rear monitors.

From Fig. 22, we see that the number of times that the participants fixated their eyes for 1 s or more on a rear monitor was large for occlusion with separated monitors. Figure 23 indicates that the peak of the histogram for the backward view was shifted to the right. The gaze time was longer overall without occlusion. For objective mode, Fig. 24 shows that there are more gaze points toward the front than toward the rear. Thus, we thought that there was less attention on vehicles behind for the objective mode.

Integrated View. Figures 25, 26, and 27 show the distribution of the gazing point of one participant for the three conditions with an integrated monitor. Figures 28, 29, and 30 show the histogram of gaze time for five people for the three conditions with an integrated monitor.

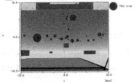

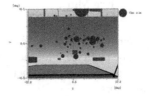

Fig. 25. Integrated subjective view with occlusion

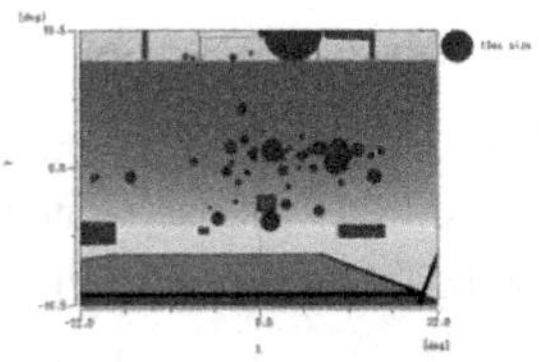

Fig. 26. Integrated subjective view

Fig. 27. Integrated objective view

Figures 25, 26, and 27 show that the gaze points under all conditions with an integrated monitor were on the trajectory extending from the center of the image to the rear monitors on the left side, right side, and above. However, the gaze points in objective mode were gathered at the center, as with separated monitors.

Figures 28, 29, and 30 show that the medians of the gaze time histograms were smaller than those for separated monitors for each condition. This suggests that gaze time tends to be shorter when the rear monitors are integrated.

Wide View. Figures 31 and 32 show the distribution of the gazing point of one participant for the two conditions with a wide-angle view. Figures 33 and 34 show

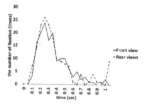

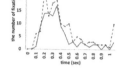

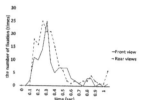

Fig. 28. Integrated subjective view with occlusion

Fig. 29. Integrated subjective view

Fig. 30. Integrated objective view

histograms of gaze time for five people in subjective and objective modes, respectively, for a wide-angle view.

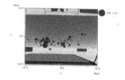

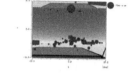

Fig. 31. Wide-angle subjective view

Fig. 32. Wide-angle objective view

Both Figs. 31 and 32 show that the gaze points were divided between the windshield and the rear monitor.

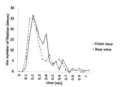

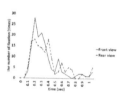

Fig. 33. Wide-angle subjective view

Fig. 34. Wide-angle objective view

Figures 33 and 34 show that gaze time was shorter when looking either to the front or backward in subjective mode and longer for only the rear monitor in objective mode.

Bird's Eye View. Figures 35 and 36 show the distribution of the gazing point of one participant for the conditions with a bird's-eye view. Figures 37 and 38 show the histogram of gaze time for five people for the conditions with a bird's-eye view.

Both Figs. 35 and 36 show that the participants turned their gaze points onto the bird's-eye monitor rather than onto the windshield.

Both Figs. 37 and 38 show that the number of times when the participants' gaze was longer than 1 s was larger than for other conditions, and the medians were larger too. Furthermore, the median of gaze time for the windshield was the smallest for the

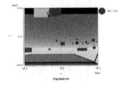

Fig. 35. Bird's-eye normal view

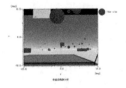

Fig. 36. Bird's-eye broad view

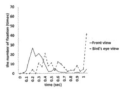

Fig. 37. Bird's-eye normal view

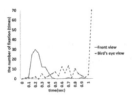

Fig. 38. Bird's-eye broad view

broad bird's-eye view. Therefore, these figures indicate that under this arrangement the participants mainly watched the bird's-eye monitor without looking forward much, especially for the broad view.

4 Conclusion

The purpose of this study is to experimentally reveal how human spatial perception is influenced by the display method, especially the presentation position and the viewpoint. The results are summarized below.

By changing the presentation of the rear vision from separated monitors to an integrated monitor, the gaze time for the rearview was shortened, the subjective evaluation was higher, and participants found it easier to avoid other vehicles approaching from behind. With a wide-angle view, the subjective evaluation and performance further improved; however, the gaze time or the rearview tended to be shorter in subjective mode and longer in objective mode. Under either condition for the bird's-eye view, the participants mainly watched the bird's-eye monitor without looking forward much. However, there was a big difference in terms of performance and subjective evaluation. The narrow field of view was not useful because the participants' reactions to other vehicles was too slow. In contrast, the best results were obtained with a broad field of view.

In conclusion, for the presentation position, the results for subjective evaluation and performance were in the order from lowest to highest for separated, integrated, and wide-angle monitors. The broad bird's-eye view is good, but we thought that this condition is undesirable when hazards that are not shown in the bird's-eye view are present because the participants' attention is on the monitor. The objective mode was less satisfactory than the subjective mode if the backward view was divided into two, as with the separated monitors and the integrated monitor because the side the viewpoints of the rear-view and side-view are different and cognitive load is high. In contrast, there are no influence by the difference in the viewpoint under the condition with the single image of the wide-angle view. We would like to implement a simpler environment and further confirm and examine the findings obtained in this study. In contrast, the objective mode was better with the single image of the wide-angle view. In addition, it is thought that the attention backward was a distraction, because the gaze time tends to be shorter than for subjective mode when looking both frontward and backward.

References

1. UNECE: Proposal for Supplement 2 to the 04 series of amendments to Regulation No. 46. http://www.unece.org/fileadmin/DAM/trans/doc/2015/wp29/ECE-TRANS-WP29-2015-084e.pdf. Accessed 24 May 2017
2. Murata T (2014) Spatial perception. https://doi.org/10.14931/bsd.4048
3. Hatada T (1993) Artificial reality and visual space perception. Japan Ergon Soc 3(29):129–134
4. Yamada M et al (1986) Definition of gazing point for picture analysis and its applications. IEICE 9(69):1335–1342

Visual Ergonomics in Control Room Environments: A Case Study from a Swedish Paper Mill

Susanne Glimne[1]([✉]), Rune Brautaset[1], and Cecilia Österman[2]

[1] Unit of Optometry, Department of Clinical Neuroscience, Karolinska Institutet,
Box 8056, 104 20 Stockholm, SE, Sweden
susanne.glimne@eyelab.se
[2] Kalmar Maritime Academy, Faculty of Technology, Linnaeus University,
391 82 Kalmar, SE, Sweden
cecilia.osterman@lnu.se

Abstract. This paper presents results from a case study at a Swedish paper mill. The study is part of a larger ongoing research project aiming to investigate visual ergonomics in control room environments. Visual conditions were measured and evaluated in five control rooms, and nine process operators answered questions about perceived workload and visual experience.

The participating operators rated their mental workload as being the highest, but also mentioned a high physical load and time pressure. Six operators indicated that they had work-related eye problems and experienced pronounced eye fatigue, as well as eye discomfort, sore eyes and blurred vision.

In one of the visited control rooms, where the operators indicated that they often experienced eye problems, the operators were clearly subjected to glare. Light was reflected within the field of view from a steel plate outside the control room windows. Compared to recommendations for office workplaces, the illuminance levels were low in all control rooms except one, causing contrast glare since computer screens generally had a higher luminance. Low contrasts were also identified in the computer screens between background and characters.

In conclusion, the measured visual conditions could be related to an increased visual load, leading to visual discomfort. This may lead to pain in the neck, shoulders, and contribute towards an unnecessary high mental load. Recommendations are suggested of how lighting conditions van be improved.

Keywords: Visual demands · Contrast glare · Lighting design

1 Introduction

The technical development has led to an increasing number of employees working in different types of control room environments, where information and media technology is used to supervise and control a large number of varying processes. Control room work is present in the industrial process and manufacturing industry, transport sector (e.g., traffic surveillance, and operation of trains, airlines and ships), as well as in the media industry such as television production and image processing.

© Springer Nature Switzerland AG 2019
S. Bagnara et al. (Eds.): IEA 2018, AISC 827, pp. 180–189, 2019.
https://doi.org/10.1007/978-3-319-96059-3_20

Several aspects are related to the visual comfort of control room operators. During work with computer screens, the design of, *e.g.*, font size, character and line spacing, contrast and colour are important. As is the lighting in the room and at the work station, and the height and placement of screens, which affect the gaze angle and viewing distance [1, 2]. When working with computers, the eyes need to converge (come together). A clear image of the screen should be projected on the retina of the eyes, either through accommodation (change in focus) or by using of correction (*e.g.* contact lenses or glasses). If the eyes must make increased efforts to adjust for a clear retinal image, the intensity of muscle activity in trapezius increases [3]. Even lower degrees of blur may cause impaired performance and visual discomfort, and people with more eye problems are more likely to suffer from musculoskeletal disorders [4]. Ensuring good visual conditions are especially important for older operators, since our vision declines with age. Reduced light level, low contrast, or glare will significantly decrease the visual functions in older individuals [5, 6].

Supervisory work control rooms generally require continuous visual assessments in combination with active interventions. Commonly, it requires a frequent change in focusing distances when monitoring different digital displays. When the light conditions differ between these visual distances, the eyes need to adapt to the luminance differences. During this adaption, the operator experiences the situation as visual impairment [7]. Frequent use of digital displays contributes to a deterioration of the so-called binocular control (collaboration between the two eyes), since most of the work occurs in a static two-dimensional (2D) visual environment – working with a flat screen [8]. If this is combined with other visually impaired factors, binocular control is further reduced resulting in an increased load on the visual system.

If a task requires a maximum load on the visual system for a longer period of time, an increase of visual demands will occur. These factors are likely to contribute to the vision and eye symptoms referred to collectively as *computer vision syndrome* (CVS). Previous research have proposed a causal link between eye-related disorders, workplace lighting, glare, dry eyes, screen intensity, readability, short reading distance, flickering light, upward gaze and reflections on surfaces [cf. 9, 10]).

Visual demanding tasks that involves sustained attention, memory and concentration, results in a lower blink frequency which may affect the tear film, and lead to eye-related disorders such as dryness, soring eyes and redness [11]. This occurs especially in the upward gaze direction while the blink frequency is further reduced. The presence of dry eyes is significantly elevated with increasing age [12].

A reduced visibility, *e.g.* from inappropriate lighting design with indirect or direct glare, will reduce the visibility of the task and/or the image on the retina. This glare will add further stress on an operators' visual system [cf. 13, 14, 15]. Glare is caused when the eyes are exposed to a stronger light than they are set for and may originate from artificial lighting, sunshine or daylight through windows, and reflections from high-gloss surfaces. Glare can be divided into two main types, *discomfort glare* (psychological) and *disability glare* (physiological). Both types may arise from direct light or indirectly reflected light. Discomfort glare is more common in the light sensitivity caused by light differences in the field of view. The glare not only directly obstruct visual ability, but also cause an increased mental strain [16].

Recently, several previous ergonomics studies have been devoted to the design and evaluation of control room environments. For example, methods for design and evaluation of control rooms [17, 18], collaborative communication [19], and effects of shift work on cognitive performance in control room operators [20], to mention a few. However, visual ergonomic conditions in control rooms has received less attention.

The study presented in this paper is part of an ongoing research project investing visual ergonomics in control room environments. In all, the project will encompass about 25 different workplaces, representing various visual conditions. The project is set out to systematically examine and evaluate visual ergonomic challenges in control rooms in relation to the design of the working environment and the tasks performed by the control room operators. The evaluation is further compared to legal requirements and guidelines available. The overall project aim is to formulate practical recommendations for physical design, and organization of work in control room environments. Specifically, this paper reports a case study performed at a Swedish paper mill, where five control rooms for various processes were examined and evaluated. The project is still in an early stage and in this paper, we wish to present and discuss our methods for collection and analysis of data, and show how our results can be used, and what knowledge that can be elicited to improve visual ergonomic conditions when designing and organizing work in control rooms.

2   Materials and Methods

Five control room environments for various processes and operations were examined and evaluated during this case study. Nine process operators (2 women and 7 men) participated in the survey with an age range of 28–58 years (average 43 years, ±10.3). Two operators participated in each control room, except for one control room with one operator. The visual ergonomic aspects were evaluated by measuring the following parameters: illuminance (lx) and luminance (cd/m^2), contrast glare and reflection glare. Further, the physical layout of the control room was described and measured in terms of visual distances and viewing angles, as well as screen/monitor visibility. Eye tracking glasses were used to analyse eye fixation changes. A subjective assessment was performed using questionnaires where the process operators responded to questions about their perceived experience of the working environment. The study adhered to the tenets of the Helsinki declaration and informed consent was obtained from all respondents.

2.1   Procedures

Subjective Assessment. Initially, a brief interview was made with participating process operators. Questions were asked about how a typical working day looks and what tasks were performed. To clarify which visual requirements exist, we also related the investigation to the participants age.

The CISS-T questionnaire on vision experience [21] (Table 1) and the NASA-TLX questionnaire on workload [22] was used to evaluate subjective experiences. Questions about vision experience were answered on a five-digit descriptive scale 0–4, corresponding to: (0) Never; (1) Rarely; (2) Sometimes; (3) Often; (4) Always. Questions on experienced workload was answered in a 10-degree Visual Analog Scale (VAS). The questions were divided into mental and physical workload, and also included questions about perceived time pressure, performance, effort and frustration.

Table 1. Questionnaire about vision experience

Tired eyes	Feel sleepy
Uncomfortable eyes	Lose concentration
Headaches	Visual objects are blurred, out of focus
Double vision	Trouble remembering tasks read
See the visual objects move, jump etc.	Read slowly
Hurting eyes	Lose place regarding the visual tasks
Sore eyes	Re-read the same information
Feeling pulling around the eyes	

Light Measurements. A portable Hagner photometer S2, LMK Mobile advanced digital luminance camera and analysis software (LMK Glare Analysis, AddOn) were used to map and measure illuminance and luminance in workplace. With the analysis software, it was possible to get quantitative renderings of luminance on critical surfaces like desks and monitors in the control rooms. The purpose of the measurements was to identify any major differences in luminance within the field of view and between the different visual environments where changes in focus occur.

Eye Tracking. An Eye Tracker (Tobii Glasses 2) together with analysis software (Tobii Pro Lab) was used to evaluate frequent changes in focusing distances due to monitoring different digital displays with different visual distances. The eye tracker was used to record eye movements when the operator performing his or her normal work tasks while wearing the glasses connected to a recording device. Each recording lasted for about 10 min per participants with a sampling frequency of 50 Hz. Prior to each recording session, the eye tracker was calibrated to fit the individual process operator.

Other Work Environment Measurements. The physical layout of the control rooms was measured and described by listing panels, displays, furniture and light sources, as well as visual distances and viewing angles for the participating process operators. The reflective properties (gloss unit) on surfaces were measured using a gloss meter Minigloss 101 N with an angle of incidence of 60° and measuring range 0–100. Visibility of the displays was mapped, character heights measured, and contrasts on screen checked.

2.2   Analysis of Data

Since this research project is at its initial stage with so far few observations and a limited number of participants, the data presented in this paper is analysed and presented using descriptive statistics. To describe and summarize data allowing for less advanced interpretations of central tendency (mean) and measures of spread (min, max, and standard deviation) are used.

3   Results

3.1   Process Operators' Estimation of Vision Experience and Workload

Eye-related problems during control room work were reported by six of the nine participants, primarily the operators in control rooms 1, 4 and 5 (Fig. 1). Five operators stated that they at least sometimes suffered from headaches during control room work.

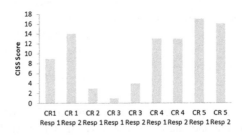

Fig. 1. Distribution of experienced eye symptoms in control room performing tasks divided per control room and respondent. Eyes symptoms included: tired eyes, uncomfortable eyes, double vision, hurting eyes, sore eyes, pulling eyes, and blurring.

Mean and standard deviation for the most scored aspects of the participants' vision experiences and perceived work load is illustrated in Tables 2 and 3. The mental workload (*e.g.* to choose, calculate, remember, search, decide) was perceived as the most prominent by the respondents. All participants indicated that they sometimes felt sleepy when performing control room tasks and four operators experienced this often.

Table 2. Mean and standard deviation (SD) for the most scored aspects of vision experience.

	Tired eyes	Uncomfortable eyes	Sore eyes	Blurred vision	Headaches	Feel sleepy	Difficult remembering
Mean	2.33	1.89	1.45	2.00	1.45	2.45	1.89
SD	±1.00	±1.27	±1.13	±1.12	±1.24	±0.45	±1.17

Note: Answered in a five-digit scale 0–4

Table 3. Mean and standard deviation (SD) for the most scored aspects of work load.

	Mental workload	Physical workload	Time pressure	Performance	Effort	Frustration
Mean	5.78	5.33	4.45	3.33	4.67	4.67
SD	±1.86	±2.24	±2.19	±3.58	±1.94	±1.80

Note: Answered in a 10-degree Visual Analog Scale (VAS) where 0 is low and 10 is high.

3.2   Lighting Measurements in the Control Rooms

The results are presented as measured values within the field of view based on the existing light level at the time of visit. Reflectivity (gloss unit) of the surface was measured in areas within the field of view that were more reflective.

Illumination levels were generally low (mean 224 lx ± 159SD) since the operators themselves preferred a subdued light level. In Control room 2, a window to the outside gave a more adapted illuminance level to the prevailing daylight.

A compilation of the measured values for illumination (lx) and luminance (cd/m²) in the five control rooms is shown in Table 4. Mean and measures of spread (min., max, and standard deviation) are used to describe and summarize the results.

Control Room 1. A screen placed within the field of view had a higher luminance (90–105 cd/m²) compared to the light condition in general (Table 4). Contrast of the screens on the table in the control room was 15.5–36% in contrast between background and characters (details on screen). Examples of measured luminance values were 4.5–10 cd/m² regarding details. The background had a luminance between 19–31 cd/m². This means that at lowest contrast was 2:1 and the highest (where it is darkest/brightest) was 6:1.

Table 4. A compilation of the measured values for illumination (lx) and luminance (cd/m²) dived per control room

	Control room 1		Control room 2		Control room 3		Control room 4		Control room 5	
	cd/m²	Lx	cd/m²	Lx	cd/m²	Lx	cd/m²	Lx	cd/m²	Lx
Mean	57	155	141	433	78	272	46	139	47	34
SD	50	74	147	172	78	121	51	181	49	9
Min	4	100	30	250	5	130	2	45	2	25
Max	130	250	460	700	230	440	120	540	130	50

Control Room 2. Measured luminance were much higher (80–105 cd/m²) in the area in front of the screens compared to the luminance of the screens (30–60 cd/m²). There was an incidence of daylight through the windows (∼260 cd/m²) resulting in high luminance within the field of view.

Control Room 3. A screen placed within the field of view had a higher luminance (∼230 cd/m²) compared to the light condition in general.

Control Room 4. A screen placed within the field of view had a higher luminance (120 cd/m²) compared to the light condition in general.

Control Room 5. Luminance differences in the field of view where caused by light reflecting from a steel plate (∼130 cd/m², gloss unit 199) through the windows (a horizontal light beam corresponding to the luminaires existed). Measured contrast was 5:1 (lowest luminance compared with the highest luminance while performing work tasks. Figure 2 illustrates the luminance differences and shows the reflected steel plate in the corridor outside the control room.

Fig. 2. Illustration of control room 5 regarding reflected glare within the field of view. A yellow/white colour corresponds to highest luminance (luminance image).

3.3   Visual Distances and Visual Objects in the Control Rooms

In order to further describe visual environments in the control rooms a summary is included in Table 5 regarding visual distances to screens, character size on used screens, and upward gaze angles to screens further away.

Table 5. Details of visual distance, character size and gaze angle

	Distance 1 (m)	Distance 2 (m)	Character size (mm)	Gaze angle[a] (°)
Room 1	1.50–1.75	2.65–2.80	3–4	10–30
Room 2	0.75–1.00	1.60–1.80	3–4	10–30
Room 3	1.25–1.50	2.30–3.60	3–4	30–50
Room 4	0.75–1.00	1.70–2.10	3–4	5–20
Room 5	0.75–1.00	1.10–1.30	3–4	30–50

[a]Gaze angle is based on office position and is dependent on individual length and angle of the chair back.

3.4   Changes in Focus and Visual Distances

The measured eye movements are presented according to changes in focus and visual distance. In Control room 1, the supervisory work involved major differences in visual distances, in combination with frequent changes between screens on the office table and mounted on the wall further away. Data from the eye-tracker recording in this control room is therefore used as an example in this paper. Analysis of the recording showed a change in focus just below 100 times and changes in major visual distances about 20 times regarding one minute of tasks performed.

4   Discussion and Concluding Remarks

The results from this case study show generally low illuminance levels since the operators preferred subdued light. Recommended illumination in office environments including computer work is minimum 500 lx [23]. Control room 2 had a more adapted lighting level to the prevailing daylight from the windows.

In Control rooms 1 and 3, contrast glare from screens resulted in higher luminance, 130 cd/m^2, and 230 cd/m^2 respectively. In Control room 1, the higher value applies to the general lighting conditions in the room. Since it was relatively dark, set luminance on the screen was too high. Equivalent conditions are found in Control room 4.

Contrast glare within the field of view in Control room 5 was caused by reflected light from a steel plate through the windows ($\sim$130 cd/m^2, gloss unit 199). The luminance contrast (the difference between the lowest and the highest luminance within the field of view). was 5:1. Previous studies show that computer users prefer a luminance between 60–85 cd/m^2 in surrounding area [7] and gloss above 60 gloss units influences visibility and visual performance [24].

The contrast between background and characters on screens in Control room 1 ranged between 15.5–36%. The luminance on screens ranged between 4.5–10 cd/m^2 and the background between 19–31 cd/m^2. This means that the lowest contrast was 2:1 and the highest was 6:1. The general recommendation regarding contrast between characters and background on displays is 3:1 as a lowest value. For optimum visibility, the contrast should be 10:1. This recommendation could be difficult to implement in an environment with a low illuminance in general, as the background would need to be in the white sense. Given that the lighting level in the control rooms is relatively low (set by workers), the background setting in whitening mode is not preferable.

Process operators in control rooms 1, 4 and 5 scored highest on eye-related problems, compared to the operators in the other control rooms. The operators' visual system is under high strain [25] since the screens are flat. When glare is present or visibility is reduced, an increase of visual demands occurs [26], and the intensity of muscle activity in Trapezius increases [27] leading to an increased risk of musculoskeletal disorders. Furthermore, an upward gaze direction with reduced blink frequency can lead to eye-related disorders such as dryness, soring eyes and redness [11]. Since the work including upward gaze, these operators are more prone to dry eye problems and may therefore have a more pronounced problem. The presence of dry eyes is significantly elevated with increasing age [12].

The mental workload (*e.g.* to choose, calculate, remember, search, decide) was perceived by the participating operators as the most prominent work load. Over time, this may lead to mental fatigue and increased stress levels [28] which could be demonstrated in this study since operators experienced difficulty remembering. All process operators also reported some sleepiness during work. This may be an effect of low illumination levels in the control rooms in addition to shift work schedule.

A previous study on sleepiness among control room operators [20] recommend improved lighting conditions as a useful ergonomic strategy, since bright light facilitates cognitive performance, *i.e.* working memory, sustained attention etc [29].

In sum, the results from this case study indicate that many supervisory tasks in the control rooms include major changes in focus distances with considerable differences in luminance, as well as sizeable luminance differences within the field of view causing visual contrast glare. Eye-related problem was present related to the visual conditions. Elevated visual requirements may contribute to pain in the neck and shoulders, and further towards an increased mental load. The research findings presented in this study support the relevance of providing an optimized visual environment that considers visual ergonomics when designing and organizing work in control rooms.

Acknowledgment. The research for this paper was financially supported by AFA Försäkring through the 160255 project "The importance of an optimized visual ergonomics for sustainable control room work".

References

1. Long J, Helland M (2012) A multidisciplinary approach to solving computer related vision problems. Ophthalmic Physiol Opt 32(5):429–435
2. Long J et al (2014) Visual ergonomics standards for contemporary office environments. Ergon Aust 10(1):7
3. Richter HO, Bänziger T, Forsman M (2011) Eye-lens accommodation load and static trapezius muscle activity. Eur J Appl Physiol 111(1):29–36
4. Hemphälä H, Eklund J (2012) A visual ergonomics intervention in mail sorting facilities: effects on eyes, muscles and productivity. Appl Ergon 43(1):217–229
5. Haegerstrom-Portnoy G, Schneck ME, Brabyn JA (1999) Seeing into old age: vision function beyond acuity. Optom Vis Sci 76(3):141–158
6. Nylén P et al (2014) Vision, light and aging: a literature overview on older-age workers. Work 47(3):399–412
7. Sheedy JE, Smith R, Hayes J (2005) Visual effects of the luminance surrounding a computer display. Ergonomics 48(9):1114–1128
8. Glimne SA, Öqvist Seimyr G, Brautaset RL (2015) Effect of 3-dimensional central stimuli on near point of convergence. Strabismus 23(3):121–125
9. Blehm C et al (2005) Computer vision syndrome: a review. Surv Ophthalmol 50(3):253–262
10. Rosenfield M (2011) Computer vision syndrome: a review of ocular causes and potential treatments. Ophthalmic Physiol Opt 31(5):502–515
11. Helland M et al (2008) Will musculoskeletal, visual and psychosocial stress change for visual display unit (VDU) operators when moving from a single-occupancy office to an office landscape? Int J Occup Saf Ergon 14(3):259–274
12. Stapleton F et al (2015) The epidemiology of dry eye disease. In: Dry Eye. Springer, pp 21–29
13. Jainta S et al (2011) Binocular coordination during reading of blurred and nonblurred text. Invest Ophthalmol Vis Sci 52(13):9416–9424
14. Jaschinski-Kruza W (1994) Dark vergence in relation to fixation disparity at different luminance and blur levels. Vis Res 34(9):1197–1204
15. Pickwell L, Yekta A, Jenkins T (1987) Effect of reading in low illumination on fixation disparity. Am. J. Optom. Physiol. Opt. 64(7):513–518

16. Garzia RP (1996) Vision and reading, vol 5. Mosby Incorporated
17. Simonsen E, Osvalder A-L (2018) Categories of measures to guide choice of human factors methods for nuclear power plant control room evaluation. Saf Sci 102:101–109
18. Stanton NA et al (2010) Human Factors in the Design and Evaluation of Central Control Room Operations. CRC Press/Taylor & Francis Group, Boca Raton
19. Kataria A et al (2015) Exploring bridge-engine control room collaborative team communication
20. Kazemi R et al (2016) Effects of shift work on cognitive performance, sleep quality, and sleepiness among petrochemical control room operators. J. Circadian Rhythm. 14
21. Hart SG, Staveland LE (1998) Development of NASA-TLX (Task Load Index): results of empirical and theoretical research. In: Advances in psychology. Elsevier, pp 139–183
22. Borsting EJ et al (2003) Validity and reliability of the revised convergence insufficiency symptom survey in children aged 9 to 18 years. Optom Vis Sci 80(12):832–838
23. IESNA (2011) The lighting handbook: reference and application. IESNA - Illuminating Engineering Society of North America, New York
24. Brunnström K et al (2008) 66.3: Visual ergonomic effects of screen gloss on LCDs. In: SID Symposium digest of technical papers. Wiley Online Library
25. Glimne S, Brautaset R, Seimyr GÖ (2015) The effect of glare on eye movements when reading. Work 50(2):213–220
26. Glimne S et al (2013) Measuring glare induced visual fatigue by fixation disparity variation. Work 45(4):431–437
27. Richter HO, Zetterlund C, Lundqvist L-O (2011) Eye-neck interactions triggered by visually deficient computer work. Work 39(1):67–78
28. Fallahi M et al (2016) Effects of mental workload on physiological and subjective responses during traffic density monitoring: a field study. Appl Ergon 52:95–103
29. Kretschmer V, Schmidt K-H, Griefahn B (2013) Bright-light effects on cognitive performance in elderly persons working simulated night shifts: psychological well-being as a mediator? Int Arch Occup Environ Health 86(8):901–914

Improving Lighting Quality by Practical Measurements of the Luminance Distribution

Thijs Kruisselbrink[1,2(✉)] ⓘ, Juliëtte van Duijnhoven[1,2] ⓘ,
Rajendra Dangol[1,2] ⓘ, and Alexander Rosemann[1,2] ⓘ

[1] Eindhoven University of Technology, 5600 MB Eindhoven, The Netherlands
t.w.kruisselbrink@tue.nl
[2] Intelligent Lighting Institute, 5600 MB Eindhoven, The Netherlands

Abstract. Light is one of the important aspects for a comfortable office environment. Too often high quality lighting is not achieved. Lighting quality can be defined by different aspects that are relevant such as the quantity, distribution, glare, spectral power distribution, daylight, directionality, and dynamics of light. The luminance distribution seems to be a suitable measure to achieve high quality lighting. The luminance distribution can be measured, with a practical accuracy, by commercially available cameras and fisheye lenses. All these aspects spectral power distribution can be measured using a camera-based luminance distribution measurement device. So, a luminance distribution measurement device is an excellent tool to measure or indicate lighting quality. It can be used to achieve a better understanding of lighting quality and potentially it can be implemented in automated building control systems.

Keywords: Luminance distribution · Lighting quality · Measurements

1 Introduction

Lighting is an important aspect of the modern office environment that largely influences the experienced comfort level. However, the lighting control strategies are often based on the energy savings criterion only, which can cause serious discomfort. Additionally, the lighting is generally designed according to illuminance requirements, not necessarily providing a high quality lighting. Lighting quality is more than meeting illuminance recommendations. It is a complicated concept that is still not fully understood [1]. To date, lighting quality doesn't have any rigorous definition, whilst many attempts [2, 3] have been made. The complicated nature of lighting quality can be fully understood only by relating the continuous measurement of photometric quantities to the subjective responses in the lab and in field studies.

In the study by Kruisselbrink et al. [4] lighting quality is described by the photometric assessment of the quantity, distribution, glare, spectral power distribution, daylight, directionality, and dynamics of light.

The quantity of light is a very important aspects related to the visual performance and satisfaction. It has been shown that the luminance is closely related to the perceived brightness, suggesting that the luminance is a more suitable metric for lighting quantity than the illuminance [5]. Moreover, the distribution of light indicated by luminance

S. Bagnara et al. (Eds.): IEA 2018, AISC 827, pp. 190–198, 2019.
https://doi.org/10.1007/978-3-319-96059-3_21

ratios or the uniformity has a large impact on the visual comfort. Glare is a discomfort or disability sensation caused by luminance values in the field of view that are much higher than the adaption luminance [6]. Multiple indicators (e.g. UGR, DGP), for different applications, have been developed that aim to quantify this subjective response.

Additionally, the color appearance [7], related to the visual comfort, and color quality [8], related to the visual performance, are relevant to lighting quality as they are linked to the spectral composition of the light and how colors are rendered. Daylight, including the view out, is an important lighting quality aspect as daylight improves the satisfaction because humans evolved under daylight [4]. View out is an import component of daylight that can influence the work performance, job satisfaction and potentially the general health by providing contact with the outside world [7]. The directionality of lighting is a quality aspect that is not used often but has a big effect on the ability to distinguish task details, surface textures and faces. Finally, the dynamics of light can help to provide a more stimulating and pleasant lit environment. Unfortunately, measuring all these individual aspects is often not feasible.

It has been shown that the luminance is closely related to the perceived brightness, suggesting that luminance based metrics are more suitable for lighting quality than the illuminance based metrics [5]. Therefore, the luminance distribution, containing a lot of information, is proposed as an indicator for lighting quality; however, the measurements are not straightforward.

In this article, firstly, we show how and with what accuracy the luminance distribution can be measured using an image-based system. Subsequently, we explore how we can measure different aspects of lighting quality using the image-based luminance distribution measurement device.

2   Measuring the Luminance Distribution

A number of studies has shown that the luminance distribution can be determined using a commercially available camera equipped with a fisheye lens [9–12]. The camera is used as the light sensor and the fisheye lens is used to capture a wide angle of view.

2.1   High Dynamic Range Imaging

The High Dynamic Range (HDR) technology is essential for measuring the luminance distribution using cameras as it allows to capture the dynamic range of real world scenarios, while standard 8-bit images have a very low dynamic range [13, 14]. Generally, HDR images are developed using a sequential exposure change technique were ordinary cameras are used to take multiple Low Dynamic Range (LDR) images by sequentially varying the shutter speed [15, 16].

Based on the measurement described in [9] it was determined what exposure values (EV), a logarithmic combination of shutter speed (t) and aperture (N) as displayed in Eq. 1, are able to capture the required dynamic range. A series of seven to nine exposures is shown to be sufficiently accurate [15].

$$EV = log_2 \frac{N^2}{t} \tag{1}$$

Subsequently, the different exposures can be merged into a HDR image by the command line HDR builder developed by Ward [16]. This HDR builder is capable of approximating the essential camera response curve using radiometric self-calibration [10, 17, 18]. The response curve is camera specific and relates the pixel values directly to the scene radiances while also accounting for corrections administered in the proprietary imaging pipeline [14, 18]. It is advised to determine the camera response curve once according to the method described by Reinhard et al. [13] and reuse this for all following measurements (Figs. 1 and 2).

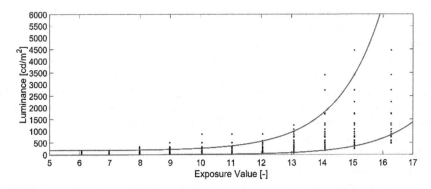

Fig. 1. Approximated relation between the exposure value and the captured luminance range. The black dots resemble the measured luminance values for each exposure, while the red lines represent the curve fitted minimum and maximum luminance captured by each exposure (Color figure online).

2.2   Luminance Calculation

Based on the floating-point RGB values of the HDR images the luminance can be determined. Initially, the RGB tristimuli are translated to the equivalent XYZ tristimuli, because the Y channel matches the sensitivity of the human eye for photopic vision, by applying a conversion matrix that is dependent to the primaries and the white point. Standardized conversion matrices can be found in [19], or can be determined based on the protocol described by Inanici [10]. Generally, the primaries and white point are assumed constant, and hence the luminance is calculated according to Eq. 2. Additionally, a linear calibration factor k is applied to relate the Y tristimulus to the photometric quantity luminance and bring the luminance into the right order of magnitude [9].

$$L = k \cdot (0.2125 \cdot R + 0.7125 \cdot G + 0.0721 \cdot B) \tag{2}$$

Finally, the vignetting effect should be accounted for. Vignetting is the effect of light fall-off at the periphery of the lens due to internal scattering [15, 20, 21], this non-linear effect along the image radius can be approximated by a polynomial function.

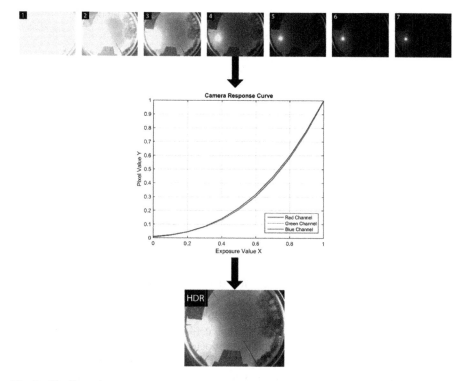

Fig. 2. Pipeline of sequential exposure HDR imaging. Seven LDR images are captured with different exposures and are merged into one HDR images by implementing the camera response curve.

Especially, fisheye lenses can exhibit significant light fall off, up to 73%, at the periphery of the lens [22]. The Vignetting effect is determined in a Ulbricht's sphere because, in theory, this integrating sphere has a uniform luminance distribution [23]. By relating the luminance values measured along the radius to the maximum luminance in the center the polynomial can be approximated. The polynomial is subsequently used to develop a correction filter for each individual pixel.

Ultimately, the calculation of the luminance based on the HDR images results in a luminance value for each individual pixel.

2.3   Accuracy

To illustrate the accuracy of camera system for luminance distribution measurements a Sony IMX2019 camera equipped with a simple fisheye lens is used. Seven LDR images, with shutter speeds according to Table 1, are captured to build the HDR image. Additionally, the aperture, light sensitivity and white balance are fixed to f/2.0, ISO 100 and [1.3 1.0 1.3], respectively. The camera response curves for R, G and B are described by Eqs. 3–5. Based on the resulting HDR images the luminance is calculated according to Eq. 6. The camera system is calibrated using a Halogen light source (2500 K).

Table 1. Shutter speeds of exposures 1 to 7, accompanied by the EV.

Exposure	Shutter Speed [µs]	EV
1	200000	4.3
2	55556	6.2
3	16129	8.0
4	4608	9.8
5	1314	11.6
6	369	13.4
7	104	15.2

$$Y_R = 0.7929 \cdot X^2 + 0.1998 \cdot X + 0.007299 \tag{3}$$

$$Y_G = 0.9397 \cdot X^2 + 0.0435 \cdot X + 0.01679 \tag{4}$$

$$Y_B = 0.9086 \cdot X^2 + 0.07674 \cdot X + 0.01466 \tag{5}$$

$$L = 3.1013 \cdot (0.2125 \cdot R + 0.7125 \cdot G + 0.0721 \cdot B) \tag{6}$$

The luminance is determined for ten samples ranging from white to dark grey placed in a light box with a LED light source (4410 K) (Fig. 3). The camera and Konica Minolta LS-100 luminance meter were focused at the sample which was placed at a distance of 1.5 m. There was no significant effect of the placement of the devices on the measured luminance values ($p = 0.998$). The measurement is repeated thrice for each sample with interchanged device positions. Ultimately, the results are averaged over the three measurements.

The measurement results are shown in Fig. 4. This measurement resulted in an average relative difference between the LS-100 luminance meter and the luminance

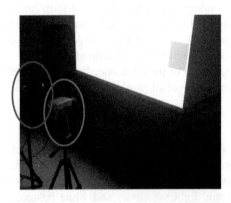

Fig. 3. Accuracy measurement setup. The camera (framed in Red) and LS-100 (framed in Blue) are placed at 1.5 meters from the sample that is placed in a light box with a LED light source. (Color figure online)

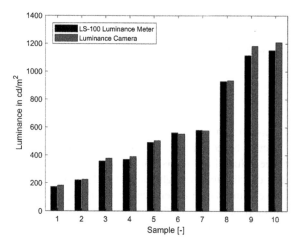

Fig. 4.

camera of 3.59% with a standard deviation of 2.24%. It should be noted that the relative difference is expected to be slightly higher for colored samples.

3 Measuring Lighting Quality

Generally, the individual aspects of lighting quality are measured independently by using expensive devices. For limited measurements, this can be acceptable; however, for continuous measurements, this is not practicable. Inexpensive measurement devices that are capable of measuring multiple lighting quality aspects instantaneously are desirable. The luminance distribution is a parameter that contains a large amount of data that can be used to indicate multiple lighting quality aspects. Previously, we already described that we can measure the luminance distribution with low cost components. With such a device we can measure or indicate the quantity, distribution, glare, daylight, directionality, and the dynamics of light. The spectral power distribution cannot be accurately measured using a luminance distribution measurement device. With only three channels (RGB) no sufficient information of the spectrum can be acquired.

The following section describes ways to measure individual lighting quality aspects using a luminance distribution measurement device.

Quantity. The quantity can be indicated by the luminance and illuminance. The luminance is easily extracted from the luminance distribution as each pixel represents a luminance value. Also the illuminance can be calculated based on the luminance distribution. It can be calculated for the surface area of the image sensor and for the surfaces of the photographed scene. The illuminance on the sensor is measured by integrating all luminance values, taking into account the cosine correction, similar to dedicated illuminance measurement devices. For measuring the illuminance of the surfaces of the photographed scene all surfaces are assumed to be Lambertian, meaning that they reflect the light diffusely; additionally, the surface reflectance is required for this measurement. The illuminance is then simply calculated according to Eq. (7), with luminance value (L) and surface reflectance (ρ).

$$E = \frac{L \cdot \pi}{\rho} \tag{7}$$

Distribution. The distribution of the light is often indicated by luminance ratios between areas of interest based on average luminance values of the areas. Generally, the average luminance value of an area is calculated by applying luminance masks. For the illuminance the distribution of light is generally indicated by the uniformity, describing the relative illuminance difference on the desktop which can be extracted according to the method described in the section Quantity. Theoretically, an illuminance distribution analogue to the luminance distribution can be developed based on this method.

Glare. Glare is a largely subjective response that is not understood entirely, however, a number of objective indicators have been developed with varying complexity. In general, these indicators consist of four quantities: the luminance, the solid angle, the displacement of the glare source and the adaptation luminance [24]. Defining these quantities can be rather complex, therefore, the glare is generally calculated using the *evalglare* software that requires a luminance distribution as input [25].

Daylight. To a large extend daylight can be indicated by the other aspects such as the luminance. However, some measures are developed especially for daylight such as the daylight factor and UDI [26]. Moreover, the view out is of importance. The view out consists of the opening in the façade providing the view and the quality of the view. The opening is generally fixed and known, unless dynamic sunscreens or brightness control is applied. The quality of view is largely subjective, however, Hellinga and Hordijk developed a methodology that is able to assess the quality of the view almost objectively [27]. This methodology makes use of images that are already available due to the HDR technology that is applied to calculate the luminance.

Directionality. Traditionally, the directionality is measured using cubic illumination, a tedious measurement using a six faced illuminance sensor vulnerable for calibration errors [28]. To measure the directionality using the luminance distribution a three dimensional object is required on which the light can fall from different directions, preferably a sphere. Subsequently, the luminance or illuminance is measured for different areas of the sphere, preferably at both sides. Based on this, the direction and the strength of the illumination vector, representing the directionality, can be determined [29]. Besides using a sphere, also a pyramid is proposed with surfaces in each relevant direction [30].

Dynamics. The variation in daylight can be described by the cumulative difference in pixel values of the luminance pictures according the method described by Rockcastle and Andersen [31]. Based on this method the degree of change, the location of change and the rhythm of the change can be described.

4   Conclusion

This study showed that the luminance distribution can be measured with a practical accuracy using low cost commercially available parts. The luminance distribution can be used to measure or indicate the different aspects of lighting quality, such as the quantity, distribution, glare, daylight, directionality and dynamics of light. Therefore, a luminance distribution measurement device is an excellent tool to measure the lighting quality.

By automating the luminance distribution measurement it is relatively easy to perform continuous measurements of lighting quality that can be related to subjective responses or integrated and automated building control systems.

References

1. Veitch JA, Newsham GR (1998) Determinants of lighting quality I: state of the science. J Illum Eng Soc 27:92–106
2. Boyce PR (2014) Human Factors in Lighting, 3rd edn. CRC Press, London
3. Veitch JA, Newsham GR (n.d.) Determinants of Lighting Quality I: State of the Science
4. Kruisselbrink T, Dangol R, Rosemann A (2018) Photometric measurements of lighting quality: an overview. Build Environ 138:42–52. https://doi.org/10.1016/j.buildenv.2018.04.028
5. Van Den Wymelenberg K, Inanici M (2014) A critical investigation of common lighting design metrics for predicting human visual comfort in offices with daylight. LEUKOS 10:145–164. https://doi.org/10.1080/15502724.2014.881720
6. Rea MS (2000) The IESNA lighting handbook. Illuminating Engineering Society of North America, New York
7. European Committee For Standardization (2009) NEN-EN 12464-1: Light and lighting - Lighting of work places - Part 1: Indoor work places
8. Bodrogi P, Brückner S, Khanh TQ (2010) Dimensions of light source colour quality. In: Conference on colour graphics, imaging and vision, pp. 155–159
9. Kruisselbrink T, Aries M, Rosemann A (2017) A practical device for measuring the luminance distribution. Int J Sustain Light 36:75–90
10. Inanici MN (2006) Evaluation of high dynamic range photography as a luminance data acquisition system. Light Res Technol 38:123–136. https://doi.org/10.1191/1365782806li164oa
11. Sarkar A, Mistrick RG (2006) A novel lighting control system integrating high dynamic range imaging and DALI. LEUKOS 2:307–322. https://doi.org/10.1080/15502724.2006.10747642
12. Tohsing K, Schrempf M, Riechelmann S, Schilke H, Seckmeyer G (2013) Measuring high-resolution sky luminance distributions with a CCD camera. Appl Opt 52:1564–1573. https://doi.org/10.1364/AO.52.001564
13. Reinhard E, Ward G, Pattanaik S, Debevec P (2006) High dynamic range imaging: acquisition, display, and image-based lighting (the Morgan Kaufmann series in computer graphics). Morgan Kaufmann Publishers Inc., San Fransisco
14. Moeck M, Anaokar S (2006) Illuminance analysis from high dynamic range images. LEUKOS 2:211–228

15. Cai H, Chung T (2011) Improving the quality of high dynamic range images. Light Res Technol 43:87–102. https://doi.org/10.1177/1477153510371356
16. Ward G (n.d.) Anyhere Software. http://www.anyhere.com/. Accessed 7 Mar 2016
17. Mead A, Mosalam K (2016) Ubiquitous luminance sensing using the Raspberry Pi and camera module system. Light Res Technol 1–18. https://doi.org/10.1177/1477153516649229
18. Mitsunaga T, Nayar SK (1999) Radiometric self calibration. In: IEEE computer society conference on computer vision and pattern recognition, vol. 1, pp. 374–380. IEEE Computer Society, Fort Collins. https://doi.org/10.1109/cvpr.1999.786966
19. Pascale D (n.d.) A Review of RGB Color Spaces
20. Inanici MN (2010) Evalution of high dynamic range image-based sky models in lighting simulation. Leukos 7:69–84. https://doi.org/10.1582/LEUKOS.2010.07.02001
21. Wüller D, Gabele H (2007) The usage of digital cameras as luminance meters. In: Electronic imaging conference, vol. 6502, San Jose, USA, pp. 1–11. https://doi.org/10.1117/12.703205
22. Cauwerts C, Bodart M, Deneyer A (2012) Comparison of the vignetting effects of two identical fisheye lenses. LEUKOS 8:181–203
23. Ulbricht R (1920) Das Kugelphotometer. Verlag Oldenburg, Berlin Und Munchen
24. Wienold J, Christoffersen J (2006) Evaluation methods and development of a new glare prediction model for daylight environments with the use of CCD cameras. Energy Build 38:743–757. https://doi.org/10.1016/j.enbuild.2006.03.017
25. Wienold J (2015) Evalglare
26. Nabil A, Mardaljevic J (2006) Useful daylight illuminances: a replacement for daylight factors. Energy Build 38:905–913. https://doi.org/10.1016/j.enbuild.2006.03.013
27. Hellinga H, Hordijk T (2014) The D&V analysis method: a method for the analysis of daylight access and view quality. Build Environ 79:101–114. https://doi.org/10.1016/j.buildenv.2014.04.032
28. Cuttle C (2014) Research note: a practical approach to cubic illuminance measurement. Light Res Technol 46:31–34
29. Dubois M-C, Gentile N, Naves David Amorim C, Geisler-Moroder D, Jakobiak R, Matusiak B, et al. Monitoring protocol for lighting and daylighting retrofits: A Technical Report of Subtask D (Case Studies), T50.D3 (2016)
30. Howlett O, Heschong L, Mchugh J (2007) Scoping study for daylight metrics from luminance maps scoping study for daylight metrics from luminance maps. LEUKOS 3:201–215. https://doi.org/10.1582/LEUKOS.2007.03.03.003
31. Rockcastle S, Andersen M (2014) Measuring the dynamics of contrast & daylight variability in architecture: a proof-of-concept methodology. Build Environ 81:320–333. https://doi.org/10.1016/j.buildenv.2014.06.012

A Comparison of Mental and Visual Load Resulting from Semi-automated and Conventional Forest Forwarding: An Experimental Machine Simulation Study

H. O. Richter[1]([⊠]), D. Domkin[1], G. H. Elcadi[2], H. W. Andersson[3],
H. Högberg[2], and M. Englund[3]

[1] Centre for Musculoskeletal Research, Department of Occupational and Public
Health Sciences, Faculty of Health and Occupational Studies,
University of Gävle, Gävle, Sweden
hrr@hig.se
[2] Department of Health and Caring Sciences, Faculty of Health
and Occupational Studies, University of Gävle, Gävle, Sweden
[3] Skogforsk, The Forestry Research Institute of Sweden, Uppsala, Sweden

Abstract. The purpose of the present study was to extend the knowledge of functional linkages between visual and mental load, performance, and prefrontal cortex (PFC) activity, during forestry forwarding work. Eleven healthy participants, range 21–51 years old, with a minimum of 1-year work experience, carried out the task of loading logs along a standardized path in a machine simulator during two counterbalanced test conditions: (i) conventional crane control, and; (ii) semi-automated crane control. Mental load was assessed by quantification of oxygenated hemoglobin (HbO_2) concentration changes over the right dorsolateral prefrontal cortex (dlPFC) via non-invasive functional near infrared spectrometry (fNIRS). Visual, autonomic, and motoric control variables were measured and analyzed in parallel along with the individual level of performance. Linear Mixed Models (LMM) analysis indicated more mental load during conventional crane work. Collectively, our data suggest that fNIRS is a viable tool which can be used in neuroergonomic research to evaluate physiological activity levels in PFC.

Keywords: Attention fatigue · Compensatory effort
Near infrared spectroscopy (NIRS) · Neuroergonomics · Time series analysis
Visual ergonomics

1 Introduction

The 4th industrial revolution can have tremendous implications on how we perceive and organize work in the future, but little is still known about the impact on human body and brain. This study contributes with new knowledge of the consequences of the current increase in automation. Work-related injuries and constant demands for a higher productivity are two of the many arguments for why forestry work must be improved. Pivotal to the current study are recent advancements in the quantification of

© Springer Nature Switzerland AG 2019
S. Bagnara et al. (Eds.): IEA 2018, AISC 827, pp. 199–208, 2019.
https://doi.org/10.1007/978-3-319-96059-3_22

brain activity by functional Near Infrared Spectroscopy, which have opened new and unique avenues for investigation (Ferrari and Quaresima 2012; Ayaz et al. 2013; Derosiere et al. 2013). The possibility of quantifying central hemodynamic changes to mental load and fatigue in subjects in an ecologically realistic setting and free gaze conditions is a particular advantage of fNIRS. The prefrontal cortex (PFC), and specifically the right dorsolateral prefrontal cortex, Brodman area 46 (dlPFC), is a brain region of key interest in the context of visual load and mental effort because of its supervisory role in behavior and its involvement in visual processing (Barceló et al. 2000; Fuster 2015; Derosiere et al. 2013; Mandrick et al. 2013). Against this background the purpose of the present study was to extend the knowledge of functional linkages between visual and mental fatigue, performance, and PFC activity, during semi-automated and conventional crane work in forestry forwarding.

2 Methods

2.1 Participants

Eleven male forwarder drivers with professional experience of more than one year in conventional forwarder crane operation, not taking any medication influencing nervous system and without neck pain of traumatic origin, Parkinson, MS or eye diseases participated in the study. The participants had the following characteristics: age 21–51 years (median 39), height 175–187 cm (median 183), weight 75–135 kg (median 93) and body mass index (BMI) 22.4–40.3 (median 28.1). The participants were informed about the study and gave their written informed consent. The regional Ethics Committee in Uppsala approved the study (Dnr 2017:049).

2.2 Experimental Setting

The study was performed in the Troëdsson Forest Technology Lab (Skogforsk, Sweden), which is a real-time forest machine simulator. The machine simulation was of a Komatsu forest forwarder, with a knuckle boom crane with a single extension and a grapple. The simulator was produced by Oryx Simulations AB (Sweden). Participants operated the simulator from a work station with an actual operator seat equipped with armrests with joystick pods and a floor-mounted accelerator. Simulated visual environment as seen from the driver's perspective - forest, road path, logs on the ground, and the machine – crane and load space – was generated and presented in real time on three back projection screens placed around the operator. The screens were at the distance of approximately 1 m from the operator, providing a field of view of the virtual visual environment of approximately 240° of visual angle horizontally and about 60° vertically. Simulator parameters describing position, movement and operation of the forwarder, crane and joystick controls were continuously logged at a sample rate of 30 Hz.

2.3   Work Task

The participants practiced the work in the simulator before taking the actual test. The task was to drive the forwarder along a long straight path, collect with the crane's grapple one by one logs that were lying on the ground by the sides of the path and put them into the forwarder's load space. The participants were instructed to perform the task as quickly as possible while maintaining accuracy. The placement and number of logs to be collected was the same for all participants and tests.

There were two test conditions, which were performed in a randomized order:

Conventional and Semi-automated Crane Operation. The main difference between the conditions was in the degree of automation of the crane operation. During conventional operation the operator controlled the crane manually by actuating the joysticks. The operator guided the crane to the location of the log, secured it with the grapple, and then guided the crane back to the load space where he placed and released the log. During the semi-automatic crane operation, the operator initiated with the press of a button an automatic crane movement to a predetermined location perpendicular to the machine on either the left or right side. Fine adjustments of the crane position and securing of the log were done under manual control. Then, after another button press, the crane with the log in the grapple automatically moved to a predetermined location within the load space, where the log was then, under manual control, placed and released. Another difference between the test conditions was in the type of manual crane control. In the conventional condition the operator separately controlled the hydraulic cylinders used to move and extend the crane arm by engaging a separate joystick function for each cylinder, which is the way most forwarders are controlled today. In the semi-automatic condition, the same joystick functions were used to control the movement of the tip of the crane in a cylindrical coordinate system originating from the base of the crane (Boom tip control). In the conventional control condition, the operator used six separate joystick functions to control the crane and the grapple. In the semi-automatic condition, the operator used five joystick functions plus one to initiate the automatic function.

2.4   Work Cycles

Each *Work cycle* consisted of collecting one log and placing it into the load space. The predefined total number of work cycles was 240 for both test conditions. The timing (start and end) of each individual work cycle and the duration thereof was determined and used in the statistical analysis.

2.5   Simulator Joystick Controls

To evaluate the usage of joystick controls during the work task for both testing condition, we computed across data samples of each work cycle the mean *number of simultaneously actuated joystick functions* (meanNrContr), the mean *number of joystick direction changes* and the mean *percent of time of joystick usage*.

2.6   Near Infrared Spectroscopy

Non-invasive measurements of regional hemodynamic changes over the dlPFC were quantified with a near-infrared spectrometer (fNIRS: PortaLite mini, Artinis Medical Systems, Zetten, The Netherlands). Mean penetration depth was ~ 12.5 mm into the cortex and the sampling frequency was 10 Hz. An increase in brain activity is generally assumed to reflect an increase in local HbO_2 as based on a mechanism known as neurovascular coupling (Ferrari and Quaresima 2012). A decrease in HbO_2 response may also reflect neuronal suppression, or alternatively the re-distribution of blood flow (Quaresima and Ferrari 2012). Neuronal activation is thought to be accompanied by the expansion (vasodilatation) of perfused blood vessels and by an increased portion of perfused vessels (recruitment) (Ferrari and Quaresima 2012). The fNIRS probe was placed on the PFC on the right dlPFC. A modified version of the 10–20 system (Oostenveld and Praamstra 2001) was used for placement of the probe. The raw data was smoothed with a moving average with a window of 5 s. Several steps were taken in the computation of baseline-subtracted local HbO_2. First, immediately prior to performance in each of the two work tasks, a baseline (rest) was recorded for 7 min during which the subjects quietly rested with their eyes opened. Secondly, baseline reference HbO_2 values were computed as the 5th percentile across a 4 min window in the middle of the rest period, which were next subtracted from all HbO_2 values obtained during the corresponding forwarding task, thus providing ΔHbO_2. Finally, the individual means of ΔHbO_2 were determined for each consecutive work cycle. The final data set consisted of Conventional crane operation (n = 7) and Semi-automated crane operation (n = 7).

2.7   Electrocardiography

Electrocardiography (ECG) was measured with Biopac MP150 system (Biopac Systems Inc., Santa Barbara, CA, USA). Data was recorded with a sampling program AcqKnowledge 4.3 (Biopac Systems Inc., Santa Barbara, CA, USA) at a sampling rate of 1000 Hz. On the sites of ECG electrode placement the skin was shaved and scrubbed with an El-Pad (Biopac) and washed with Alsol (skin alcohol solution 10 mg/ml). A ground electrode (Ambu Neuroline Ground, Ambu A/S, Ballerup, Denmark) was placed on the skin on the processus spinosus of the cervical vertebra C7. An ECG electrode (Ambu Blue Sensor VLC, Ambu A/S, Ballerup, Denmark) was placed directly under the right and left clavicle. The impedance for the electrodes within a pair was about 2–8 kΩ in all cases. To evaluate heart rate variability, we computed the standard deviation of the successive differences in duration of periods between heart beats, i.e. between normal-to-normal intervals (SDNN). A normal-to-normal interval is the interval between adjacent QRS complexes resulting from sinus node depolarization. The SDNN was computed for three sequential time periods of the work task, each period lasted about 10–15 min dependent on the total task duration (SDNN). The SDNN from these periods was also averaged to get a value per test for analyses. In the LMM analyzes outlined below SDNN for three sequential time periods of the work task constituted a control variable (covariate).

2.8   Eye-Head Rotation and Gaze

To evaluate processing of visual information during forest forwarding work, head and eye movements were measured with a SmartEye Pro (software version 6.0) eight-camera eye-tracking and head-tracking system (Smart Eye AB, Sweden) at a sample rate of 30 Hz. The following parameters were calculated for each work cycle: *maximal lateral head rotation angle* (MaxHeadAngle), *maximal lateral eye rotation angle* and *maximal lateral gaze angle* during the phase of gripping the logs, *number of blinks*, *number of gaze fixations* (NrFix) and *mean duration of blinks* and *gaze fixations*. To account for individual differences in duration of work cycle, the number of blinks and gaze fixations in a work cycle was divided by the duration of the corresponding work cycle in order to obtain a standardized measure of the number of blinks and fixations per second. Before the computations of the head rotation angle and gaze angle, data points corresponding to moments of loss of eye and head tracking were detected based on the tracking quality estimates provided by the SmartEye sampling software and were replaced by values obtained in linear interpolation between the adjacent good samples. Gaze and head rotation angles data were smoothed with a moving average with a window width of 1 s.

2.9   Evaluation of Mental Fatigue and Work Process

Evaluation of *mental fatigue* was performed after each test on a Borg CR10 scale ranging from 0 ("no fatigue at all") to 10 ("extreme fatigue"), with value 3 corresponding to "moderate fatigue". Several parameters describing work perception in the simulator in the semi-automatic test condition in comparison to conventional test condition as a reference were evaluated with a modified NASA Task Load Index (NASA TLX) after all tests were completed. The parameters described the perception of: (1) *physical load* (in the hands, arms and neck), (2) *mental load*, (3) *time pressure*, (4) *performance level*, (5) *physical and mental effort to achieve the desired performance level*, and (6) *frustration level*. The evaluation scale which was used to compare the semi-automatic work with the conventional work (reference) had the following rating alternatives: "much less", "slightly less", "equal", "slightly more" and "much more". Participants were, in addition, asked to comment on the perceptual experiences during work in the simulator as opposed to real forest work, in relation to e.g. the lack of depth perception.

2.10   Data Analysis and Statistics

Data processing and computation of variables were performed with Matlab v.9 (The MathWorks, Inc., USA). Statistical analyses were performed with SPSS v.24 (IBM SPSS Statistics, IBM Corporation, USA). Statistical significance level was set at $p < 0.05$. Performance related differences between the conventional and semi-automatic test conditions were tested with Repeated Measures ANOVA. To test the effect of independent variables on the ΔHbO_2 concentration a regression analysis by Linear Mixed Models (LMM) was performed. One representative variable from each main category of variables described in the method section was chosen to be included in the

LMM analysis as a covariate. In the subsequent analytic step, if one or more of the selected variable did not make a significant contribution to the model ($p \geq 0.10$), this/these variable(s) was removed, and the model was re-estimated. The LMM was specified as a random coefficient model with both fixed and random effects. The random effects were allowed to vary across individuals. The variance/covariance between the random effects was set as diagonal.

3 Results

3.1 Performance Related Differences Between Conventional and Semi-automated Work

The averaged across participants and work cycles ΔHbO_2 concentration was higher during the conventional work condition (average 3.8 µMol, SD: 2.5) than in the semi-automatic work condition (average 2.1 µMol, SD: 2.7). The difference was on the border of significance ($p = 0.06$). The group averaged baseline subtracted oxygenated hemoglobin concentration change over time and work type is shown in Fig. 1.

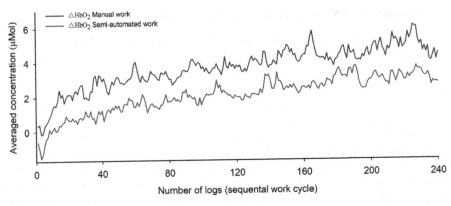

Fig. 1. Average ΔHbO_2 concentration change over time in the two test conditions (n = 7).

The work cycle duration was on average slightly longer in semi-automatic test condition with an average of 11.9 s (SD: 2.6) compared to the conventional test condition with an average of 10.6 s (SD: 1.4). The difference was on the border of significance ($p = 0.05$). The total work time was not significantly different between test conditions ($p = 0.17$), being on average 41 min (SD: 4) for conventional and 44 min (SD: 4) for semi-automatic condition.

The average number of simultaneously actuated joystick functions was significantly higher in the conventional (average 4.2, SD: 0.2) than in semi-automatic condition (average 3, SD: 0.2) ($p < 0.001$).

The heart rate variability measure SDNN decreased in both test conditions between the baseline and work period, for conventional condition from an average of 53.6 ms (SD: 25.7) to an average of 39 ms (SD:13) ($p = 0.017$) and for semi-automatic from an average of 52.4 ms (SD: 25) to an average of 41.3 ms (SD: 15) ($p = 0.052$). There was no significant difference in SDNN between test conditions during either the baseline or the work period ($p = 0.8$ and $p = 0.25$, respectively). On average SDNN tended to increase more in magnitude over successive work cycles during semi-automatic work relative to manual work. Throughout the last third of the work period SDNN was consequently slightly lower during manual work in comparison to semi-automated work.

There was no significant difference in eye rotation angle, being on average 18.5° (SD: 4.3) for conventional and 18.6° (SD: 5) for semi-automatic condition ($p = 0.8$). Maximal lateral head rotation angle during gripping of logs was not significantly different between testing conditions (average 38°, SD: 5 for both, $p = 0.9$). Neither lateral gaze angle (average 57°, SD: 5 for conventional and average 57°, SD: 2 for semi-automatic) was significantly different for the two testing conditions ($p = 0.9$).

The number of gaze fixations per second was significantly higher during the conventional (average 1.34, SD: 0.23 per sec) test condition in comparison to semi-automatic test (average 1.27, SD: 0.21) ($p < 0.01$). The number of gaze fixations per second ranged from 0.43 to 2.69 in the conventional test and from 0.37 to 2.51 in the semi-automatic test condition. The average duration of gaze fixations was significantly longer during the semi-automatic test condition (0.48, SD: 0.1 s) relative to conventional test condition (average 0.44 s, SD: 0.1) ($p = 0.01$). The duration of gaze fixations ranged from 0.3 to 11 s during the conventional test condition and from 0.3 to 9.5 s in the semi-automatic test. The total average number of fixations for the whole test was 3190 (SD: 502) for conventional work and 3495 (SD: 957) for semi-automatic work.

The number of blinks per second was not significantly different between two testing conditions (both average 0.25, SD: 0.09 per sec, $p = 0.38$). Neither the average duration of blinks was significantly different ($p = 0.9$) being on average 0.5 (SD: 0.1) sec for both conditions.

The mental fatigue ratings showed higher values for conventional test condition (median 3) than for semi-automatic condition (median 2.8). Wilcoxon test showed that this difference was significant ($p = 0.047$). Perception of physical load, was rated "slightly less" on median during semi-automatic work when compared to conventional work (Wilcoxon test, $p = 0.005$). All other subjective parameters of work perception were rated equal in both work conditions. Seven of eleven drivers reported difficulties with the lack of depth perception during work in the simulator in comparison to real forest machine work.

3.2   Linear Mixed Models Analysis

The results from the LMM analysis is presented in Table 1 below. This analysis revealed that dlPFC activity differed between work tasks ($p < 0.05$). On average activity in the *Manual task* was larger than the *Semi automated task*. Activity increased systematically over each successive *Cycle* ($p < 0.005$) in both test conditions. The interaction *Worktype × Cycle* indicated more activity over successive work cycles during *manual work* ($p < 0.001$). The *mean number of controls direction changes* showed a trend towards significance ($p < 0.10$), the more direction changes the larger magnitude oxygenated blood volume. There was also a small significant effect from the *mean maximal lateral head rotation* (p < 0.005). SDNN impacted with a main effect (p < 0.001) and as a *Worktype* × SDNN interaction effect (p < 0.01). The larger magnitude of SDNN the lower predicted dlPFC activity. During manual work this relationship was more pronounced in that SDNN and ΔHbO_2 tended to exhibit a more distinct negative correlation than was the case during semi-automated work.

Table 1. Summary of the main effects from the LMM analyses on prefrontal (dlPFC) activity during Conventional and Semi-automated crane operation work.*

Parameter	Estimate	Significance	95% CI
Intercept	1.245	0.156	−0.538, 3.028
Conventional	2.587	0.016	0.526, 4.647
Semi-automated	0ᵃ		
Work cycle	0.012	0.005	0.005, 0.019
MaxHeadAngle	0.008	0.002	0.003, 0.014
SDNN	−0.026	0.001	−0.041,−0.011
Conventional × meanNrContr	−0.040	0.599	−0.191, 0.110
Semi-automated × meanNrContr	0.109	0.086	−0.015, 0.234
Conventional × Cycle	0.003	<0.000	0.002, 0.005
Semi-automated × Cycle	0ᵃ		
Conventional × NrFix	0.022	0.004	0.007, 0.036
Semi-automated × NrFix	−0.004	0.455	−0.016, 0.007
Conventional × SDNN	−0.036	0.007	−0.062, −0.010
Semi-automated × SDNN	0ᵃ		

*Information criteria, smaller is better, Akaike's Information Criterion (AIC) = 13834. a = reference. For abbreviations see Methods.

4   Discussion

The results from this study indicate that hemodynamic activity recorded over dlPFC with fNIRS increase in magnitude over time, is impacted by test condition, and differ with respect to the contents of work, which impacts on mental load. These main findings, discussed more in detail below, indicate that fNIRS can be used to assess the degree to which the PFC is strained during mentally and physically demanding forestry crane operation work.

4.1   Predictors to DlPFC Activity

There was a fixed effect of work type on dlPFC activity. On average PFC activity was lower during semi-automated crane operation work. The effect of the number of work cycles (number of logs that was loaded) on PFC activity was presumably caused by tonic dlPFC neuronal discharge as related to the active maintenance of task-related attention templates in working memory. Cortical resources related to the inhibition of undesired behavioral responses may also have played a role (Hockey 1997; Faber et al. 2012; Wang et al. 2016). In this scenario, dlPFC neurons systematically increased their tonic discharge over time to stay focused on the task and to counteract mental fatigue and/or boredom and to inhibit urges to quit.

During conventional crane operation work participants had to process a larger flow of visual input relative to semi-automatic work, indicated by more frequent gaze shifts: higher number of gaze fixations with shorter duration. This finding, together with higher Borg ratings of mental fatigue, indicated a higher mental load in the conventional test condition.

The effect of maximal lateral head rotation angle during the phase of gripping the logs on dlPFC activity likely was related to mental effort to move the head to extreme and strenuous outer positions. A hallmark sign of prefrontal processes is to be enlisted during non-routine demanding situations.

The interaction *mean number of simultaneously actuated joystick functions across data samples of each work cycle × semi-automated work* showed a trend on mental load, as measured by dlPFC activity. The more actuated joystick functions the more dlPFC activity during semi-automated crane operation work. On average, a larger number of simultaneously actuated joystick functions occurred during conventional forestry work. According to the adaptive coding model of PFC function, the population of cortical neurons in dlPFC adjust their function to match the requirements of any given task undertaken and the more tasks, in this case number of simultaneously used controls, the more recruitment of cortical cells and the more intensive engagement in the task is expected (Duncan 2001).

dlPFC activity decreased in magnitude when SDNN increased in magnitude and this was probably so because dlPFC disengaged as task demands became less stressful. The interaction effect *work type × SDNN* indicated less experience of stress during semi-automated crane operation work, especially during the last third of the work period.

The individual level of productivity (work cycle duration) did not impact upon PFC, i.e., dlPFC activity did not increased more in magnitude over time among subjects that were working at a faster pace or vice versa. The participants likely allocated similar central resources to execute the task. A limitation in this respect may have been the relatively short test time duration in comparison to real forestry work.

4.2   Conclusions

Automation in forest machines can reduce the level of intensity of visual processing and mental load in forest machine operators. fNIRS technique is suitable for laboratory applications simulating forestry machine work. Depth perception was a noteworthy problem in forestry machine simulator with 2D screens in comparison to real (3D) work. Field studies are required to further evaluate applicability of fNIRS method in forestry work and to evaluate differences in visual processing and mental/physical load in real work situation.

Funding. This study was in part supported by grants from the Swedish Council for Working Life, Social Research Grant 2009-1761 and grants from *Södra Skogsägarna* and *Norrskog*.

We acknowledge Research Engineer N. G. Larson for excellent engineering assistance. The authors declare that this research was conducted in the absence of any commercial or financial relationships that could be construed as a potential conflict of interest.

References

Ayaz H, Onaral B, Izzetoglu K, Shewokis PA, McKendrick R, Parasuraman R (2013) Continuous monitoring of brain dynamics with functional near infrared spectroscopy as a tool for neuroergonomic research: empirical examples and a technological development. Front Hum Neurosci 7:871

Barceló F, Suwazono S, Knight RT (2000) Prefrontal modulation of visual processing in humans. Nat Neurosci 3(4):399–403

Derosière G, Mandrick K, Dray G, Ward TE, Perrey S (2013) NIRS-measured prefrontal cortex activity in neuroergonomics: strengths and weaknesses. Front Hum Neurosci 7:583

Duncan J (2001) An adaptive coding model of neural function in prefrontal cortex. Nat Rev Neurosci 2(11):820–829

Faber LG, Maurits NM, Lorist MM (2012) Mental fatigue affects visual selective attention. PLoS ONE 7(10):e48073

Ferrari M, Quaresima V (2012) A brief review on the history of human functional near-infrared spectroscopy (fNIRS) development and fields of application. Neuroimage 63(2):921–935

Fuster JM, Bressler SL (2015) Past makes future: role of pFC in prediction. J Cogn Neurosci 27 (4):639–654

Hockey GR (1997) Compensatory control in the regulation of human performance under stress and high workload; a cognitive-energetical framework. Biol Psychol 45:73–93

Mandrick K, Derosiere G, Dray G, Coulon D, Micallef JP, Perrey S (2013) Prefrontal cortex activity during motor tasks with additional mental load requiring attentional demand: a near-infrared spectroscopy study. Neurosci Res 76(3):156–162

Oostenveld R, Praamstra P (2001) The five percent electrode system for high-resolution EEG and ERP measurements. Clin Neurophysiol 112(4):713–719

Wang C, Trongnetrpunya A, Samuel IB, Ding M, Kluger BM (2016) Compensatory neural activity in response to cognitive fatigue. J Neurosci 36(14):3919–3924

Human Gaze-Parameters as an Indicator of Mental Workload

Frode Volden[1]([⊠]), Viveka De Alwis Edirisinghe[1],
and Knut-Inge Fostervold[2]

[1] Norwegian University of Science and Technology, Gjøvik, Norway
frodv@ntnu.no
[2] University of Oslo, Oslo, Norway

Abstract. In this study we have investigated which eye-parameters that most reliably can indicate increased mental workload. Being able to detect high mental workload in individuals, allows for early detection of potentially dangerous situations, and possibly adjustment of the information flow that creates the high workload. N-back memory tasks with four difficulty levels were designed to induce mental workload for a sample of 21 university students. 17 eye parameters were measured using an Eye Tracker at a sampling rate of 250 Hz. Data indicate that peak fixation duration is the most suitable eye parameter to estimate mental workload. It has a negative relationship with mental workload, where higher peak fixation duration can be observed at lower mental workload and lower peak fixation duration at higher mental workload. Moreover, blink frequency, blink count, peak blink duration, and pupil diameter show a significant positive relationship to mental workload. Most of the saccade parameters failed to show a significant relationship, while fixation frequency, fixation duration, fixation count, blink duration, saccade velocity, and peak saccade amplitude showed a partial relationship with mental workload.

Keywords: Eye-tracking · Mental workload

1 Introduction

High mental workload can lead to physical, psychological and social issues. Performing demanding tasks for a long period can cause stress and fatigue [1], this can be a source for both health and performance issues on people. Poor performance can bring severe consequences for critical jobs such as driving, aviation, and surgical operations. For example, in the driving context higher mental workload and poor performance can be a cause of accidents [2].

Physical measures, such as heart rate, Electroencephalogram (EEG), respiration rate, alertness monitoring, skin conductance level have been explored to find a relationship with mental workload [2]. Eye parameters is another physiological measure that has been used in the context of mental workload estimation. The most significant feature about eye parameter measurement compared to the other physiological measures mentioned above, is the possibility to use non-wearable and non-intrusive equipment to measure them. The purpose of this study is to find out the relationship

© Springer Nature Switzerland AG 2019
S. Bagnara et al. (Eds.): IEA 2018, AISC 827, pp. 209–215, 2019.
https://doi.org/10.1007/978-3-319-96059-3_23

between different eye parameters and mental workload and hence report the most suitable eye parameter(s) for mental workload estimation.

Based on existing research and the possible parameters that may be measured with eye-trackers, 17 different gaze-parameters where identified and included in the study. Holmqvist et al. [3] is one of several sources for the identification of interesting parameters.

Eye parameters can be categorized into four: blinks, saccades, fixations, and other. Eye parameters belong to these categories have all been used to estimate mental workload (Table 1).

Table 1. Gaze parameters used in the study.

Category	Parameter
Blinks	Count, Frequency, Duration, Peak duration
Fixations	Count, Frequency, Duration, Peak duration
Saccades	Count, Frequency, Duration, Peak duration, Amplitude, Peak amplitude, Velocity, Peak velocity
Other	Pupil diameter

Visual N-back tasks from 1 to 4 was used in this study both to induce mental workload, and as the situation where gaze parameters where recorded. Jaeggi et al. [4]. Consider N-back tasks as an excellent method to control mental workload in experimental settings. More mental workload is induced by increasing the complexity level of the task. N-back task can be designed in two different ways: (a) as an auditory n-back task or (b) as a visual n-back task. An auditory N-back task let the users hear different sounds repeatedly, and then they have to remember and recall the previous sounds. In a visual N-back task which we used in this experiment, participants are shown pictures one at a time, and are than asked to recall previously shown pictures. N-back 1 means recalling the previous picture, N-back 4 means recalling a picture shown 4 slots back. The first induces little mental load, the last puts a lot of load on the memory system (working memory).

In general, it is not always easy to identify the amount of workload in a given situation for a specific person since the mental workload induced by the task vary from person to person. However, it is possible to find out the relative level of mental workload, for example, whether a person experiences high mental workload or low mental workload.

One of the techniques used for this purpose, and the one used in this study, is the NASA-TLX form. The NASA-TLX form developed by Hart and Staveland [5]. creates a total score for workload, and consists of 6 questions; *Mental Demand, Physical Demand, Temporal Demand, Performance, Effort, and Frustration.* The question on physical demand was not used in this study. Each question is answered on a 21-point scale.

2   Method

The experiment was automated by being programed as a website. In the experiment, four levels of n-back tasks were used as a means of inducing workload on the participants, and different eye parameters were measured under each N-back task. Each level of workload (N-back level) consisted of a sequence of 40 images which were shown with 4 s intervals. When recognizing an image, the participant pressed the spacebar on the keyboard. Images were either characters (50%), fruits (25%) or vegetables (25%). Images appeared randomly at 4 different screen positions (Fig. 1).

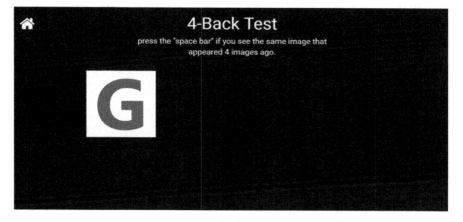

Fig. 1. Example screen from 4-back condition

At the end of each task, participants reported their mental workload by submitting the online NASA-TLX form presented on the website (Fig. 2).

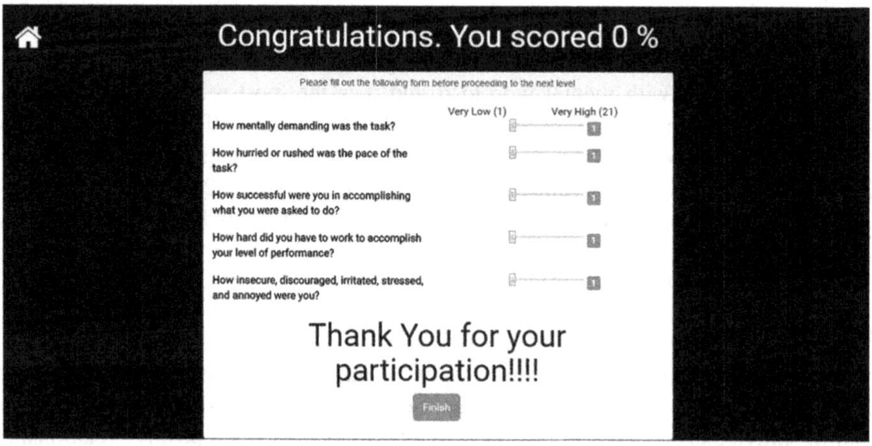

Fig. 2. NASA-TLX form after each level.

2.1 Participants

21 students (12 males) with a mean age of 25 years participated in the study. Participants were exposed to increasingly difficult and challenging mental workload situations, allowing for within subject comparisons of behaviors and performance.

2.2 Apparature

The recording and detection of the dependent variables (eye parameters) was carried out using a remote eye tracker connected to a laptop computer, that also were running the eye-tracker software. A SMI RED250mobile Eye Tracker was used in the experiment. SMI Experiment Center 3.6 software was used to perform the experiment, and parameter detection from eye-tracking data was done by SMI BeGaze 3.6. Data analyses was done using IBM SPSS v.21.

3 Results and Discussion

3.1 N-Back

As many other studies have suggested, visual n-back tasks can be used to increase a person's mental workload systematically, and it was observed among the students during the study. Therefore, using the n-back task to induce the mental workload can be considered a valid method, and therefore it will work as a good platform for further studies of gaze-behaviors as dependent of increased workload. The higher n-back, the more demanding the task was scored by the participants. There was a significant difference between the four n-back levels regarding the mental demand ratings, $F(1.97, 37.35) = 88.22$, $p < .001$ (Fig. 3).

3.2 Fixations

One of the principal purposes of this research was to find out the most significant eye parameter that can be used to estimate mental workload of university students. Results suggest that peak fixation duration has a strong negative and highly statistically significant relationship with mental workload and it is the most reliable eye parameter among the 17 eye parameters used in this study. The mean difference between low mental workload ($M = 2658.23$, $SD = 448.56$) and high mental workload ($M = 2290.81$, $SD = 428.02$) was statistically significant, $t(17) = 3.32$, $p = .004$, $d = .86$. This means that when users were highly mentally loaded they tend not to fixate a point for a long time (Fig. 4).

3.3 Blinks

Both blink count and blink frequency have a significant positive relationship with mental workload, and blink frequency has the strongest relationship of the two. This confirms previous research findings. Peak blink duration shows a positive relationship with mental workload. In other words, peak blink duration increases with mental

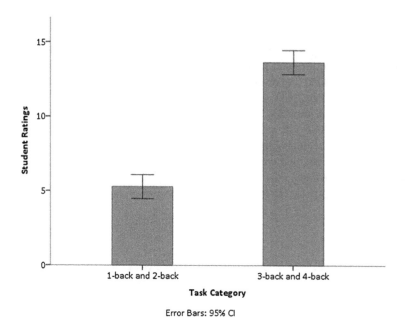

Fig. 3. Scores of experienced mental demands of the task

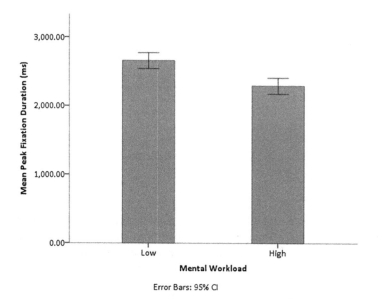

Fig. 4. Peak fixation duration

workload. This is a new finding related to estimating mental workload. For the paired-samples t-test, a normal distribution for mean differences of peak blink duration was obtained by removing four outliers (N = 17). The results showed that there was a significant difference in mean peak blink duration, t(16) = −3.08, p = .007, d = 1.37 between low workload condition (M = 565.94, SD = 296.74) and high workload condition (M = 973.21, SD = 637.66). In other words, when mental workload increased, peak blink duration would also increase, and it decreased under the low mental workload. ANOVA show a significant relationship with mental workload. A relatively strong relationship can be observed between 3-back and 4-back tasks. However, there is a positive correlation where peak blink duration increases with the mental workload. This is a new finding, and no research has previously focused on the relationship between peak blink duration and mental workload (Fig. 5).

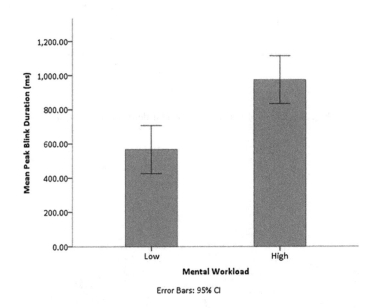

Fig. 5. Peak blink duration

3.4   Pupil Diameter

Pupil diameter show a significant positive relationship with mental workload and confirms earlier findings. Though lighting conditions can increase pupil diameter, in this experiment, it was not affected because of the same lighting conditions used for every user. Therefore, the increased value of pupil diameter due to lighting conditions can be kept as constant for every user, and it does not have any negative impact on our conclusions. Pupil diameter analysis using paired-samples t-test was done by filtering the sample removing two outliers and the new sample of 19 participants. There was a significant difference in means of pupil diameter, t(18) = −4.568, p < .001, d = 0.24 between low mental workload group (M = 4.09, SD = .55) and high mental workload group (M = 4.22, SD = .61). In general, pupil diameter increases with increased mental workload. This effect was expected and is in accordance with existing research.

3.5 Saccades

Of the 17 different eye parameters used in the study, none of the saccade measures showed any strong relationship to mental workload. Only some very small effects on peak saccade velocity and mean saccade velocity were found. This is in accordance with previous research on the area. Saccadic intrusions were not included in the study but is a well-known parameter to estimate mental workload. We therefore suggest this parameter to be included in future research.

4 Conclusion

The N-back task seems like a good way to induce mental workload in this study and can be used for further studies on the area.

In general findings support and validates existing research on which gaze parameters is affected by increased mental workload, with the addition of a number of peak values that show great promise. Results suggest that *peak fixation duration* has a strong negative significant relationship with increased mental workload and among the 17 parameters included in this study, it is the most reliable predictor of mental workload. Both *blink count* and *blink frequency* have a significant positive relationship with the mental workload, and blink frequency show the strongest relationship of the two. Peak blink duration shows a positive relationship with the mental workload.

References

1. Brookhuis KA, De Waard D, Kraaij JH, Bekiaris E (2003) How important is driver fatigue and what can we do about it. Human Factors in the Age of Virtual Reality, 191. Shaker Publishing, Maastricht, p 207
2. Gable TM, Kun AL, Walker BN, Winton RJ (2015) Comparing heart rate and pupil size as objective measures of workload in the driving context: initial look. In: Adjunct Proceedings of the 7th International Conference on Automotive User Interfaces and Interactive Vehicular Applications. ACM, pp 20–25
3. Holmqvist K, Nyström M, Andersson R, Dewhurst R, Jarodzka H, Van de Weijer J (2011) Eye tracking: A comprehensive guide to methods and measures. OUP, Oxford
4. Jaeggi SM, Buschkuehl M, Perrig WJ, Meier B (2010) The concurrent validity of the n-back task as a working memory measure. Memory 18(4):394–412
5. Hart SG, Staveland LE (1988) Development of nasa-tlx (task load index): Results of empirical and theoretical research. Adv Psychol 52:139–183

Colour in Glossolalia: "As Long as It's Black" (In Western Culture)

Leonor Ferrão[1,2(✉)]

[1] CIAUD, Lisbon School of Architecture, Universidade de Lisboa,
Rua Sá Nogueira, Polo Universitário do alto da Ajuda,
1349-063 Lisbon, Portugal
lferrao@fa.ulisboa.pt
[2] CITAD, Universidade Lusíada,
Rua da Junqueira 188-198, 1349-001 Lisbon, Portugal

Abstract. The aim of this article is to discuss some aspects that deal with colour in Western culture. From typography to fashion and product design, black suggests, among many other things, normality, simplicity, distinction, sophistication, discretion, sadness, grief, death, renunciation, iniquity, secrecy. The word BLACK also circulates in common discourse, sometimes as a noun, sometimes as an adjective, and sometimes in idiomatic expressions. Given the vastness of contexts in which it is used and circulates, I will focus on the meanings of this colour in artefacts that form a significant part of Western material culture, as it is impossible to remove from them the symbolic force that results from their being BLACK.

Keywords: Human factors · Western culture · Material culture
Semantics · Black (Colour)

1 Opening Note

This article is part of the Glossolalia project: an alphabet of critical keywords on design which has been in development since 2013 [1–3]. It is a subsidiary of Michel Pastoureau's studies on the symbolic dimension of colour in Western culture [4–6].

Before being a characteristic of a thing, it recalls the absence of light and therefore refers to the time preceding the divine *fiat lux*. Even though the night was sung by poets like Orpheus, men have always been terrified of the dark because it recalls the beginnings of their existence, when they did not know the reasons that explained the alternation between day and night and they did not know how to make light.

2 Black in Old Narratives

Black has been associated with funeral rites since the Neolithic age. In Ancient Egypt, Anubis, the jackal-god, is black. However, black in Ancient Egyptian culture did not have a negative connotation, but was instead associated with fertility, fecundity and divinity.

© Springer Nature Switzerland AG 2019
S. Bagnara et al. (Eds.): IEA 2018, AISC 827, pp. 216–226, 2019.
https://doi.org/10.1007/978-3-319-96059-3_24

In Norse mythology, the raven – a bird that is totally black – is a positive symbol, an attribute of Odin, god of war. For the Germans, the raven and the crow, among other animals, were sacred and eaten after the sacrifices. But such practices, deeply rooted among the peoples of the North, could not be tolerated by Christianity. Hence, in 751AC, Saint Boniface complained to the pope about the difficulty of applying the long list of prohibitions he was supposed to enforce. The pope responded by prioritizing the raven as the first animal to interdict, because a Christian could not eat black animals. This bird's bad name comes from Genesis: Noah sent a raven to inquire about the retreat of the waters of the flood. According to medieval exegesis, instead of returning to the ark with the awaited news, the raven stayed on the ground eating corpses; Noah then sent a dove (Genesis VIII, 6-11). It is from this that we get the negativity of black (raven) and the positive value of white (dove).

In ancient mythologies, hell is either black or red. In Greek mythology, besides black, hell is frozen and guarded by a fearsome black dog with three heads, sometimes depicted with two (see Fig. 1).

Fig. 1. Black-figured amphora, c. 490BC, height 43.8 cm, British Museum, London. Source: http://www.britishmuseum.org/research/collection_online/collection_object_details.aspx? assetId=336885001&objectId=459079&partId=1. (Color figure online)

Almost all the peoples of Antiquity observed the flight and plumage of ravens to probe the presages. According to Pliny the Elder, "ravens are the only birds that seem to have any comprehension of the meaning of their auspices" (Nat. Hist., X, 15). Pastoureau therefore wonders whether ravens' alleged clairvoyance, so admired by the Romans and the Germans, is the true reason behind the mistrust of the bird by Christian theologians of the Middle Ages (and not just because it is entirely black).

At this time in the history of the West, two systems seem to have coexisted: the polarity between black and white on the one hand, and the triad of black, white and red on the other. From the year 1000 onwards, black began a long symbolic journey towards discretion and humility, but for centuries its infernal connotation persisted, with occasional exceptions, as well as its association with death and related rituals. Among Catholics and Protestants, black continues to be the colour of mourning (Fig. 2a) although it can be "relieved" progressively, becoming replaced or mixed with other colours, such as grey and purple, or perhaps not: in some communities, mourning cannot be withdrawn and black is therefore worn until death. In Andalusia (southern

Spain), there is a very special type of mourning: women wear black garments to participate in the Holy Week festivities, celebrating the death of Christ (see Fig. 2b). It is a Catholic celebration that mixes spirituality with sensuality.

(a)   (b)

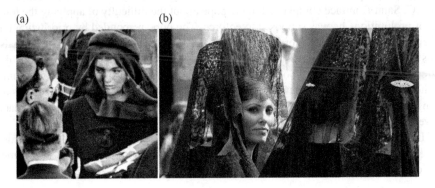

Fig. 2. (a) Jackie Kennedy at JFK's funeral (November 25, 1963). Source: http://www. allposters.fr/-sp/Jacqueline-Kennedy-at-President-John-Kennedy-s-Funeral-Affiches_i9361068_; (b) Andalusian women in black at Holy Week. Source: http://sevilla.abc.es/pasionensevilla/ actualidad/noticias/guia-de-la-semana-santa-2014-icomo-hay-que-vestirse.html

3   Black, Symbol of Evil

Black is the colour of the Devil and his entourage of fallen angels. It is the colour of sin, hatred, anarchy, Nazi and fascist paramilitary troops (the German *Schutzstaffel* and the Italian *Camicie nere*), totalitarianism and violence. It is also the colour of Nihilism, sadness, melancholy and depression, and therefore the colour of the punk subculture (as opposed to hippie culture).

Black unfolds into various shades of negativity, as in the expressions film noir, roman *noir*, describing diseases that kill, like the black plague. When it refers the colour of the skin it is a symptom of ugliness or ferocity. The Queen of Sheba (who despite her dark skin was considered a beautiful woman), the Magus King Balthazar, the mythical Prester John (believed to be a Christian king) and St. Maurice, the first black Christian martyr (the patron saint of dyers) are exceptions because, for Europeans, beauty and goodness were associated with white skin and blond hair. The crusading spirit, in its multiple facets and in its multiple locations (from the Iberian Peninsula, from which the Muslim kingdoms were expelled by Portuguese and Spanish Christian kings, to the Holy Land), centuries of human trafficking and colonization and, recently, successive waves of migrants and Islamic fundamentalism have helped keep mistrust and fear of the other alive. For this reason, there has been no shortage of arguments to update and legitimize segregation and violence, feeding ancestral stereotypes instilled since childhood (see Fig. 3) [7, 8].

The negativity of the black colour extends to flags. The first black banners were used by the Roman legions; whether or not they had that purpose, they were part of the strategy to intimidate enemies. The black standard followed, adopted by the Abbasid

Fig. 3. The fascination of indigenous people with the blonde hair of Mariazinha: illustration by Ofélia Marques (1902-1952) for *Mariazinha em África* (1947), a Portuguese children's literature classic.

caliphate (from 747AD). After that, pirates, corsairs, anarchists and jihadists used or still use black flags for the same reasons, that is, to cause terror and, if possible, to sow chaos.

4   From Symbol of Austerity and Belonging to a Social Group to Symbol of Social Distinction

Colour in monastic habits started to be considered a luxury, and therefore despicable. This attitude would change in the Carolingian era, when the importance of colour to identify the state and build a feeling of belonging to a group was gradually recognised. Disputes involving white monks and black monks [9] became well known. The arrival of heraldry dictated the end of black's leading position above all the other colours, along with white and red. This transformation coincided with a symbolic change: black ceased to be only the colour of mourning or of the devil (and his creatures) [10]. Coats of arms emerged in the middle of the twelfth century, providing another form of social distinction and affirmation of belonging to a group. It began by being an identification code used in battles and battle preparation activities (jousts and tournaments), used by the most diverse social groups (families, professional groups, cities, states). The only rule seems to have been not to usurp others' coats of arms. Pastoureau studied the complex web of figures and the limited colour code, among which black punctuates, as well as yellow (gold), white (silver), red and green, which were the main colours of Western culture in use during the Middle Ages. The colour code applies to knights, as seen in the thirteenth-century chivalric romances. The black knight stands out, whose colour was an attribute of a figure in the foreground (like Lancelot or Gauvain) who hides his identity for the sake of modesty, but shows his value. The high moral standards of the black knights have been echoed in romantic literature (and still are today [11]). According to a recent study, this seems to have been the meaning behind the epithet Black Prince given to Edward of Woodstock (1330-1376) rather than his allegedly black soul [12].

Recalling the practical aspects of obtaining black dyes, it should be noted that the first dyes used only produced long-lasting browns or ephemeral blacks because they fade after being exposed to air and natural light. Until the fortuitous discovery in 1856 of Perkin's mauve, black was first made using blue (*Indigofera tinctoria L.* or *Isitatis tinctoria L.*) and then red (*Rubbia tinctoria L.*). Exposure to the air (oxidation) concluded a time-consuming and an expensive process. Quality (and durability) in black dyes was only achieved in the late 14th century. In order to obtain a true black, a substance extracted from the chestnut tree had to be used. In certain climates (namely in Eastern Europe, Middle East and North Africa), several species of insects lay their eggs on the chestnut leaves, which react, enveloping the cocoons with a substance that encloses the larvae in a new cocoon. This was collected before the beginning of summer. These cocoons then needed to be dried slowly and exported to the European market, where prices reached very high levels. The high demand thus dictated and pressed for the techniques to become more and more sophisticated.

4.1 Black, the Colour Compatible with the Dignity and Sobriety That Christians Should Have

From the end of the Middle Ages to the seventeenth century, black and white were not considered colours: the printing press, engravings, the development of calligraphy, the Protestant Reformation (followed by the Catholic Reformation), and scientific progress placed black in a very particular position, different from other colours, although maintaining the former dialogue with white (used mostly as a background or in details).

During this long period, black became the colour of choice for the elites' dress, possibly in response to laws on luxury control. However, the high demand for precious fabrics dyed in black and black furs (to be used in clothing) seem to contradict such a reaction if we look only at the price and do not consider the symbolic shift of the black colour (Fig. 4a) – Italians and Spaniards defined a standard of (good) taste (Fig. 4b and c) that contaminated other European states [14], both Protestant (Fig. 5) and Catholic – including children (Fig. 6a and b) and other social groups (Fig. 6c).

In Protestant cultures, black symbolizes modesty, temperance, humility, reserve and asceticism. In this sense, it works as a resistance against worldly vanity, which is interesting because for other social groups it has been, and continues to be, a symbol of social differentiation, status, and even a statement of pride (as in the case of male gypsies, who wear black all their lives).

(a)   (b)   (c)

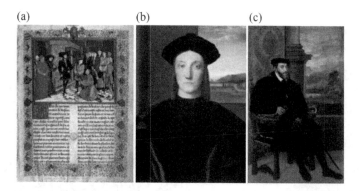

Fig. 4. (a) Philip the Good, third Duke of Burgundy (of the house of Valois) receiving the Chroniques of Hainaut painted by Rogier van der Weyden, 1448, Manuscript (Ms 9244), 44 × 31,2 cm, Bibliothèque Royale de Belgique, Brussels. Source: www.wga.hu; (b) Portrait of Guidobaldo da Montefeltro by Raffaello Sanzio, c. 1507, oil on wood, dim. 71 × 50 cm, Galleria degli Uffizi, Florence. Source: www.wga.hu, (c) Charles V by Tiziano, 1548, oil on canvas, 205 × 122 cm, Alte Pinakothek, Munich. Source: www.wga.hu.

Fig. 5. Portrait of Regents of the St. Elisabeth Hospital of Haarlem by Frans Hals, 1641, oil on canvas, dim. 153 × 252 cm, Frans Hals Museum, Haarlem. Source: www.wga.hu.

In Catholic states, the black's symbolic journey in female and male dress was suspended during the seventeenth and eighteenth centuries, a period that coincided with an explosion of colour due to baroque culture (although black continued to be used as the colour of mourning). In the middle of the nineteenth century, black returned to men's clothing, coinciding with the development of industry. It was the colour used by the bourgeoisie and various social groups: doctors and magistrates dress in black as it is a colour that attests to the wearer's credibility.

(a)   (b)   (c)

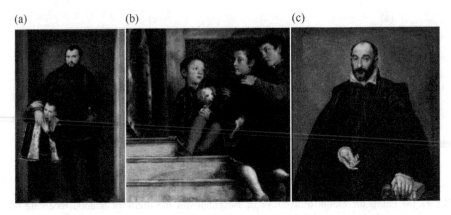

Fig. 6. (a) Children of Vendramin family (detail) by Tiziano, 1540-45, oil on canvas, dim. 206 × 289 cm, National Gallery, London. Source: www.wga.hu; (b) Count Giuseppe da Porto with his son Adriano by Paolo Veronese, 1551-52, oil on canvas, dim. 207 × 137 cm, Galleria degli Uffizi, Florence. Source: www.wga.hu; (c) Andrea Palladio by El Greco, 1575, oil on canvas, dim. 116 × 98 cm, Statens Museum for Kunst, Copenhagen. Source: https://www.wikiart.org/en/el-greco/portrait-of-a-man-andrea-palladio-1575.

5   Seeing the World in Black and White

Visual perception of colours depends on the type of light, whether it is natural or artificial (electric light gives a visual experience that is very different from the light candles or oil lamps provide). Representation of the world in black and white after the invention of photography, and then cinema, creates a sensory experience that is very different from that which results from the visual perception of the real.

Until the democratization of colour in photography, which is much more recent than usually thought, images were circulated by engravings and then by black and white pictures (according to Pastoureau, this explains the lack of interest that the historiography of art attributes to colour). However, the cultural meaning of colours is much more important than light (and the many natural and artificial possibilities it offers) or black and white or colour reproduction. The technological question is, therefore, only one aspect of a complex equation that inexorably refers to the prevailing symbolic dimension. Many directors continued to prefer black and white movies, such as Roberto Rosselini (1906-1977) and Michelangelo Antonioni (1912-2007), for strictly poetic reasons, or used both features, such as Wim Wenders (b.1945) in *Der Himmel über Berlin* (1987): the black and white sequences show the angels' point of view and the colour ones the human's.

Fig. 7. The black prayer book of Galleazzo Maria Sforza by a Flemish miniaturist, possibly Philippe de Mazerolles, 1466-76, Manuscript (Codex Vindobonensis 1856), dim. 25,0 × 17,6 cm, Österreichische Nationalbibliothek, Vienna. Source: www.wga.hu. (Color figure online)

6   Black in Writing

The increase in literacy, as it became progressively essential in the education of social elites, dictated the dissemination of writing (in black) for the most diverse uses; at the same time, the first books made using movable type printing presses (also with black ink) were made. Both types of writing helped make black into a kind of non-colour. [15] However, richly illuminated manuscripts continued to be produced. There remain six illuminated manuscripts on pages dyed in black, dating from the third quarter of the fifteenth century. The following figure shows two pages of one of them, possibly commissioned by King Charles the Bold (1433-1477), the fourth Duke of Burgundy – for unknown reasons, the book came into the possession of Galleazzo Maria Sforza (1444-1476), Duke of Milan (see Fig. 7 below). It is a work sumptuously illuminated in gold and silver, which highlights the other colours (red, brown, blue and green). The black background emphasizes the delicacy of the borders, profusely decorated with acanthus leaves, flowers, fruits and birds that frame miniatures (which represent the space in three dimensions) or border the text, in white using Gothic script (the Textus Semiquadratus variant) [16].

7   *Nigredo*

In Alchemy, *nigredo* is the first of three steps on a journey to the *magnum opus*, the transmutation of metals into gold and the creation of the philosopher's stone (*lapis philosophorum*).

Alchemy was a very old practice, fiercely persecuted in Catholic states in early modern times. In addition to its alchemical meaning, *nigredo* seems to apply to any experience that challenges the established knowledge, as Yourcenar points out in her novel on Zeno of Bruges [17]. In this sense, the *nigredo* formula (despite the allusion to black) is opposed to the idea of "darkness" because it represents the search for new knowledge and openness what can result from it (Fig. 8). Although alchemy is not directly related to the culture of the Enlightenment (both Catholic and Protestant), it symbolizes the struggle to overcome human limits and the human desire for freedom of thought and action, inseparable features of the modern condition [18].

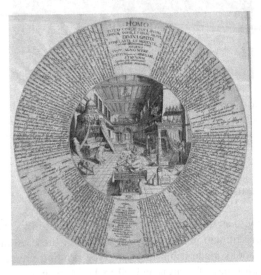

Fig. 8. The Alchemist's Lab by Hans Vredeman De Vries (?), engraving from a circular plate, diameter 41 cm, in *Amphitheatrum Sapientiae Aeternae* (*Amphitheater of Eternal Wisdom*) published in Hamburg (1595), Metropolitan Museum of Art, New York. Source: https://www. metmuseum.org/art/collection/search/397573.

8 "As Long as It's Black": Towards Standardization in Production and Consumption

"The Chanel Ford – the frock that all the world will wear – is model '817' of black crêpe de Chine [...] It will become a standard wardrobe component for all women of taste". This sentence from the catalogue of MoMu's exhibition entirely dedicated to black in fashion design [19] was written by an anonymous journalist and was published in *Vogue* magazine (October, 1926). Irony aside, it recognizes the similarities between "la petite robe noir" ("the little black dress") (Fig. 9a) and the Model T, produced by the American automobile manufacturer Henry Ford. Curiously, the former emerged from a Catholic context, as a symbol of simplicity and affordable elegance; the latter – which could only be black – was inspired by Puritan morality.

(a)   (b)   (c)

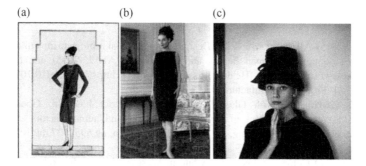

Fig. 9. (a) "La petite robe noir" by Coco Chanel (1926). Source: https://www.pinterest.com/pin/396387204696773879/; (b) and (c) Audrey Hepburn dressed by Givenchy, 1950s. Sources: https://www.fq.co.nz/fashion/celebrity-fashion/audrey-hepburns-givenchy-lovestory; https://style.over.net/kateri-kos-nakita-najveckrat-iscemo-na-spletu/; http://www.liverpoolmuseums.org.uk/walker/exhibitions/cecilbeaton/audrey_hepburn.aspx.

The model T (see Fig. 10a and b) was also the first affordable car and the first to be mass-produced. Ford's comments on its colour confirms the symbolic dimension of black in Protestant spirituality, on the one hand, and opens a new semantic feature, symbolizing mass production and mass consumption, disguising social differences (which continued to exist) [20].

(a)   (b)   (c)

Fig. 10. (a) and (b) The assembly line of Model T and Henry Ford near a Model T (1908–1926). Source: http://www.bookofdaystales.com/tag/model-t/; https://iconicphotos.wordpress.com/2009/05/02/henry-ford-model-t/; (c) The assembly line in Modern times by Charlie Chaplin (1936). Source: http://chaplin.bfi.org.uk/resources/bfi/filmog/film_large.php?fid=59441&enlargement=bfi-00n-8qx.jpg.

Chaplin's satire on the alleged failures of scientific management (Fig. 10c) reminds us that the promises of liberation through the machine paradoxically result in new forms of subjection and dehumanization. The "petite robe noir", like the use of black in modern architecture and design, went in another direction: maximum luxury, maximum discretion, maximum sophistication (the best are not at all affordable, it must be said).

References

1. Ferrão L (2015) All about words [on Design]: on Glossolalia project. In: 6th international conference on applied human factors and ergonomics (AHFE 2015) and the affiliated conferences, vol 3. Springer, Heidelberg, pp 6253–6257. https://doi.org/10.1016/j.promfg. 2015.07.774. Procedia Manufacturing
2. Ferrão L (2016) Color on Glossolalia: much ado about blue. In: Goonetilleke R, Karwowski W (eds) advances in physical ergonomics and human factors. Advances in intelligent systems and computing, vol 489. Springer, Cham, pp 595–607. https://doi.org/10. 1007/978-3-319-41694-6_58
3. Ferrão L (2016) Color in Glossolalia: white in Western Culture. In: Rebelo F, Soares M (eds) Advances in ergonomics in design, AHFE 2017. Advances in intelligent systems and computing, vol 588. Springer, Cham, pp. 411–420. https://doi.org/10.1007/978-3-319-60582-1_41
4. Pastoureau M (2007) Dictionnaire des couleurs de notre temps: symbolique et société. Christine Bonneton, Paris
5. Pastoureau M (2008) Noir: histoire d'une couleur. Édition Seuil, Paris
6. Pastoureau M, Simonnet D (2008) Couleurs, le grand livre. Éd. du Panama, Paris
7. Castro F (1947) Marques, Mariazinha em África: romance infantil (Mariazinha in Africa: a children's novel). Ática, Lisbon
8. Carvalho R (2017) A representação do negro na ilustração para a infância e juventude em Portugal (1926-1961) (The representation of the black in the illustration for childhood and youth in Portugal). PhD thesis. Universidade de Lisboa
9. Pastoureau M (1997) Jésus chez le teinturier. Couleur et teinture dans l'Ocident médiéval. Éditions du Léopard d'or, Paris
10. Pastoureau M (1993) Traité d'héraldique. 5ième ed., forward by de Jean Huber. Picard, Paris
11. Scott W (1920) Ivanhoe, or the Knight Templar, London
12. Was Edward the Black Prince really a nasty piece of work? http://www.bbc.com/news/magazine-28161434. Accessed 25 May 2018
13. Lupton E (2004) Thinking with type: a critical guide for designers, writers, editors & students. Princeton architectural Press, New York
14. Quondam A (2007) Tutti i colori del nero: moda e cultura del gentiluomo nel Rinascimento, Angello Colla Editore, Vicenza, p 75
15. Calvet L-J (1996) Histoire de l'écriture. Hachette Littératures, Paris
16. Vienna, Österreichische Nationalbibliothek, Codex Vindobonensis 1856. https://www.facsimilefinder.com/facsimiles/black-prayer-book-galeazzo-maria-sforza-facsimile. Accessed 28 May 2018
17. Yourcenar M (1968) L'œuvre au noir. Gallimard, Paris
18. Cassirer E (1951) Philosophy of the Enlightenment, transl. by F. Koelln and J.P. Pettegrore. Princeton University Press, Princeton
19. Dirix E (2010) La petite robe noir, the little black dress, the status of black in 1920s fashion. In: Debo K (eds) Black, masters of black in fashion & costume. Lannoo Publishers, Tielt, p 163
20. Taylor FW (1911) Scientific management. Harper, New York

Psychophysiology in Ergonomics

Comparison Between Gender Evaluation of Perfume Designed by Providers Commercially and that Evaluated by Participants

Akihisa Takemura[✉] and Fuma Mori

Setsunan University, 17-8 Ikeda Naka-machi, Neyagawa, Osaka, Japan
a-takemu@led.setsunan.ac.jp

Abstract. It is useful the gender of perfumes which users feel with sniffing is revealed. We conducted the sensory evaluation experiment of sniffing perfumes. Seven perfumes those were selected from fourteen perfumes based on pre-experiment were diluted four conditions of concentration. Twenty eight samples were evaluated about the intensity, the comfort, and five impressions (those were "Prefer/Not prefer", "Sweet/Not sweet", "Thin/Thick", "Exotic/Sober" and "Masculine/Feminine") in random order by fifty participants. It was revealed sensory evaluations of perfumes of various concentrations. For example, there were differences in "Masculine/Feminine" evaluation depending on concentration, and even at the same concentration, the variance of "Masculine" or "Feminine" evaluation was very large depending on perfume. Furthermore, "Masculine" or "Feminine" evaluations by participants were compared with gender recommended by the sellers. Some perfumes for men or unisex were evaluated as "Feminine". Moreover, impressions of perfumes were compared between female and male participants for each concentration. In some perfumes, there were large differences about evaluations of some impressions.

Keywords: Perfume · Sensory evaluation · Concentration

1 Introduction

Recently the use of smells has been often reported to cause others' discomfort and unhealthy. In Japan, the new word "Koh-gai" was created, which means harm caused by fragrance of perfume, softener, air freshener and so on. Perfumes have been used for a long time in the living environment. It is easy to imagine that the influence on others who are in the same space as the user is great. However, for sensory evaluation of perfume, there are few studies using concentration of aroma as a parameter. In recent, it has been widely allowed the values for clothing of neutral or reverse sex, and it is also generally accepted that there are perfume users of gender different from the assumption of seller. It is interesting to clarify what kind of gender evaluation the impression of perfume is appreciated. It will also help to understand the mechanism of impression evaluation of general fragrances. In this study, we aim to grasp the sensory evaluation of commercial perfume with concentration as parameter, and to compare the impression of gender by the seller and the user.

© Springer Nature Switzerland AG 2019
S. Bagnara et al. (Eds.): IEA 2018, AISC 827, pp. 229–238, 2019.
https://doi.org/10.1007/978-3-319-96059-3_25

2 Method

Sensory evaluation experiments were conducted at the laboratory under conditions of
the room temperature 18.0–23.5 °C and the humidity 20–67%. Seven perfumes were
selected based on preliminary examination (Table 1). The Un jardin sur le nil (NL) was
unisex, the Incense Ultramarine (UM) was men's, and the other five (RS, CL, CO, N5
GS) were women's. For each perfume, the basic bag (made of polyethylene tereph-
thalate) in which 0.1 mL of liquid perfume was volatilized in 10 L of odorless air was
prepared. In addition, the aroma in the basic bag was diluted to the concentration of the
four conditions shown in Table 1 to prepare sample bags, and the participants evaluated samples of 28 conditions in total in order based on Latin square method.

Table 1. Sample perfumes and their concentration conditions

Note	Fragrance	Supplier	For	Symbol	Dilution ratio
Citrus	Un jardin sur le nil	HERMES	Unisex	NL	10^3, $3 * 10^2$, 10^2, 1
Single Floral	Rosarium Eau de Parfum RX	Shiseido	Women	RS	10^4, 10^3, 10^2, 1
Floral bouquet	Eau de Classy	Vasilisa	Women	CL	10^4, 10^3, 10^2, 1
Floral green	Eau de Toilette	COACH	Women	CO	10^4, 10^3, 10^2, 1
Floral aldehyde	N°5 PARFUM	CHANEL	Women	N5	10^6, 10^5, 10^2, 1
Chypre	Incense Ultramarine	Givenchy	Men	UM	10^5, 10^3, 10^2, 1
Oriental	Guipure & Silk	Jeanne Arthes	Women	GS	10^3, $3 * 10^2$, 10^2, 1

The participants were 50 students who passed the T&T Olfactometer (that was the
olfactory test in Japan), the average age was 21.6 years old, 25 males and 25 females
were included. They included 17 perfume users and 33 non-users. Before evaluations,
the participants were not informed of the product name of the sample and the setting
gender by the sellers.

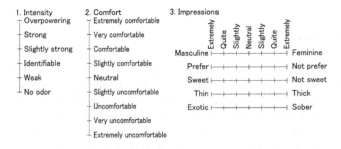

Fig. 1. Rating scales

As shown in Fig. 1, the evaluation items were the intensity of aroma, the comfort
and five items of impression. For the items of impression, four factors of the "Beautiful", the "Sweet", the "Rich" and the "Flamboyant" were extracted in the factor
analysis of the preliminary study.

Therefore, we selected "Prefer/Not prefer", "Sweet/Not sweet", "Thin/Thick", "Exotic/Sober" as representative scales of each factor. In addition, "Masculine/Feminine" for the "Gender" impression was added to them, and a total of five items was prepared.

3   Results and Discussion

3.1   Relationship Between Concentration and Evaluations

In order to grasp the tendency of basic perfume evaluation, the relationship between concentration and evaluation was investigated.

Figures 2 and 3 show the relationship between the concentration and each evaluation. The horizontal axis represents the reciprocal of the dilution factor, and the vertical axis represents each evaluation scale. The plot shows the average evaluation, and the area of circle shows the frequency of evaluation. In the relationship between the concentration and the intensity, the regression lines of four average evaluations are also described, assuming that Weber-Fechner's law adapted for this relationship. In the relationship between concentration and intensity, plots well matched the regression lines, and it could be said that the Weber-Fechner's law was also well adapted in the case of perfume evaluation. The five kinds of perfumes except the N5 and UM were almost the same gradient.

In terms of the concentration and the comfort, the NL, RS, CL, and CO were evaluated comfort as the concentration was higher, but were neutral in the highest concentration (dilution factor 1). On the other hand, the N5, UM and GS tended to decrease monotonously, the basic condition was evaluation on the unpleasant side. However, the deviations of evaluations were very large in any perfumes on the high concentration.

In the relationship between the concentration and the gender impression, there were a lot of "Feminine" evaluations on high concentration in many perfumes. However, the NL of unisex perfume and the N5 of women's perfume were evaluated "Masculine" as many as "Feminine" evaluation on all concentrations. The UM of men's perfume was evaluated as "Masculine" on the medium concentration, and participants evaluated "Feminine" on the high concentration. Even in the case of women's perfume RS and CL, there were many "Masculine" evaluations on the high concentration, the average value was not "Feminine" but neutral. It was found that the user evaluation of high concentration perfume was generally regarded as "Feminine". However, even if it is a women's perfume, depending on the type of perfume it may be categorized as "Masculine", sometimes caught in both "Masculine" and "Feminine". From these, it was found that the effect of the type of perfume was great for the gender impression by users. Furthermore, even men's perfume was found to be "Feminine" depending on the concentration. Thus it was found that the effect of the concentration on gender impression of perfume was large.

In the relationship between the concentration and the "Beautiful", participants evaluated "Not prefer" on the high concentrations in all perfumes. The NL, RS, CL and CO were evaluated "Prefer" on the higher concentration. The N5, UM and GS decreased monotonously with respect to the concentration.

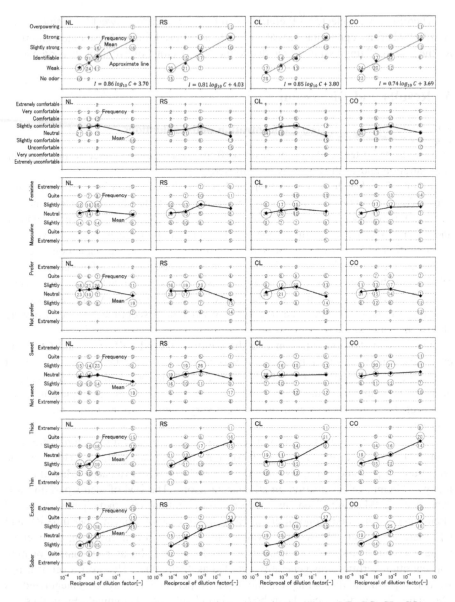

Fig. 2. Relationships among concentration and evaluation items (NL, RS, CL, CO)

In the relationship between the concentration and the "Sweet", the NL, RS and N5 monotonically increased on the low concentration and decreased on the high concentration. The CO and GS also monotonically increased with respect to the concentration. In this relationship, the tendencies of perfumes were different from each other.

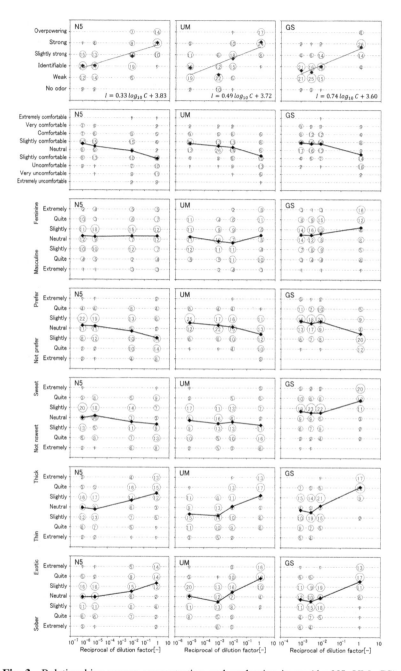

Fig. 3. Relationships among concentration and evaluation items (the N5, UM, GS)

In the relationship between the concentration and the "Rich", in all perfumes as higher concentration the participants evaluated "Thick". Since the relationships between the logarithm of the concentration and the "Rich" were linear, it was suggested that the correlation between the intensity of aroma and the "Rich" was high.

The relationship between the concentration and the "Flamboyant" showed a similar tendency to the relationship between the concentration and the "Rich".

3.2   Relationship Between Gender Impression and Other Evaluations

In order to estimate what kind of background the user's "Gender" evaluation is determined, the relationships among "gender" evaluation and other evaluations were examined. Figure 4 shows the relationships among the "Gender" evaluation and other evaluations. The horizontal axis shows the aroma intensity and the four items of impression, and the vertical axis is the "Gender" scale. The area of the circle shows the frequency, and the correlation coefficient is also shown in the figure.

In the relationship to the aroma intensity and the "Gender" evaluation, focusing on the correlation coefficient, there was almost no correlation in many perfumes. However, the GS only showed a slight positive correlation. In the view of the distribution tendency of the evaluation, in all perfumes, there were many neutral "gender" evaluation at low intensity, and there was a wide deviation at high intensity. In the UL of men's perfume, there were a lot of "Masculine" evaluations at high intensity. In the NL of unisex perfume and five women's perfumes, there were many "Feminine" evaluations at all intensity evaluations. In other than the NL, it can be said that it was a "gender" evaluation according to the "gender" set by the sellers.

In the relation between the "Beautiful" and the "Gender", weak positive correlation was found from the correlation coefficient in the NL, RS, and N5. There was a lot of "Feminine" evaluation in "Prefer" evaluations in many perfumes. However, for the UM of men's perfume, there were "Masculine" in "Prefer" as many as "Feminine".

In the relationship between the "Sweet" and the "Gender", the correlation coefficient revealed a relatively high positive correlation in perfume other than the CL. In the CL, it was considered that the correlation coefficient was calculated to be low because there was a relatively large number of "Masculine" evaluations on "Sweet" evaluations. From this, it could be said that the characteristic of the CL was the strong relevance between "Sweet" evaluations and "Masculine" evaluations. In other perfumes participants evaluated as "Feminine" on "Sweet" evaluations and as "Masculine" on "Not sweet" evaluations. On the other hand, some perfumes, such as the RS, CL, N5, UM, were evaluated "Feminine" on "Not sweet" evaluations. Especially, it is interesting that the relationship between "Not sweet" and "Feminine" evaluations was strong in the UM of men's perfume.

In the relation between the "Rich" and the "Gender", a weak positive correlation was found in all perfumes from the correlation coefficient. As a distribution tendency of evaluation, many "Feminine" evaluations were seen on "Rich" evaluations. On the other hand, there were many perfumes that were positive quadratic relationships since many "Feminine" evaluations were seen also on "Thin" evaluations. The deviation of the NL of unisex perfume was large regardless of the "Rich" and the "Gender". In addition, the positive correlation with the GS for women's perfume was stronger than

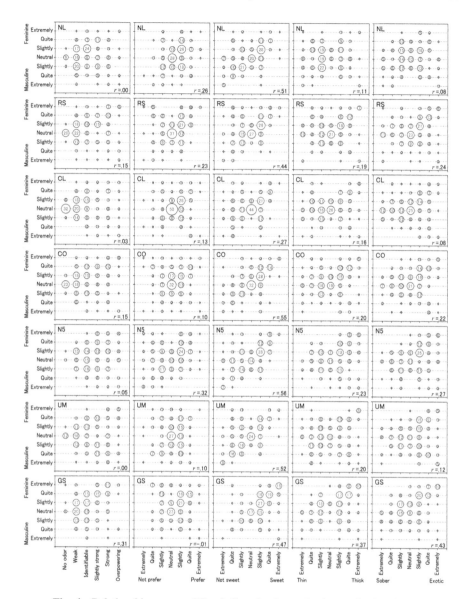

Fig. 4. Relationships among "Gender" evaluation and other evaluation items

the others. Thus it could be said that "Thick" perfume was "Feminine" perfume. Even in the UM of men's perfume, the "Feminine" evaluation was high on "Thick" evaluations.

In the relation between the "Flamboyant" and the "Gender", the correlation coefficient showed a relatively high positive correlation with the GS. A weak positive correlation was observed in the RS, CO and N5. There was almost no correlation in the NL, CL, UM. One of the reasons was that "Exotic" evaluations of the NL of unisex perfume was concentrated around the neutral. Another reason might be because the "Gender" evaluations were evenly distributed irrespective of "Exotic" evaluations.

Although the number of evaluations was small, "Masculine" evaluations were seen on not only "Exotic" evaluations but "Sober" evaluations, indicating that the individual differences in relation between the "Flamboyant" and the "Gender" were large. The cause is guessed to be because the variance of the "Gender" was very large on "Exotic" evaluations. Even for women's perfumes of the RS, CO, N5, GS, "Masculine" evaluations were seen on "Exotic" evaluations.

3.3   Comparison Impression Evaluation Between the Male and Female Participants

Figures 5 and 6 shows a comparison between impression evaluations of 25 men and women. The results of the t test are also shown in the figure. In the NL and GS, there was hardly any evaluation difference between men and women in any impression item.

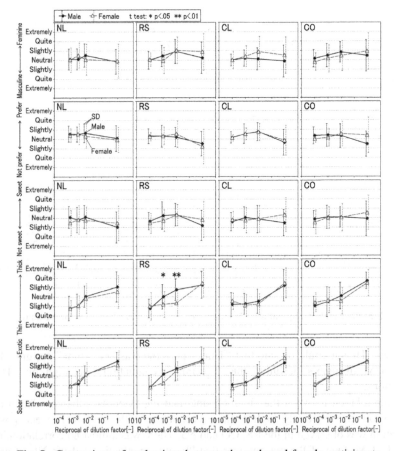

Fig. 5. Comparison of evaluations between the male and female participants

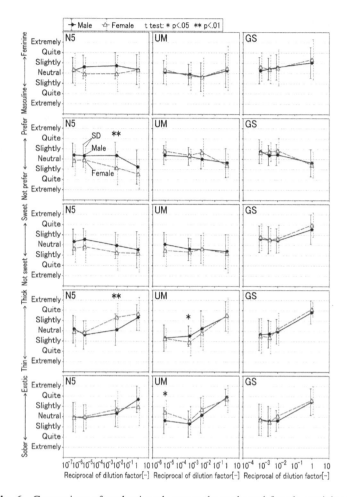

Fig. 6. Comparison of evaluations between the male and female participants

In the RS, the male participants evaluated significantly higher "Thin" evaluation on the moderate concentration. In the CL, female participants evaluated more "Feminine" than male participants, although there was no significant difference.

In the CO at a high concentration, the male participants rated lower "Beautiful" evaluation than the female participants, whereas the females evaluated "Prefer". In the N5, there were large differences in the four items other than the "Flamboyant". At all concentration, the males evaluated "Prefer" than the females. The males evaluated the N5 as "Sweet" and "Thin" than the females. In the "Gender", the males evaluated more "Feminine" in spite of neutral averaged evaluation by the females.

In the UM, the males evaluated significantly higher "Thin" on middle concentrations. In addition, the females evaluated significantly higher "Exotic" on lower concentrations. As in the NL and GS, it was quantitatively shown that there were perfumes with no difference between males and females and perfumes with large differences as in the N5.

4 Conclusion

We examined the intensity, the comfort, and the impressions of perfume, especially the "Gender" impression, using concentration as a parameter. Consequently, it was found that the type and concentration of perfume had a great effect on "Gender" evaluation. Moreover, it was found that the "Gender" was not necessarily evaluated according to the gender set for the sellers depending on the concentration of perfume. From the relationship between the "Gender" and the other evaluation items, the characteristics of perfume seemed to be the case, and there were cases those the "Gender" were consistent with the gender set by the sellers. Moreover, from the investigation of the impression evaluation differences between men and women, it was found that there was no difference in the NL and GS, whereas a large difference in the N5.

Relationships Between Autonomic Nervous System Indices Derived from ECG Signals

Chié Kurosaka[✉], Hiroyuki Kuraoka, Shimpei Yamada,
and Shinji Miyake

University of Occupational and Environmental Health, Japan, Kitakyushu, Japan
chie-k@health.uoeh-u.ac.jp

Abstract. We investigated the validity of heart rate variability (HRV) as an autonomic nervous system (ANS) assessment method and also investigated the correlations between all pairs of several indices obtained from ECG signals during a task performance. We recorded ECGs from 18 healthy participants during two 30-min mental tasks, i.e., a word processor typing task and a mental arithmetic task, and calculated five indices of HRV spectral analysis and five parameters from the Poincaré plot. In the results, significant correlations were found among the participants between LF and SD2 and between HF and SD1. There were no significant differences in the correlation coefficients between the two tasks, and no subjective scores correlated with physiological indices. The correlation coefficients between the sympathetic nervous system activity indices such as LF/HF ratio and CSI, however, were not very high, and these results were different from previous studies. Based on these results, we investigated the correlation coefficient time series in each participant and found that there were fluctuations in correlation coefficients between two indices which had been reported in other studies to indicate the same ANS activity. It was clarified that weak correlations among participants occurred by the fluctuations of correlation coefficients obtained in each participant.

Keywords: Heart rate variability · Poincaré plot · Mental workload

1 Introduction

It has been reported that spectral analysis of heart rate variability (HRV) and Poincaré plot analysis make it possible to obtain indices of autonomic nervous system (ANS) activity. It is also widely known that the high frequency (HF) component and the logarithm of the product short-term (SD1) and long-term (SD2) are related to the cardiac vagal tone, and the ratio of the low frequency (LF) component to HF, i.e., LF/HF ratio, and the ratio of SD2 to SD1 are used as a marker of sympathetic nervous system activity. The validity of HRV as a method of evaluating ANS activity has been investigated in many studies and is now being used as a stress evaluation method.

HRV becomes an unsuitable index of ANS in the case of a state of deep breathing and utterance. We must also consider the effect of body motion on HRV when HRV parameters are used for the evaluation of mental stress, in which case multidimensional assessments using several physiological measures with HRV parameters are necessary. In practical application, however, it may not be easy to obtain many physiological

© Springer Nature Switzerland AG 2019
S. Bagnara et al. (Eds.): IEA 2018, AISC 827, pp. 239–244, 2019.
https://doi.org/10.1007/978-3-319-96059-3_26

signals. In this paper, we investigated the validity of HRV as an ANS assessment method during task performance, and also investigated the correlations between all pairs of several indices obtained from ECG signals.

2 Method

2.1 Participants

Eighteen healthy male graduate students aged 22–27 years (mean: 24.2 yrs.) participated in this study. All the participants provided written informed consent. This study was approved by the Ethics Committee of the University of Occupational and Environmental Health, Japan.

2.2 Physiological Measurement

R-R intervals (RRI) were obtained from ECG signals that were recorded from the CM_5 lead. We calculated five indices of HRV (LF, HF, LF/HF ratio, coefficient of variation of RRI (CV-RR), and pNN50), and five parameters from the Poincaré plot (length of the transverse axis (SD1), length of the longitudinal axis (SD2), CVI (=log (SD1*SD2)), CSI (=SD2/SD1), and area of the ellipse (S = π*SD1*SD2/4).

2.3 Mental Workload

The participants performed two mental tasks: word processor typing (TYPING) and mental arithmetic (MA). In the TYPING task, a report of an accident at the Fukushima nuclear power plant [1] and a blank document are displayed side by side on a 27-inch computer screen. The participants are required to type in the blank document the same sentences that appeared in the report. The number of characters typed in the document are used as the task performance, regardless of typographical errors.

The MA task is based on the MATH algorithm proposed by Turner et al. [2]. A numerical calculation is displayed on a computer screen for 2-s, after which the target number is displayed following the word "EQUALS" for 1.5-s. Participants are required to press the left mouse button if the target number is correct and to click the right mouse button if not. The MA problems are self-paced, and after participants respond to a question the next one is displayed. The MA task contained five levels of difficulty, as shown in Table 1. The initial question is always level 3. The next question is at a higher level when the correct answer is given and lower when the response is incorrect. The number of questions answered, the average level of them, the rate of correct answers, and time series of all the responses are calculated.

2.4 Subjective Assessment

Subjective fatigue was evaluated using the Subjective Feelings of Fatigue (SFF) questionnaire [3], which consists of 25 items about five factors: feeling of drowsiness (I), instability (II), uneasiness (III), local pain or dullness (IV), and eyestrain (V).

Table 1. MA task level.

Level	Formula
1: easy	2-digit + 1-digit
2	2-digit +/− 1-digit
3	2-digit +/− 2-digit
4	3-digit + 2-digit
5: difficult	3-digit − 2-digit

2.5   Procedure

The experiment contained two 30-min task periods (TYPING and MA) and a 5-min resting period between tasks. The subjective assessments were recorded four times, before and after each task.

2.6   Statistical Analysis

The correlation coefficients among the participants in all pairs of the ten indices and the subjective scores were derived for each task (TYPING and MA). A strong correlation criterion was set as $r >= .70$, based on a general knowledge of statistics. Paired t-test was used to compare the averages of correlation coefficients between the two tasks.

The RRI time series during the mental tasks was divided into sixty 30-s blocks and was used for analysis. Correlation coefficients of all pairs in all ten indices were obtained from sixty points in each task. The correlation coefficients of fifteen plots (7.5-min data) were calculated again to investigate the tendencies in each participant. The analysis segment was shifted by thirty seconds, and the correlation coefficient of time series was obtained by repeating this procedure forty-six times. This analysis was carried out only during the MA task to investigate the relationship between physiological indices and task performance.

3   Results

3.1   Correlation Coefficients Among Participants

There were significant correlations between LF/HF ratio and CSI (TYPING: mean $r = .523$, MA: mean $r = .503$), and between HF and CVI (TYPING: mean $r = .675$, MA: mean $r = .649$), but none of them were very high. Significant correlations were found between LF and SD2 (TYPING: mean $r = 0.845$, MA: mean $r = .800$), and between HF and SD1 (TYPING: mean $r = .723$, MA: mean $r = .709$). The correlation coefficients between CVRR and LF were high (TYPING: mean $r = .840$, MA: mean $r = .781$), which was reasonable due to the calculation formula. There were no significant differences in correlation coefficients between the TYPING and the MA task. No subjective scores correlated with physiological indices.

3.2   Individual Analysis During the MA Task

Figure 1 shows a sample of time series of the correlation coefficients during the MA task. The r_{HF-SD1} and r_{LF-SD2}, which show a strong correlation, fluctuate in this figure. Four correlation coefficients fluctuated independently of task performance. These fluctuating patterns were different in each participant, but all of the participants had them in common. In Fig. 2(a), in the same participant as shown in Fig. 1, we can see the Poincaré plot drawn during the period in which all of the four correlation coefficients were high. Figure 2(b) shows the ellipse of RRI in the section where the correlation coefficient between indices related to sympathetic nervous system activity, i.e. the LF/HF ratio and CSI, were low. No one had a constantly high correlation during the 30-min MA task. Especially, fluctuations of the correlation coefficients between LF/HF ratio and CSI were remarkable, as shown in Fig. 1.

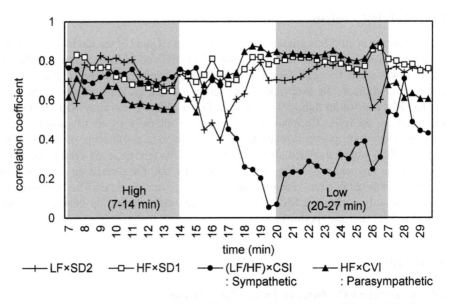

Fig. 1. Time series of correlation coefficients between indices during the MA task. This shows a sample of one participant.

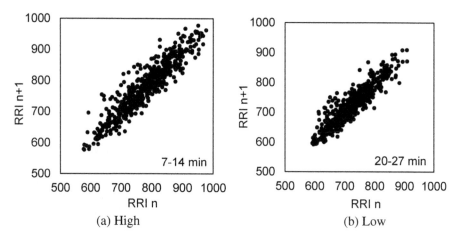

Fig. 2. Poincaré plots in the sections where correlation coefficients between indices of sympathetic nervous system activity are High and Low.

4 Conclusion

It has been reported that the correlation between LF/HF ratio and CSI (= SD2/SD1) is extremely strong [4, 5]. Those studies investigated the relationships among participants in several indices obtained from ECG signals during a resting period. In our study, we attempted to show the tendencies in an individual during mental workloads. It is necessary to verify the validity of this analysis because indices from spectral analysis of HRV, i.e., LF, HF, and LF/HF ratio, were obtained from ECG signals in a very short term (30-s) in this study, but even if it was not long enough, the requirement of the frequency resolution for calculating LF and HF was fulfilled. Smith et al. investigated the validity of short-term analysis using 30 beats of ECG signals [6, 7]. In their report, spectral analysis using the Lomb-Scargle algorithm could differentiate between sleep and meditation, although the sensitivity of the analysis was not assessed. We must continue our investigations, but the analysis method in this study may be useful to indicate individual tendencies.

We indicated that correlation coefficients between LF/HF ratio and CSI among participants were not very high, and that these results were different from previously reported results. It was clarified that these results occurred because of the fluctuations of correlation coefficients obtained in each participant. The size of ellipse was slightly small in the section where the correlations were low (see Fig. 2), the cause of which may have been due to the reduction of width in RRI fluctuation. The physiological meanings of these tendencies may be clarified by the investigation of other physiological signals, such as respiratory and blood pressure.

Acknowledgements. The data we analyzed was obtained in a collaborative study with Zheng Yi, DAIKIN Industries Ltd.

Conflict of Interest. The authors have no conflicts of interest directly relevant to the content of this article.

References

1. Cabinet Secretariat Website. http://www.cas.go.jp/jp/seisaku/icanps/SaishyuGaiyou.pdf. Accessed 7 Mar 2018
2. Turner JR, Heiwitt JK, Morgan RK, Sims J, Carrol D, Kelly KA (1986) Graded mental arithmetic as an active psychological challenge. Int J Psychophysiol 3(4):307–309
3. Sasaki T, Matsumoto S (2005) Actual conditions of work, fatigue and sleep in non-employed, home-based female information technology workers with preschool children. Ind Health 43:142–150
4. Hsu C, Tsai M, Huang G, Lin T, Chen K, Ho S, Shyu L, Li Y (2012) Poincaré plot indexes of heart rate variability detect dynamic autonomic modulation during general anesthesia induction. Acta Anaesthesiol Taiwan 50:12–18
5. Hoshi RA, Pastre CM, Vanderlei LCM, Godoy MF (2013) Poincaré plot indexes of heart rate variability: relationships with other nonlinear variables. Auton Neurosci Basic Clin 177:271–274
6. Smith AJ, Owen H, Reynolds KJ (2013) Heart rate variability indices for very short-term (30 beat) analysis. Part 1: survey and toolbox. J Clin Monit Comput 27:569–576. https://doi.org/10.1007/s10877-013-9471-4
7. Smith AJ, Owen H, Reynolds KJ (2013) Heart rate variability indices for very short-term (30 beat) analysis. Part 2: validation. J Clin Monit Comput 27:577–585. https://doi.org/10.1007/s10877-013-9473-2

The Effects of Playing Music During Surgery on the Performance of the Surgical Team: A Systematic Review on Published Studies

Pegah Rastipisheh[1], Shirin Taheri[1], Ahmad Maghsoudi[2(✉)],
Mohsen Razeghi[3], Alireza Choobineh[4], and Reza Kazemi[1]

[1] Department of Ergonomics, School of Health,
Shiraz University of Medical Sciences, Shiraz, Iran
[2] Student Research Committee,
Shiraz University of Medical Sciences, Shiraz, Iran
maghsoudiomid@gmail.com
[3] Physiotherapy Department, School of Rehabilitation Sciences,
Shiraz University of Medical Sciences, Shiraz, Iran
[4] Research Center for Health Sciences, Institute of Health,
Shiraz University of Medical Sciences, Shiraz, Iran

Abstract. Objectives: The present study aims to evaluate the effects of playing music during surgery on the performance of surgical team (including surgeons, nurses and staffs) through a systematic review of published studies.

Methods: Relevant database including Medline/PubMed, Scopus, and Science direct were searched up to November 2017 to find related articles.

Results: after removing duplicates and irrelevant articles, our comprehensive literature search found 23 articles that met inclusion criteria. Among these studies, 17 studies reported positive effects of music on the performance of the surgical team included providing more relaxing and more pleasant environment, making them calmer, performing tasks more accurately and precisely, decreasing mental workload and task completion time, increasing situation awareness, reducing stress and anxiety and improving memory consolidation and making them to enjoy their work. On the contrary, 4 studies reported negative effects of music during surgery including negative impact on task completion task, poor auditory performance and more likely repeated requests. Two Remaining articles didn't report any significant differences between the comparative groups. Because of different scales that were used in the studies, conducting meta-analysis on the data was not possible.

Conclusion: Most studies have shown the positive effects of music on the surgical team during operation, however it should not be neglected for its adverse effects. Therefore, it is possible to improve the performance of the surgical team during operation by playing controlled music.

Keywords: Music · Operating room technicians · Operating room personnel
Surgical team

© Springer Nature Switzerland AG 2019
S. Bagnara et al. (Eds.): IEA 2018, AISC 827, pp. 245–253, 2019.
https://doi.org/10.1007/978-3-319-96059-3_27

1   Introduction

Music is an art of producing sound through vocal or instrumental techniques to impress listener's mental status [1]. Music has gone beyond the barriers, so it has achieved a position as a universal language that motivates aged people and cheer depressed and bored people [2].

It has been reported that music can reduce the patient's pain and anxiety during and after the surgery. A few studies showed that music may affect the performance of the surgical team [3]. Listening to music is possible in operating rooms around the world so that surgeons listen to various types of music from classic to rock. A survey that conducted in UK showed that 90% of the surgeons, especially plastic surgeons, listen to music during surgery [4].

By listening to music, improvement of the surgeon's performance has been proved especially in laparoscopic suture tying, mesh alignment, and layered pigskin closures [5, 6]. Moreover, by listening to their favorite music, surgeons have better mental and physical reactivity [7].

Therefore, music is an economical, advantageous and reliable means that liked by everyone and can be used to reduce stress or stress-related problems [8].

Although music helps soothing difficulties but still moral and pragmatic limitations are obstacles on the way of experiencing its effects on stress reduction and improving surgeon's performance [9]. By creating comfort, relaxing and comfortable atmosphere, music supports medical tasks. Surgeons have different feedback about playing music during surgery. Some believe that it is a useful tool in operating rooms but some others note that it can have negative and adverse effects [3].

This study is a systematic review of available recent researches that shows how playing music during surgery can affect surgical staff's performances.

2   Methods

2.1   Search Strategy

This systematic review was designed and conducted in 2017. Only English language articles were evaluated in the present study. An extensive literature search was done in relevant English database including Medline/PubMed, Scopus, Cochrane library and Science Direct. Using appropriate Medical Subject Headings (MeSH) terms, we applied the following search strategy in each of the databases up to November 2017; (music) AND (operating room OR operating room technicians OR operating room personnel OR surgical team OR surgeon).

2.2   Study Selection and Eligibility Criteria

All studies that evaluated the effects of playing music on the performance of the surgical team during surgery were included in this study.

2.3   Data Collection Process

After obtaining relevant articles, two researchers independently evaluated the quality of the papers. These two researchers were blinded to names of the authors and their affiliations as well as the journals in which the articles were published. Any discrepancies between the two researchers were resolved by a third reviewer. After identifying eligible articles, the pre-designed forms were used for data gathering.

3   Results

3.1   Study Selection

A comprehensive literature search in the databases resulted in 176 potentially related articles. After evaluating these papers, we found 23 eligible articles and included in the study. The flowchart of the study selection is displayed in Fig. 1.

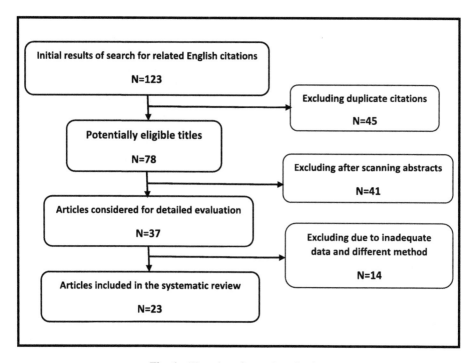

Fig. 1. Flowchart for study selection

3.2   Study Characteristics

In these 23 eligible articles, the effects of playing music during surgery on the performance of the surgical team have been evaluated. Among these studies, 18 studies reported positive effects of music on the performance of the surgical team. In contrast,

4 studies reported negative effects of music during surgery. Two remaining articles didn't report any significant differences between the comparative groups [10, 11]. Because of different scales that were used in the studies, conducting meta-analysis on the data was not possible. Table 1 shows the main characteristics and results of the studies included in this study.

Table 1.

Study	Study population	Sample size	Results
Corhan et al. [12]	Undergraduates students	18	– Better performance with the rock than with the easy-listening music – Greater accuracy with the fixed interval schedule and random interval schedule than the continuous music – More correct detections were made with the random interval schedule than the fixed interval
Fontaine et al. [13]	Undergraduates students	35	– Subjects who listened to familiar music displayed higher levels of arousal than those who listened to non-familiar music and no-familiar music group showed higher levels of arousal than subjects listening to no music at all – Over-all percent detections were significantly higher for subjects listening to familiar music compared the other groups
Allen et al. [6]	Male surgeons	50	– Less autonomic reactivity in all psychological measures – Better speed and accuracy in task performance – Lower stress level
Hawksworth et al. [14]	Anesthetists	200	– Reduce the vigilance of anesthetists – Lmpaired the communication with other staff – Distract their attention from alarms
Miskovic et al. [15]	Junior surgeons	45	– Worse performance in the group that listened to activating music compared to control groups
Ullmann et al. [16]	Physicians and nurses	171	– Positive effect on communications among the staff – Making staff calmer and more efficient
Conard et al. [17]	Laparoscopic surgery experts	8	– Dichaotic music has a negative impact on time until task completion – Classical music makes the task performance more accurately

(continued)

(continued)

Study	Study population	Sample size	Results
Makama et al. [18]	Surgical team	151	– Reduction stress and anxiety, masking worrying sounds, prevention of distraction, minimizes annoyance, positive effect on speed and accuracy, reduction of analgesic requirements – Enhancement of performance
Siu et al. [5]	Surgical team	10	– Decreased time to task completion and total travel distance – Reduced muscle activations and increased median muscle frequency – Improved their performance
Zeinaly et al. [19]	Surgical team	300	– Positive effect on communications among the staff – Improving concentration
George et al. [20]	Surgeons, anesthesiologists and nurses	100	– Improved their concentration – Reducing their autonomic reactivity in stressful surgeries – Make them calmer
Conard et al. [21]	Surgeons	31	– Classical music improved memory consolidation – Classical music improved their performance
Way et al. [22]	Surgeons	15	– Poorer auditory performance
Faraj et al. [23]	Surses and medical staff	52	– Enjoyed their work more – Better performance
Lies et al. [24]	Plastic surgery residents	12	– Improvement in repair quality – Decreased repair time
Weldon et al. [25]	Surgeons	20	– More repeated requests and increased operation time
Chung et al. [26]	Medical students	53	– Classical music group experienced a decrease in mean arterial pressure compared to the no music and the death metal music groups – No music and classical music environments were more relaxing, more pleasant, and better at boosting their performances compared to the death metal group
Fancourt et al. [27]	Aspiring surgeons	352	– Rock music impairs the performance of men but not women – Rock music increasing time to task completion and surgical mistakes

(continued)

(continued)

Study	Study population	Sample size	Results
Lopez et al. [28]	Surgeons, anesthetists and nurses	67	– Better performance – The ability to concentrate more easily – Ability to perform the job skilfully and with more precision
Shakir et al. [7]	Residents and plastic surgery fellows	12	– Preferred music have a positive effect on trainees' microsurgical performance
Tseng et al. [29]	Nurses	80	– Mozart music decreased their mental workload – Increased situation awareness – Relieved stress to achieve positive reaction

4 Discussion

Based on the included studies, it seems that music can have more positive effects on performance of the surgical team than other decreasing stress methods such as education [23] and chewing gum [24].

Ten eligible articles included in the present study, demonstrated that playing music can improve the performance of the surgical team [9, 10, 18, 19] due to decreasing stress and increasing surgeon's efficacy [8], providing warm and satisfactory atmosphere for patients and medical staff [11], increasing learning, preventing of worse initially function, memory consolidation [6], creating pleasure [15], and decreasing autonomic reactivity [13, 20].

Five studies showed that because of decreasing constant surgery time duration [9] and cardiovascular activities, music can increase the speed and precision and concentration of the surgeons and other staffs [10, 12, 13, 19].

The results of the three studies indicated that music can reduce stress by improving the mental states of the staffs and creating satisfaction and pleasure in the working environment [5, 7, 10, 21]. Moreover, in a research we observed that music can decreases mental workload among nurses [21]. The results of other three researches showed that music might enhance peace and calmness in the surgical team [8, 18] by decreasing autonomic reactivity [13].

It should be noted that one of the most important issues during surgery is the effective conversation among surgical team. The results of two studies showed that music can improve staff communications [12] by decreasing their stress [8].

Some studies showed that music could mask worrying sounds [10] and increase situation awareness [21], while the results of some other researches demonstrated that music might decrease concentration of anesthetists on the alarms [6] and mask the surgeon's voice resulting in more repeated requests [17].

The results of the two studies suggested that music might decrease muscular fatigue due to decreasing in muscular activity [13] and enhance relaxation in the working environment resulting in reduction in arterial pressure [18].

5 Limitations

The research methodology of the most previous studies conducted on this issue was subjective and few of them had objective methodology. Furthermore, the differences between tools and methods caused that the results of these studies were not quite comparable. For this reason, in the present study we were unable to perform meta-analysis on the collected data.

6 Conclusion

Most studies have revealed positive effects of music on surgeon team during operation, although its negative effects should not be neglected. It seems that using objective evaluation methods during music playing can provide more tangible and clearer results.

References

1. Miskovic D, Rosenthal R, Zingg U, Oertli D, Metzger U, Jancke L (2008) Randomized controlled trial investigating the effect of music on the virtual reality laparoscopic learning performance of novice surgeons. Surg Endosc 22(11):2416–2420
2. George S, Ahmed S, Mammen KJ, John GM (2011) Influence of music on operation theatre staff. J Anaesthesiol Clin Pharmacol 27(3):354
3. Ahmad M (2017) Role of music in operating room. J Anesth Crit Care Open Access 7 (5):00279
4. Lies SR, Zhang AY (2015) Prospective randomized study of the effect of music on the efficiency of surgical closures. Aesthet Surg J 35(7):858–863
5. Siu K-C, Suh IH, Mukherjee M, Oleynikov D, Stergiou N (2010) The effect of music on robot-assisted laparoscopic surgical performance. Surg Innov 17(4):306–311
6. Allen K, Blascovich J (1994) Effects of music on cardiovascular reactivity among surgeons. Jama 272(11):882–884
7. Shakir A, Chattopadhyay A, Paek LS, McGoldrick RB, Chetta MD, Hui K, et al (2017) the effects of music on microsurgical technique and performance: a motion analysis study. Ann Plast Surg 78(5):S243–S247
8. Thoma MV, La Marca R, Brönnimann R, Finkel L, Ehlert U, Nater UM (2013) The effect of music on the human stress response. PLoS ONE 8(8):e70156
9. Fratianne RB, Prensner JD, Huston MJ, Super DM, Yowler CJ, Standley JM (2001) The effect of music-based imagery and musical alternate engagement on the burn debridement process. J Burn Care Rehabil 22(1):47–53
10. Moorthy K, Munz Y, Undre S, Darzi A (2004) Objective evaluation of the effect of noise on the performance of a complex laparoscopic task. Surgery 136(1):25–30
11. Sármány J, Kálmán R, Staud D, Salacz G (2006) Role of the music in the operating theatre. Orvosi Hetilap 147(20):931–936

12. Corhan CM, Gounard BR (1976) Types of music, schedules of background stimulation, and visual vigilance performance. Percept Motor Skills 42(2):662
13. Fontaine CW, Schwalm ND (1979) Effects of familiarity of music on vigilant performance. Percept Motor Skills 49(1):71–74
14. Hawksworth C, Asbury AJ, Millar K (1997) Music in theatre: not so harmonious. A survey of attitudes to music played in the operating theatre. Anaesthesia 52(1):79–83
15. Miskovic D, Rosenthal R, Zingg U, Oertli D, Metzger U, Jancke L (2008) Randomized controlled trial investigating the effect of music on the virtual reality laparoscopic learning performance of novice surgeons. Surg Endosc. [Internet] 22(11):2416–2420. http://onlinelibrary.wiley.com/o/cochrane/clcentral/articles/888/CN-00666888/frame.html, https://link.springer.com/content/pdf/10.1007%2Fs00464-008-0040-8.pdf
16. Ullmann Y, Fodor L, Schwarzberg I, Carmi N, Ullmann A, Ramon Y (2008) The sounds of music in the operating room. Injury 39(5):592–597
17. Conrad C, Konuk Y, Werner P, Cao CG, Warshaw A, Rattner D, et al (2010) The effect of defined auditory conditions versus mental loading on the laparoscopic motor skill performance of experts. Surg Endos [Internet]. 24(6):1347–1352. http://onlinelibrary.wiley.com/o/cochrane/clcentral/articles/248/CN-00759248/frame.html, https://link.springer.com/content/pdf/10.1007%2Fs00464-009-0772-0.pdf
18. Makama J, Ameh E, Eguma S (2010) Music in the operating theatre: opinions of staff and patients of a Nigerian teaching hospital. Afr Health Sci 10(4):386–389
19. Zeinaly M, Aghdashi M, Afshar A, Masoudi S (2010) Music and work process in the operating room. Urmia Med J 20(4):324–327
20. George S, Ahmed S, Mammen KJ, John GM (2011) Influence of music on operation theatre staff. J Anaesthesiol Clin Pharmacol 27(3):354–357
21. Conrad C, Konuk Y, Werner PD, Cao CG, Warshaw AL, Rattner DW, et al (2012) A quality improvement study on avoidable stressors and countermeasures affecting surgical motor performance and learning. Ann Surg [Internet] 255(6), 1190–1194. http://onlinelibrary.wiley.com/o/cochrane/clcentral/articles/692/CN-01019692/frame.html, http://ovidsp.tx.ovid.com/ovftpdfs/FPDDNCFBGHNMLI00/fs046/ovft/live/gv025/00000658/00000658-201206000-00028.pdf
22. Way TJ, Long A, Weihing J, Ritchie R, Jones R, Bush M, et al (2013) Effect of noise on auditory processing in the operating room. J Am Coll Surg 216(5), 933–938
23. Faraj AA, Wright P, Haneef JH, Jones A (2015) Listen while you work? the attitude of healthcare professionals to music in the OR. ORNAC J 33(2):31–32, 34–50
24. Lies SR, Zhang AY (2015) Prospective randomized study of the effect of music on the efficiency of surgical closures. Aesthet Surg J [Internet]. 2015; 35(7):858–863. http://onlinelibrary.wiley.com/o/cochrane/clcentral/articles/179/CN-01215179/frame.html, https://watermark.silverchair.com/sju161.pdf?token=AQECAHi208BE49Ooan9kkhW_Ercy7Dm3
ZL_9Cf3qfKAc485ysgAAAaEwggGdBgkqhkiG9w0BBwagggGOMIIBigIBADCCAYMG
CSqGSIb3DQEHATAeBglghkgBZQMEAS4wEQQMYRwyNbPddWk8GwTfAgEQgIIBV
IC8lKBhHOrV2CdnBnTYHE7AqaFDoTW8tUzpQ9XIyIi5pbRPhBQJbmg7H90mdDa0AZ
0cO1s0y_JUOBrcdy8DyELxMXt7drE9Im9cE6qtUpw8UX6hxm3Hnxb6ii9rQOjs426SgiTT
nubHs6GZfAoHXhz_QRMIxkSgPWQ7E7H8tO8gHElW3T1W956-PDDjf40Asp1ff1FOeP
LOdrNC5v32WRKDC4mp6UZ29y0JdcBwEY3jvcEU5BJ1J7SkxEK0wUaiyR0AS-9Zo4P
UK97jeXVoSu1JkyB0loFF4v6ueR-hbu2ptSZ2OTA-thQumn0UkqWCHkWENlf9Wb5uM
ALtKT6SqGe-dHeG843HAs3GB5VoS5XkCjWp-AMAqfjvUF6XBhpDVWMkpUf_OIdna
CWDIwqehjVm-5606meKC2CmsKUbtaNVV-MpFtrhASvolsM5z2FfdL9_n1g
25. Weldon SM, Korkiakangas T, Bezemer J, Kneebone R (2015) Music and communication in the operating theatre. J Adv Nurs 1(12):2763–2774

26. Chung B, Shen J, Yang P, Keheila M, Abourbih S, Khater N, et al (2016) Comparison of three different auditory environments and their effect upon training in novice robotic surgeons. Journal of endourology Conference: 34th world congress of endourology, WCE 2016 South Africa Conference start: 20161108 Conference end: 20161112 [Internet] 30:A66 p. http://onlinelibrary.wiley.com/o/cochrane/clcentral/articles/499/CN-01293499/frame.html
27. Fancourt D, Burton TMW, Williamon A (2016) The razor's edge: Australian rock music impairs men's performance when pretending to be a surgeon. Med J Aust 205(11):515–518
28. López Ibáñez N, Ruiz De Casas A, Morales Conde M, Moreno Ramírez D, Camacho Martínez FM (2016) Sociocultural aspects of the influence of music on healthcare staff working in the operating room *(eng abstruct)*. Piel 31(4):236–241
29. Tseng LP, Liu YC (2018) Effects of noises and music on nurses' mental workload and situation awareness in the operating room. Advances in Intelligent Systems and Computing, pp 450–454

Delicious-Looking Package Color of Bottled Black Tea

Shino Okuda[(✉)]

Doshisha Women's College of Liberal Arts, Kyoto 6020893, Japan
sokuda@dwc.doshisha.ac.jp

Abstract. Three kinds of bottled black tea were prepared, and each black tea was poured into a plastic bottled wrapped with eleven kinds of colored labels. In this study, participants observed the black tea in the plastic bottle with one of the colored labels, and evaluated "predicted sweetness", "predicted sourness", "predicted bitterness", "predicted roasted-flavor" and "predicted deliciousness". Subsequently, they evaluated the impression using the semantic differential method with seventeen pairs of adjective. According to the results, the red label could make the bottled black tea visually sweet, and the black label could make it visually bitter. Also, the yellow and yellow-green labels made the black tea more familiar and healthy, and the blue-green and blue labels made it more refreshing and sophisticated, Consequently, the package color of the plastic bottle can control the deliciousness of the black tea, and can determine the impression of the bottled black tea.

Keywords: Package color · Bottled black tea · Predicted deliciousness

1 Introduction

The perceived taste, deliciousness and preference of beverage are affected by the package design of the bottle. The color is most effective contribution in the properties on appearance [1]. It was reported that the colors of food and beverage have a significant effect on the preference and the perceived tastes [2]. It was also reported that sweetness is a strong promoter of overall pleasantness of soft drinks [3]. In addition, we found that yellow and yellow-red colors make the dishes more palatable, but red-blue colors make the dishes less palatable [4]. Also, we revealed that visual tastes and palatability are determined by the color of green tea and the mug color, but impression is determined by mug color [5]. From the above background, this study aims at revealing which packaging color of bottled black tea looks delicious.

2 Methods

2.1 Visual Stimuli

First, we prepared three kinds of bottled black tea, the common black tea (Kocha-no-jikan tea with lemon, UCC Ueshima Coffee Co., Ltd.), reddish black tea (Java-tea straight red, Otsuka foods Co., Ltd.) and yellowish black tea (lemon tea blended Earl Grey tea, Yamazaki Baking Co., Ltd.). Each black tea was poured into a plastic bottled wrapped with eleven kinds of colored labels. Table 1 shows the chromaticity values of each kind of black tea and each colored label of plastic bottle (Fig. 1).

Table 1. Chromaticity values of visual stimuli.

Visual Stimuli	L^*(D65)	a^*(D65)	b^*(D65)	C^*(D65)
Black tea-A (common black tea)	82.7	6.9	57.2	57.6
Black tea-B (reddish black tea)	59.9	28.8	79.0	84.1
Black tea-C (yellowish black tea)	90.8	−3.1	37.4	37.5
Package Red (R)	62.7	37.3	8.3	38.2
Package Yellow-Red (YR)	70.0	33.6	30.1	45.1
Package Yellow (Y)	81.6	6.3	54.9	55.2
Package Green-Yellow (GY)	77.8	−19.2	49.1	52.8
Package Green (G)	67.1	−34.7	4.8	35.0
Package Blue-Green (BG)	63.0	−17.4	−20.7	27.0
Package Blue (B)	59.4	4.0	−29.8	30.0
Package Red-Blue (RB)	59.1	24.2	−18.7	30.6
Package Black (Bk)	40.8	−0.2	0.5	0.6
Package Gray (Gr)	70.4	−0.7	−2.1	2.3
Package White (Wh)	94.4	−0.6	0.2	0.7

2.2 Procedure

Participants observed the black tea in the bottle, and evaluated four kinds of tastes, "visual sweetness", "visual sourness", "visual bitterness", "visual roasted flavor" and "visual deliciousness" with a numerical scale from 0 to 10. They also evaluated the impression using a semantic differential method with seventeen pairs of adjectives, familiar-unfamiliar, healthy-unhealthy, relaxing-unrelaxing, elegant-vulgar, unique-common, warm-cold, natural-artificial, soft-hard, gorgeous-vulgar, European-Japanese, modern-classical, interesting-uninteresting, female-male, brisk-depressing, adultlike-childish, valuable-cheap.

2.3 Participants

Twenty undergraduate students voluntarily participated in this experiment. They belonged to DWCLA, and they are all female in their twenty. All participants had normal color vision confirmed by screening with the Ishihara color perception test.

Fig. 1. Visual stimuli.

3   Results

3.1   Visual Deliciousness of the Black Tea in the Different Color Packaged Bottle

Figure 2 shows the mean values for the visual deliciousness of all kinds of black tea in the plastic bottle with different colored labels. It was shown that the predicted deliciousness of the reddish black tea is lower than the others ($P < .001$). Also, the common black tea and yellowish black tea are visually delicious in the Red, Yellow and Yellow-Red and Green-Yellow bottles. Whereas, the visual deliciousness of all kinds of black tea are lower in Gray, Blue and Blue-Green bottles.

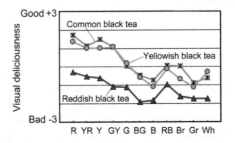

Fig. 2. Evaluation results of Visual deliciousness

3.2   Impression of the Black Tea in the Different Color Packaged Bottle

Figure 3 shows the factor scores acquired from the results of the factor analysis. According to these results, "Familiarity", "Evaluation" and "Refreshment" were extracted as factor 1, 2 and 3, respectively. It was shown that the black tea in Yellow or Yellow-Red bottle are higher familiarity. Also, the black tea in Red or Red-Blue bottle are higher evaluation, and that in Blue-Green and Green-Yellow are higher refreshment. Moreover, the reddish black tea was lower in familiarity than other kinds of black tea.

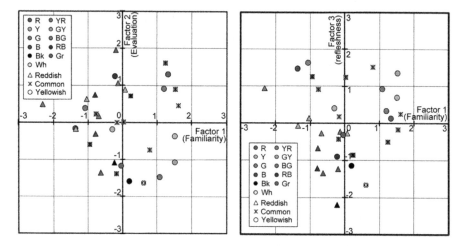

Fig. 3. Factor scores acquired from results of factor analysis.

4 Conclusion

Above mentioned results of this study are summarized as follows.

(1) Predicted deliciousness is determined by the color of black tea and the mug color.
(2) Reddish black tea does not look visually delicious, whereas, Red, Yellow and Yellow-Red and Green-Yellow bottles can make black tea visually delicious.
(3) Yellow-Red bottle can make black tea familiar, and Blue-Green and Green-Yellow bottles can make it refreshing.

Acknowledgment. I would like to thank Junna Miyamoto, an undergraduate student for helping our subjective experiment.

References

1. Jaros D, Rohm H, Strobl M (2000) Appearance properties – a significant contribution to sensory food quality? Lebensmittel-Wissenschaft Technol 30:320–326
2. Johnson J, Clydesdale F (1982) Perceived sweetness and redness in coloured sucrose solutions. J Food Sci 47:747–752
3. Tang X, Kalviainen N, Tuorila H (2001) Sensory and hedonic characteristics of juice of sea buckthorn (Hippophae rhmnoides L.) origins and hybrids. Lebensmittel-Wissenshaft und Technol 34-2:102–110
4. Okuda S, Okajima K (2013) Effect of plate color on visual palatability of food. Proc AIC 2013:1248–1252
5. Okuda S, Okajima K (2015) Color design of mug with green tea for visual palatability. Int Assoc Soc Des Res:1551–1563

A Holistic Approach to Operator System Comfort

Susanne Frohriep[✉]

Grammer AG, 92224 Amberg, Germany
susanne.frohriep@grammer.com

Abstract. The operator system consists of all elements of the user interface to the vehicle: seats, controls, and further components in the operator's bodyspace. To ensure operator efficiency, well-being and health, all elements must be analyzed and developed on a system level. The product development process at Grammer AG has defined "perceived quality" as its guiding theme. Its foundation is deducting design requirements from understanding users and use cases for the respective product applications. The developed product range encompasses driver and passenger seats for trucks, agriculture and construction equipment, fork lift trucks, bus, rail, multifunctional armrests, and automotive seat components such as head restraints, armrests, center consoles and interior components. The Grammer AG Ergo-Innovationlab provides the scope of perceived quality testing, and performs innovation verification.

Keywords: Product comfort · Operator system · Perceived quality

1 Introduction

Why comfort is relevant in product development: It is the combination measure of all human interfaces with the product, entailing acting forces, physical effects, human sensory input and perception.

1.1 Operator System

The operator system consists of all elements of the user interface to the vehicle: the seat that positions the operator, takes up operator load, and provides comfort, controls and actuators that allow for vehicle and implement operation, and further components of the operator's bodyspace [1]. As such, the operator system consists of many elements, but is used and perceived as a unit. Thus, the components must be analyzed and developed on a system level in a holistic approach to ensure operator well-being and health, and thus, efficiency [2].

The product development process at Grammer AG has defined *"perceived quality"* as its guiding theme, which is defined as "a positive user interaction with the product in all relevant use cases". Its foundation is deducting design requirements from understanding users and use cases for the respective product applications. The relevance of *perceived quality* in product development is to build up sufficient knowledge on

S. Bagnara et al. (Eds.): IEA 2018, AISC 827, pp. 258–267, 2019.
https://doi.org/10.1007/978-3-319-96059-3_29

physical and mental user interaction within the operator system in order to enable such positive interaction for the greatest possible number of users.

1.2 Comfort Definition

Comfort is, by its nature, always subjective. It is the co-occurrence of *affinity*, which entails a positive psychological reaction, and *accommodation*, which refers to a physiological fit of product to user [3]. Thus, the comfort experience is composed of the multitude of user psychological and physiological experience with the product, influenced by their state and trait characteristics, as well as the environment and their expectations concerning the product.

The experience multitude is consistently computed into a holistic comfort impression by the user through a fluid process over interaction time. Human beings compute this interaction rating with low conscious effort [4]. Even though different factors comprise the holistic comfort impression, it is perceived as a unified experience. Only if single factors become apparent, they start to dominate the impression. These can influence the rating positively or negatively. When an aspect comes to the attention of the user and becomes prevalent within the overall impression, it will dominate the comfort rating. This can occur in either direction: The negative occurrence has been named "limiting comfort factor", entailing that the holistic comfort experience cannot be better than its weakest aspect. The positive occurrence can be referred to as the "wow-factor" of a product, exceeding the expectation of the user [5].

Once a negative factor becomes dominant, comfort cannot be experienced any longer: While comfort is not merely the absence of discomfort, discomfort needs to be limited to a non-significant level in order allow for a comfort experience [6]. Only working on limiting discomfort will not automatically increase comfort, affinity factors must be enhanced as well. Thus, experiencing comfort can be defined as *"an overall positive user interaction experience with a product"*, where positive experience refers to mental and physical experience in affinity and accommodation of users to products. A positive interaction is furthermore characterized by the experience of flow or seamlessness, terms describing activities that are characterized as effortless and uninterrupted. In such types of activities in operator systems, user concentration can be upheld and work execution is less exhausting.

1.3 Ergo-Innovation Lab

The field of ergonomics provides the compendium for analyzing and rating the interaction between user and product. Its goal is to understand the interaction system fully and optimize this interaction. In order to be able to do this, human beings in their variation are the foundation. Grammer Ergonomics is the hub for planning and distributing all ergonomic content as applicable. Internal specifications have been developed for the parameters of perceived quality and associated test methods, describing test methodology, data interpretation and aggregation.

The Ergo-Innovationlab provides the full range of perceived quality testing in its six domains: Anthropometrics, ergomechanics, perception, thermal comfort, seat comfort testing and variable environments. The scope of the Ergo-Innovationlab goes beyond offering comfort assessment and analyses by providing basic knowledge, performing innovation verification, and expanding internal and external capabilities. The developed product range encompasses operator systems for trucks, agriculture and construction equipment, fork lift trucks, bus and rail, and automotive interior and seat components. Recent topics have been supporting the human spine in rotated positions, product layout for global user populations, seat and vehicle operation concepts, and benefits of haptic feedback. The development of associated ergonomic tools and methods is a continuous process. It is connected to permanently changing parameters ("moving targets"), such as new technologies that influence vehicles and workplaces, e.g.: autonomous driving, connectivity and smart products.

2   Assessing Comfort

The experience multitude of comfort is comprised of an array of factors from the realms of affinity and accommodation. Human beings compute their comfort rating with low awareness for these factors, because comfort is perceived as a unified experience. Only in the case that single positive or negative factors become dominant in the holistic impression, they become apparent to the user as relevant sub-factors of comfort. In rating comfort of products, isolating factors in testing can provide the basis for understanding reasons for rating differences.

2.1   Comfort Factors

The aspect of affinity summarizes factors of perception, such as visual appearance, acoustic feedback and haptics. The aspect of accommodation entails factors of the physical interface such as body posture, seat and controls fit, microclimate and transmitted vibration. A user will perceive comfort only if all affinity and accommodation factors offer at least acceptable comfort, or if a marginal rating in one aspect is compensated by a clearly positive one in the other. If one factor is rated unacceptable, it becomes dominant and comfort can not be reached any longer.

2.2   Comfort Rating and Evaluation

In order to evaluate a product, factors which are relevant to human interaction are isolated. These differ according to the possible ways of human-product interaction. If a product is visible, tangible and audible during interaction, visual, haptic and acoustic ratings need to be included. If a product takes up load and positions the users, the interaction parameters pressure, posture and thermo-physiology are to be assessed. If usage occurs in motile, dynamic environments, biomechanics and vibration is to be taken into account. The comfort rating is performed by scaling the parameters according to their perceived quality and defining the thresholds of all relevant comfort factors to at least the level "acceptable" or "marginal".

2.3   Comfort Assessment Tools: KIT (Kinematic Innovator Tool)

Employing tools in comfort assessment can enable rating different comfort parameters without having to build specific prototypes for each iteration. The Ergo-Innovation Lab makes use of the specifically developed kinematic innovator tool (KIT), which allows for the investigation of seat contours and kinematics.

Movements that seat users perform depend on their activities and their environment. Passengers have more liberty of action than drivers or operators, who are bound to controls and actuators to perform their tasks and control the machine. Their movements should be supported by seat kinematics, and different functions can call for different requirements. KIT enables the research of seat kinematics with test subjects by providing seat motion experience and making it assessable (see Figs. 1 and 2).

Fig. 1. Kinematic innovator tool KIT [2]

KIT consists of contour elements in a support structure with variable centerline and side support. Further functions are the coupled motion of seat elements and the variation of the seatback pivot point. This functionality allows for a free contouring of the seat interface and the determination of the optimal kinematics for specific applications [1].

Fig. 2. Kinematic innovator tool KIT [2]

3 Configuration of Operator Systems According to Comfort Criteria

Users and usability in the relevant use cases provide the basis for configuring operator systems on a system level in a holistic approach. Product users in their variability determine the product configuration to allow for successful interactions.

3.1 User Research

User research encompasses interaction data acquisition with available products as well as rating of innovations. In order to assess how well seats in commercial vehicles and

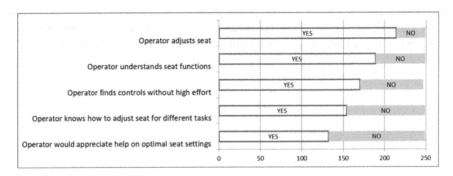

Fig. 3. User research on seat functionality [7]

their controls are understood and used, an online questionnaire with 249 users was performed in spring 2017 [7]. Respondents were operators of tractors (65.9%), trucks (49.4%), bulldozers (30.5%), forklifts (29.3%), excavators (23.7%), cranes (17.3%), combine harvesters (14.9%) and others (4.0%, multiple selections possible). Their age was between 15 and 70 years, operating experience between one and 51 years, 11.8 years on average, and the group was predominantly male (N = 234).

Seats were predominantly rated as user friendly, with a rating of 3.7 on a scale of 1 to 5. Only 5% of the votes went to "low" and "very low". To understand the prerequisites for this rating, questions were posed to situations of adjusting the seats. 86% of respondents stated they adjust their seats, and 68% find all controls with low effort. 76% find the functions to be clear.

The sub-group of users that adjust their seats (N = 214) were asked in which situations they do so. The highest numbers mentioned the situation when starting to operate the machine after someone else had used it (84%). 61% stated they would change position after a while to avoid discomfort.

These statistics show a high satisfaction and usability of professional operators with functions and adjustments of available seats. However, also development potential for seat functionality can also be deducted, since 14% never adjust their seats, 25% does not comprehend all functions, and one third would like support in finding the right settings. Overall, over half of the respondents stated they would like to get advice on the best seat position. This request was stated over all age groups and vehicle categories.

3.2   Usability Criteria

One aspect of perceived quality is usability, which refers to the human-product physical and mental interaction. ISO 9241-11 defines usability as the ability of people to use a product, at what cost and to what extent they are satisfied with using the product [8]. Thus, exactness and completion of goal attainment under consideration of the needed exertion, the satisfaction and freedom from impairment form criteria for rating product usability.

In configuring operator systems these usability criteria are applied in parameters such as the ones defined by Norman: affordance, reference, constraints, mapping, feedback and mental models with the goal of reaching an intuitive operation [9]. This type of operation unburdens the users, so that they need less exertion for the operator system and have more energy for task completion.

3.3   Understanding Interaction: Seat Control Concept Development

Applying usability criteria in seat adjustment refers to the arrangement of controls, their shape and feedback. When adjusting to the seat position, users of commercial vehicles have the concrete goal of optimal positioning in the operator system for vehicle and implement control. Therefore, effectivity and efficiency are in their main focus, and technology is the means of attaining that goal. Frustration by difficult to find or use controls is to be avoided (Fig. 4).

Planning control arrays is determined by the importance and frequency of use of the functions. For control shape, target group body measures, operation force and direction

Fig. 4. Expected position of controls on agriculture vehicle seats [2]

are to be respected. Feedback should be designed under consideration of the interaction criteria in order to reach a logical operation concept.

User study output defines the positions in which users expect functions. For example, the swivel function, which turns the entire operator system relative to the cabin environment, is expected in the position front of left seat cushion or armrest (see Fig. 3). For dimensioning and shape of control elements, user hand measures, operating force and direction is respected. Mature positioning, contouring, orientation and feedback of controls according to interaction criteria yields an intuitive operation concept.

4   Understanding and Optimizing Interaction

The complexity of human product interaction is currently not completely understood by science. In order to develop products for human use according to comfort and health principles, it is necessary to further invest efforts into advancing this knowledge. To expand ergonomics content and capacity of the Ergo-Innovation Lab, there are strategic partnerships in the research partner network, e.g. in spine research and user experience. This spine research spans a wide range and furthers understanding in methods development, alleviation of spinal disease, and healing, advancing a deeper understanding of the multi-factorial topic back pain. Topics of the past years were a better understanding of the human spinal system, spine modelling, and spinal therapies.

4.1   Spine Science and Physiological Optimization

Risk factors for low back pain numerous and complex and include the individual, the work, the environment and the society [10]. In the vehicle work place, the human spine is challenged by static loading and enforced postures due to work tasks and long-term exposure. Operators underlie a multitude of mandatory links to their environment: visual control, physical touch, positioning and load uptake. These spinal challenges are exacerbated under vibration conditions. Newell et al. showed in a laboratory experiment with a commercially available air suspended seat that reaction time become significantly longer under vibration influence and the error rate rises [11]. A seat configuration that supports the body in working conditions can minimize these negative effects. Also, the usage of armrests has a positive effect on reaction time. Even though enforced postures generally have a negative influence on reaction time, Newell found faster reaction times under enforced postures with armrest use than forward facing postures without.

While it is not possible to determine the exact load on the spine for living human beings, relief on its structures can be determined by instrumented implants [12]. At the Charité Berlin, occurring forces between vertebrae were measured with patients with spinal implants. Load uptake by armrest usage reduced implant loads by one fifth, leaning on the seatback by one third on average (see Fig. 5).

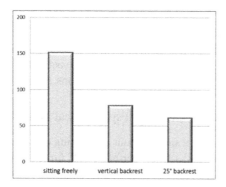

Fig. 5. Effect of posture on resultant force [in %] on spinal structures under vibration conditions

Research under vibration exposure shows that the maximal force on the implant rises with the intensity of the vibration. However, leaning on the seatback in vibration conditions reduced the occurring loads to a lower level than those measured for free, upright sitting without vibration exposure [12]. This points out that a seat configuration allowing the highest possible percentage of the user population a good support by seat elements plays a major role in relieving body structures and upkeeping performance.

Only few studies research spinal movements of healthy persons during the day. A study on this topic was performed with the measurement system Epionics Spine by the Carité Berlin with 208 healthy individuals over 24 h observation time [12]. The test subjects performed their regular activities, the only restriction was showering and bathing due to the measurement system (Fig. 6).

The results showed a very large number of spinal movements over the day, up to more than 13.000, with higher values by 29% for female than male persons. In the high frequencies, lumber spine movements were smaller than 10°. On average, 63% of the day was spent in the lumber flexion region with a lordosis angle change of 10–40° related to standing.

Research of the *Institute of Orthopaedic Research and Biomechanics* at the Trauma Research Center Ulm has shown that the slouch posture alternated with upright posture positively influences all spinal structures. This insight has been incorporated into seats

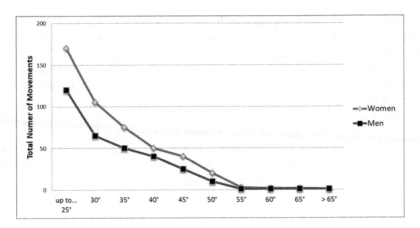

Fig. 6. Change of lordosis angle for men and women [N = 208, Median]

by Grammer AG. They enable body position change according to the ergomechanics® principle, i.e. the spine can assume different positions in seat contact.

4.2   Mental Load Management and Psychological Optimization

Mental work load results from simultaneous processing of numerous stimuli, especially under time constraints of tasks to be performed. If furthermore potential hazards are present, such as in the operation of large work machines in environments with living organisms, mental load is exacerbated. Professional, long-time users of work machines ("experts") experience much less stress and frustration than novices. Subjective stress can be measured by standardized procedures such as the NASA Task Load Index, which assesses perceived stress and strain in six categories (Fig. 7).

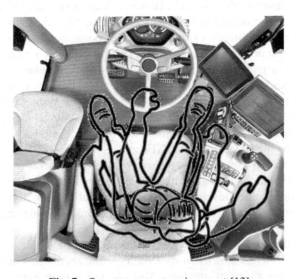

Fig. 7. Operator system environment [13]

For the execution of necessary tasks with work machines, the operation of the machine should only be a background activity without high mental load for the most beneficial setting. In order to meet this target, controls are categorized according to their importance and their frequency of use and then positioned accordingly. Thus, finding the respective functions becomes unambiguous and intuitive, and this lowers operator mental load.

Operator system configuration according to ergonomic criteria also has a positive effect on operator mental load. In both forward and rearward facing work positions, the use of armrest lowers the subjectively perceived stress and strain [11].

5   Conclusions

The holistic approach to operator system comfort takes users with their characteristics and capabilities as the foundation to product design. Products are configured to their specific targeted user groups and use cases. This approach minimizes driver load and stress by accounting for all aspects of interaction, such as the visual and haptic comfort, adjustment ranges, vibration comfort and posture and thus achieving perceived quality and its associated effects of operator health and efficiency through adequate product design.

References

1. Pheasant S (2003) Bodyspace, 2nd edn. Taylor & Francis, London
2. Frohriep S (2018) Perceived Quality als Leitbild in der Entwicklung von Operatorsystemen. In: Proceedings Arbeitswissenschaftliches Kolloquium, Wieselburg, pp 215–221
3. Frohriep S (2015) Comfort into seat design. In: Proceedings of innovative seating conference, 25 February, Düsseldorf
4. Kahnemann D (2011) Thinking, fast and slow. Penguin, London
5. Frohriep S (2017) From measurement to meaning: methods of comfort assessment and corresponding ergo-lab functions. In: Proceedings international comfort congress (ICC), Salerno
6. Knoll C (2006) Einfluss des visuellen Urteils auf den physisch erlebten Komfort am Beispiel von Sitzen. Dissertation. TU Munich, Garching
7. Rotte T (2017) eSai – electronic seat adjustment interface, Graduation thesis. TU Delft
8. DIN (1999) Ergonomische Anforderungen für Bürotätigkeiten mit Bildschirmgeräten, Teil 11: Anforderungen an die Gebrauchstauglichkeit – Leitsätze," DIN EN ISO 9241-11, Deutsches Institut für Normung, Berlin
9. Norman D (2013) The design of everyday things. Basic Books, New York
10. Nordin M (in print) The relevance of back pain. Ergomechanics 3
11. Newell G, Mansfield N (2008) Evaluation of reaction time performance and subjective workload during whole-body vibration exposure while seated in upright and twisted postures with and without armrests. Int J Ind Ergon 38:499–508
12. Rohlmann A, Wilke H-J (in print) Messung der Wirbelsäulenbelastung und der Anzahl der Rückenbewegungen. Ergomechanics 3
13. Boving M (2017) Best practices user experience GUX^{TM} – the operating concept behind the operator system. In: Proceedings 3. VDI-Fachkonferenz HMI und unterstützende Systeme in mobilen Arbeitsmaschinen, VDI-Verlag, Düsseldorf

Psychophysiological Cueing and the Vigilance Decrement Function

G. M. Hancock[(✉)]

California State University - Long Beach, Long Beach, CA 90840, USA
gabriella.hancock@csulb.edu

Abstract. Vigilance is the mental capacity by which observers maintain their attention across time. It is most commonly operationalized as the ability to detect rare and critical signals. Due to inherent constraints in human processing capacities, the longer one expects an observer to keep watch, the more likely it is that a searched-for signal will be missed. Consequently, without some means of computer or automation-based assistance, failures in operator vigilance are likely to occur. Mackworth called this characteristic decline in performance over time the 'vigilance decrement.' Many modern-day operational tasks that entail life-or-death consequences require vigilant monitoring, mostly of visual stimuli. Consequently, it is unsurprising that research has sought to establish and validate methods of counteracting such an adverse behavioral trend. One such strategy is cueing. Cuing provides the operator with a reliable prompt concerning signal onset probability. Traditional protocols have based such cues on task-related or environmental factors. The present work addresses the methodological concerns of using cues based on these factors in the design of effective vigilance cueing systems. This present work proposes an alternative perspective supported by empirical research to explore and validate this novel approach. This study examines the efficacy of cueing when based on an operator's psychophysiological state (i.e., cortical blood oxygenation) in a novel vigilance task incorporating dynamic rather than static visual displays. Results pertaining to performance outcomes, physiological measures (heart rate variability), and perceived workload and stress are interpreted via Signal Detection Theory and the Resource Theory of vigilance.

Keywords: Vigilance · Cueing · Psychophysiology

1 Introduction

1.1 Vigilance

Vigilance is the mental capacity to maintain attention over extended periods of time (Parasuraman 1979; Teichner 1974; Mackworth 1948). Humans, however, have difficulty sustaining attention and, without some sort of countermeasure or intervention, failures in vigilance are likely to occur. This behavioral pattern of performance showing a diminishing capacity to detect, as a function of time on watch, is referred to as the vigilance decrement function (Mackworth 1948; Hancock 2013). Research attempting to mitigate this decrement is vital to many real-world vigilance situations

© Springer Nature Switzerland AG 2019
S. Bagnara et al. (Eds.): IEA 2018, AISC 827, pp. 268–281, 2019.
https://doi.org/10.1007/978-3-319-96059-3_30

that require the identification of rare but critical signals, and seek to minimize the adverse effects that are associated with missing such targets of interest (Warm et al. 2009a, b). Such applications include, but are not limited to, air traffic control (Hitchcock et al. 1999), baggage screening (Harris 2002), diagnostic medical screening (Gill 1996), driving (Hancock 2014) and, in the scope of this work, military surveillance in the form of improvised explosive device (IED) detection (Szalma et al. 2018).

1.2 Decrement Mitigation Practices: Cueing and Knowledge of Results

Several practices have been examined as potential avenues for attenuating the decrement function and thereby safeguarding effective monitoring performance. Two such techniques that have proved reliable in these efforts are knowledge of results (KR) and cueing. KR is the post-hoc process of providing participants with feedback regarding the speed (McCormack 1959) and/or accuracy of their performance (Szalma et al. 2006a; b). Cueing, on the other hand, is the a priori presentation of a reliable prompt that provides the observer with an indication that a critical signal is to be forthcoming (Hitchcock et al. 1999). While both methods have demonstrated efficacy as decrement mitigation strategies (Annett and Paterson 1967), cueing has been further promoted by some researchers given its greater utility under strenuous temporal demand (Aiken and Lau 1967) and its tendency to refrain from inflating false alarms rates in the way that KR does (Annett and Patterson 1967).

Cue type was consequently manipulated in the present work to investigate its effects on cueing's efficacy as a vigilance decrement mitigation technique. Vigilance studies concerning cueing in the existing literature almost exclusively have employed cues based on environmental indicators of signal onset probability. That is cues, based on what is currently known of the state of the environment, a critical signal is therefore more likely in the next epoch of time, distance, or during the next x number of trials. Given that the state of the environment is notoriously difficult to predict, the basis of these cues may therefore be more reliable if predicated upon on the physiological state of the operator's readiness, which is more easily and reliably observable.

1.3 Neurofeedback Cueing and Vigilance

The efficacy of cueing as a mitigation technique has been established since the 1960s. However, Metzger and Parasuraman (2001) rightfully observed that the effectiveness of a cue hinges on its reliability. Hitchcock et al. empirically supported this assertion as their data revealed a direct relationship between cue reliability and vigilance performance. That is, as cue reliability declined, so too did performance efficiency, and completely reliable cues (100%) afforded the ability to maintain detection rates higher than 90% over the course of a vigil (Hitchcock et al. 2003).

As noted earlier, traditional experimental protocols have used cues based on the environment or task factors. This practice can prove problematic when trying to integrate cueing into real-world, operational tasks where predicting the state of the environment, even in the near future, is difficult. Cues generated as the result of an operator's cognitive or neurological functioning may therefore prove a more reliable foundation when projecting an imminent lapse in vigilance.

Neurofeedback is the awareness of one's own bodily functions via the continual, real-time apperception of biological dynamics such as electrical activity of the outer layers of the cerebral cortex (Raymond et al. 2005). While much of the existing literature concerning the physiological underpinnings of vigilance have concentrated on electroencephalography (EEG: Belyavin and Wright 1987; Kamzanova et al. 2014), a developing line of research is now using functional near-infrared spectroscopy (fNIRS; De Joux et al. 2015). The high spatial resolution afforded by fNIRS in combination with the fact that it assesses metabolic rather than transient electrical activity (as EEG does), renders fNIRS a more reliable (i.e., vigil-length) candidate for cue formation when compared to EEG measures like brain waveforms or even event-related potentials.

1.4 Purpose

The purpose of the current study was therefore to ascertain if the established mitigation effects of cueing maintains its efficacy when the nature of the cue shifts from one that is externally driven (i.e., based on the state of the environment) to one that is internally driven measure (i.e., based on the state of the operator) such as neurofeedback. This neurofeedback was nominally based on cortical blood oxygenation in the prefrontal cortex as measured by fNIRS. To address this question, the experiment investigated vigilance performance, the physiological state of the operator (as evaluated by psychophysiological arousal of the autonomic nervous system and mobilization of metabolic resources), as well as the cognitive and affective state of the observer with respect to workload, distress, and task engagement.

2 Methods

2.1 Participants

Twenty-seven participants (15 females, 12 males) were recruited from the Psychology Department of a large university in the southeastern United States to take part in this study. Participants' ages ranged between 18 and 31 years old, with an average age of 20.4 years (SD = 2.90 years). Observers received extra credit in exchange for their participation, reported normal or corrected-to-normal vision, and were asked to refrain from ingesting caffeine for the 24 h preceding their experimental session.

2.2 Equipment

All visual stimuli were developed using Virtual Battlespace 2 (VBS2; Bohemia Interactive, Prague, Czech Republic). The software depicted simulated representations of a typical foot patrol route through an Afghan village in daylight conditions complete with interspersed, ecologically valid critical signals. These included: a yellow fuel can, a black trash bag, a black motorcycle battery, and a wooden box top which has been sunk into the ground (and see Szalma et al. 2014). The dynamic video clips were derived by excising files from the full-version virtual of the vigil. The 5 s duration of each clip was empirically determined via data yielded from a psychophysical equivalency two alternative forced choice detection task (and see Hancock, in press).

Detection performance measures were recorded for off-line analyses using a custom Qualtrics program (Qualtrics, Provo, UT, US). All relevant variables for these analyses were measured, synchronized with each other, and time-locked in relation to stimulus-onset. All questionnaires were administered electronically using a Qualtrics software platform. All raw data were tabulated and exported for analysis using a standard statistical package (e.g., IBM Statistical Package for the Social Sciences, Software Version 21.0, IBM Corporation, Armonk, NY, US).

The Dundee Stress State Questionnaire (DSSQ; Matthews et al. 1999; Matthews et al. 2002; Matthews et al. 2013) is an assessment tool designed to gauge subjective state in performance contexts. A validated, short version of this questionnaire, consisting of 30 items, was used in this study (Matthews et al. 2013). The NASA Task Load Index (TLX; Hart and Staveland 1988) was also used to assess weighted global perceived workload.

Electrocardiogram (EKG) data were collected using a BioPac MP150 data collection system and Acknowledge software 3.9.1 (BioPac Systems Inc., Aero Camino Goleta, CA, US). All cardiac data were collected with a gain of 500 and at a rate of 1,000 Hz. Electrodes were BioPac EL503 disposable, silver-silver chloride electrodes with a 1 cm diameter (BioPac Systems Inc., Aero Camino Goleta, CA, US). Cardiac data were collected via a triangulated electrode configuration. The ground electrode was placed on the tenth rib of the left thorax. Data collection electrodes were placed on the tenth rib of the right thorax and on the left side of the chest (5 cm to the left of the jugular notch and inferior to the left clavicle). Heart rate variability (HRV), defined as the variation in time (milliseconds) between sequential heartbeats, was used as an objective measure of autonomic nervous system arousal accompanying mental workload.

Changes in the concentration of oxyhemoglobin (oxy-Hb) in cerebral bloodflow were measured via a 16-optode continuous wave fNIR Imager 1000 system (fNIR Devices LLC, Potomac, MD, USA). LED-based sensors were set flush across the participant's forehead and secured using a tie-strap. Once activated, the sensors recorded neural responses in the prefrontal cortex (PFC) via blood oxygen level dependent signals (i.e., the BOLD response) in tissue reflectivity. A number of experimental studies have observed a robust effect of right cerebral hemispheric dominance during vigilance tasks (Warm and Parasuraman 2007; Helton et al. 2007; Warm et al. 2009a, b). Data were collected via Cognitive Optical Brain Imaging Studio software (Version 1.4.0.25; Ayaz 2005) and reduced via fNIRSoft software (Version 4.3; Ayaz 2010).

2.3   Physiological Data Reduction

Heart Rate Variability. Cardiac data were reduced by measuring the amount of time between subsequent R spikes of the electrocardiogram. R spikes typically constitute the greatest change in amplitude accompanying the contractions of the cardiac ventricles (Jevon 2010). A standard deviation was calculated for the two-minute period of rest prior to task engagement, which served as the baseline value. Phasic values were computed by averaging the standard deviations of RR intervals (HRV activity) two to four seconds following stimulus onset of each trial. Data processing was limited to the specified two-second time window as this represents enough time to capture any reaction to the trial, while not being so excessive as to average out any observable effect. Also, two seconds appears to be the largest average time between subsequent heartbeats in the normal population (Grajales and Nicolaescu 2006). HRV values associated with correct detections were averaged.

Cortical Blood Oxygenation. All optode data were analyzed provided that raw oxygenation levels (i.e., the ratio of oxygenated blood to blood volume) occupied the range spanning 400–4000 mV once the appropriate filters (i.e., low pass, 60 Hz notch filter, ambient light filter, and Motion Artifact Rejection (SMAR)) were applied. Any optodes with signal values falling outside this range were excluded via channel rejection. The refined light intensity data were then used to calculate oxygenation. Raw oxygenation data were subjected to another low-pass filter, and detrending was applied to expel drifts in the signal. The performance blocks were defined in relation to the manual markers that indicated the beginning of each block. The local hemoglobin maximum value recorded by each optode and for each block was then identified, and averages of these maxima values computed according to condition, the early (first ten trials) or late (last ten trials) position of the block within the vigil, and for each of the five periods on watch.

2.4   Experimental Task and Procedure

Participants read and signed the IRB's informed consent form. Electronic versions of the demographics questions and the pre-task DSSQ were then administered. The EKG electrodes and the fNIRS strap were then put in place. Raw oxygenation was checked to ensure that levels fell within the 400–4000 mV range. If not, adjustments were made as far as possible to the strap and its placement to confirm that non-normal values were not due to some physical barrier (e.g., hair placement). Physiological baseline data were collected during a two-minute rest period before beginning the task. A reminder was then issued to participants to remain as still as possible in order not to introduce movement artifacts into the physiological data. A static image depicting the critical signals to be detected was then shown so as to familiarize observers with them. Participants were told that they would be asked to watch a series of short video clips wherein they were expected to identify one of the four critical signals. If they happened to see one of these signals, they were to respond by clicking on a box labeled 'Signal Detected' located immediately below the video clip. The response box was placed below the video clip as opposed to superimposed over it so that the presence of the mouse would not obscure any portion of the visual display. Participants were told that if they did not see any of the critical signals in the video, that they were to refrain from clicking, and that the subsequent clip would begin automatically.

Each participant then underwent a vigilance task comprised of 120 total trials. Each trial comprised the presentation of one five-second video clip. Trials were arranged in 20 total blocks; each block consisting of six trials with five, one-second inter-stimulus intervals. Each block was therefore 35 s in duration. In order to accommodate the re-setting of the BOLD response in consideration of the fNIRS data, an interval of 20–30 s occurred between each block. To ensure that participants remained vigilant throughout this interval, its duration was randomized (between 20 and 30 s) so that the participant was unsure of the onset of each subsequent block. As a result, the vigil was approximately 21 min in total duration. For analyses, this vigil was parsed into five periods on watch, each period comprising four blocks. The randomization of which trial within the block would contain the signal and which signal would be presented in that trial was determined a priori, and all participants experienced the same 'canned' vigil.

Each participant was randomly assigned to one of two groups: a cueing group vs. a control group. One of the blocks in each period on watch was randomly selected to be the block wherein a tone was presented upon the presentation of the first trial therein. Individuals in the cueing group were told (prior to beginning the task) that should they hear this tone at any time during the vigil, it would be indicative that their physiological signals had declined and that they should therefore re-orient themselves to the task. Control participants were told that the sound of the tone was an indication that the computer system was saving their data. The same tone was presented to both groups at the same volume, and at the same pre-specified junctures during the vigil so that any changes in performance could be attributed to the content of the cue, rather than the presence of additional auditory stimuli.

Following the completion of all five periods on watch, participants provided responses to electronic versions of the post-task DSSQ and the NASA-TLX, which were counter-balanced. Finally, the experimenter debriefed and answered any applicable questions for the participants following the administration of these measures.

3 Results

3.1 Vigilance Performance: Correct Detections

There was no significant main effect for condition (cueing versus no cueing) on vigilance performance in terms of correct detections (F $(1, 25) = 1.843$, $p = 0.187$). Consequently, the stated hypothesis was not supported. The condition by period on watch interaction also proved to be non-significant (F $(3.061, 76.523) = 0.080$, $p = 0.972$).

Mauchly's test indicated a violation of the sphericity assumption with regards to period on watch ($\chi^2 (9) = 20.122$, $p = .017$). Therefore, the degrees of freedom were corrected using the Greenhouse-Geisser estimates of sphericity ($\varepsilon = .765$). Figure 1 illustrates the subsequent statistically significant main effect of period on watch on vigilance performance in terms of correct detections (F $(3.061, 76.523) = 9.819$, $p < .001$, $\eta_p^2 = .282$). Pairwise comparisons revealed that there was a statistically significant greater percentage of correct detections in Period 4 relative to Period 1 (mean difference = 24.176, $p = 0.001$), Period 4 relative to Period 3 (mean difference = 15.797, $p = .004$), and in Period 5 relative to Period 1 (mean difference = 25.962, $p < 0.001$). The percentage of correct detections therefore appears to have increased over time.

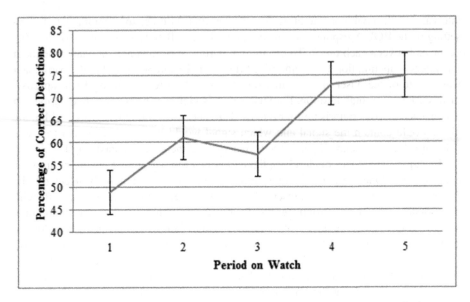

Fig. 1. Main effect of period on watch on correct detections. Error bars are standard errors.

3.2 Vigilance Performance: False Alarms

There was no statistically significant main effect for condition (cueing versus no cueing) on vigilance performance in terms of false alarms (F (1, 25) = 0.642, p = 0.431). The period on watch by condition interaction was also not significant (F (3.072, 76.811) = 0.545, p = 0.657).

Mauchly's test indicated a violation of the sphericity assumption with regards to period on watch (χ^2 (9) = 19.490, p = .022). Therefore, the degrees of freedom were corrected using the Greenhouse-Geisser estimates of sphericity (ε = .768). With this adjustment there was a statistically significant main effect of period on watch on vigilance performance in terms of false alarms (F (3.072, 76.811) = 3.783, p = .013, η_p^2 = .131). Pairwise comparisons disclosed that there was a significantly higher average number of false alarms in Period 2 relative to Period 5 (mean difference = 1.827, p = 0.024), and this main effect is illustrated in Fig. 2.

3.3 Heart Rate Variability

Phasic heart rate variability for all trials in which participants correctly detected the critical signals were analyzed via a 2 (CONDITION: Cueing vs. Control) X 5 (PERIODS ON WATCH) mixed ANOVA with repeated measures on the second factor. Analyses revealed no main effect for condition on heart rate variability (F (1, 23) = 0.028, p = 0.868).

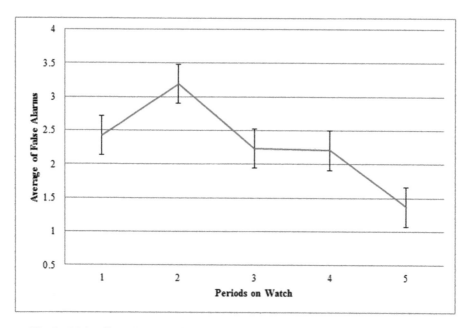

Fig. 2. Main effect of period on watch on false alarms. Error bars are standard errors.

Mauchly's test indicated a violation of the sphericity assumption with regards to period on watch (χ^2 (9) = 24.420, p = .004). Therefore, the degrees of freedom were corrected using the Greenhouse-Geisser estimates of sphericity (ε = .630). With this adjustment, there was a significant main effect of period on watch on HRV (F (2.522, 57.995) = 5.460, p = .004, η_p^2 = .192). Pairwise comparisons disclosed that there was a significantly higher HRV in Period 1 relative to Period 5 (mean difference = 0.050, p = 0.006), and in Period 4 relative to Period 5 (mean difference = 0.027, p = 0.022). This main effect is illustrated in Fig. 3.

Analyses also specified that the period on watch by condition interaction was also statistically significant (F (2.522, 57.995) = 3.830, p = .019, η_p^2 = .143). As shown in Fig. 4, both conditions seem to experience a decrease in HRV as a function of time on task. The general linear decline appears to be steeper for the control group relative to the cued group.

3.4   Cognitive and Affective States

Univariate analyses conducted on the DSSQ differences (post – pre) scores revealed no statistically significant main effect of condition on any of the subscales of: task engagement (F (1, 25) = 0.398, p = .534), distress (F (1, 25) = 0.428, p = .519), or worry (F (1, 25) = 0.671, p = .420). As a result, the prediction that cueing would differentially impact cognitive and affective states (as measured by the DSSQ) was not supported. Two-tailed T-tests were conducted to test changes on DSSQ difference scores as a result of participating in the vigil. Analyses revealed no statistically

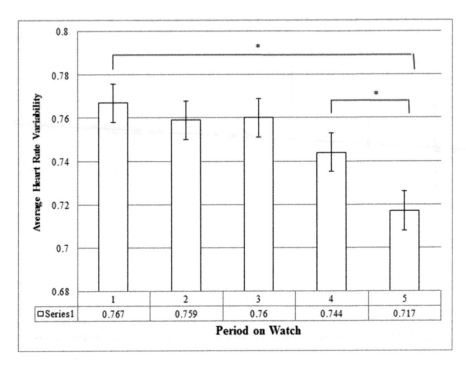

Fig. 3. Main effect of period on watch on heart rate variability. Error bars are standard errors. Asterisks denote statistically significant differences.

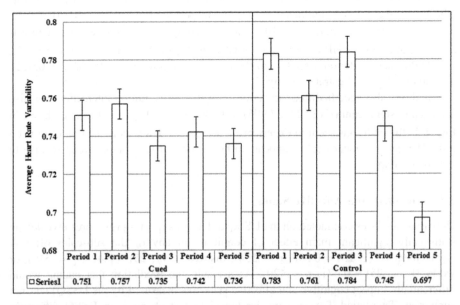

Fig. 4. Significant Condition by Period on Watch interaction in heart rate variability. Error bars are standard errors.

significant change in worry (t (26) = –0.875, p = .390). Results did, however, reveal statistically significant changes in both task engagement (t (26) = –5.685, p < 0.001) and distress (t (26) = 6.452, p < .001). Participants reported significantly lower task engagement and significantly higher distress as a result of undergoing the vigil.

There was no statistically significant main effect for condition on weighted NASA-TLX global workload averages (F (1, 225) = 0.040, p = .843). Thus, the notion that cueing would differentially influence workload was not supported. A two-tailed t-test did, however, reveal that the vigil imposed significant workload (t (26) = 22.428, p < .001). None of the weighted subscale scores (Physical Demand, Mental Demand, Temporal Demand, Performance, Effort, or Frustration) proved significant.

4 Discussion

For almost seventy years, empirical research has studied factors that contribute to the vigilance decrement function (Mackworth 1948; Scerbo 1998a), such as event rate (Parasuraman 1979), signal discriminability and task type (Parasuraman and Mouloua 1987). When characteristics of these elements combine with inherent constraints in information processing capabilities, failures in vigilance often occur. Given that these failures frequently occur in performance domains that entail significant risk to human health and well-being, the consequences of such lapses in vigilance can be dire. Consequently, researchers have sought to establish, validate, and implement effective countermeasures to such declines in order to maintain and improve vigilance performance. The two most successful countermeasures to date have been knowledge of results (providing post hoc feedback concerning performance) and cueing (presenting reliable prompts regarding single onset probability) (Wiener and Attwood 1968).

While experimental studies have identified certain influences underpinning the decrement, and have tested the cogency of different methods of contravening the maladaptive performance trend, one important consideration that has not been addressed is cue type. This study therefore focused on cue type (cues based nominally on operator state rather than environmental factors), and their potential effects on vigilance performance. A greater understanding of this relationship is necessary as cueing based on an operator's physical state could prove to be a more effectual countermeasure for the vigilance decrement given that it can be more accurately assessed in performance domains where information relating to signal onset probability is difficult to assess, predict, and provide.

Typical vigilance performance is characterized by a decline in both correct detections and false alarms as a function of time on watch. The characteristic vigilance decrement was not observed in the present experimental circumstances. Instead, the pattern of results appears to resemble a learning curve with correct detections by and large increasing, and false alarms generally decreasing over time. While previous studies have infrequently reported a marked increase in correct detections during the final period on watch, a phenomenon known as the end spurt effect (Bergum and Lehr 1963; O'Hanlon 1965), the pattern of results herein observed does not resemble the characteristic vigilance performance decrement (Mackworth 1948) as correct detections broadly rose with time on task. These results, again, could be due to insufficient sensitivity to distinguish the critical signals from their surroundings.

Some research may suggest that the vigil was of insufficient duration (at 21 total minutes) to provoke the performance decrement. Many experimental studies have elicited the decrement by employing vigils of significantly greater duration such as 90 min (Pattyn et al. 2008; O'Hanlon 1965), two hours (Mackworth 1948), or more (Johnson and Merullo 1996). However, more recently, certain researchers have suggested that it is not so much the length of the vigil as it is the task demands it imposes upon the participant that should theoretically drive the decrement. In fact, several studies with sufficiently challenging task demands have reliably observed the decrement function in a matter of minutes (Neuchterlein et al. 1983) or even within a few trials (Jerison 1963).

Two task characteristics that could therefore have led to the absence of the performance decrement are the event rate/target rate and inadvertent cognitive breaks. There appears to be an inverse relationship between event rate and hit rate, wherein correct detections decrease as the event rate increases (Jerison and Pickett 1964). Event rate has been widely manipulated in the literature with researchers using 10 events per minute (Valentino et al. 1993), 30 events per minute (Arruda et al. 2007), and 40 events per minute (Gunn et al. 2005). The current 6 events per minute therefore constitutes an uncommonly low event rate in vigilance. Pattyn et al. (2008) utilized a similarly low event rate (2 or 3 events per minute) and saw their error rate decrease over time, which corresponds to an increase in correct detections as was observed in the present study. A lower than normal event and target rate may therefore explain the fact that the vigilance decrement was not observed in the current dataset and also to other domains such as maintenance in which such low event rates apparently ameliorate the decrement.

Moreover, in consideration of the fNIRS measure, the block design incorporated intervals (with a duration of 20–30 s) between each block wherein no stimuli were presented in order to allow for participants' BOLD response to return to normal levels. While participants were explicitly told to maintain their level of attention throughout these intervals (as the stimuli could appear at any time), the fact remains that participants could have used these periods of time as cognitive breaks during which their mental resources were replenished. Experiencing short intervals without any tasks demands, or even switching between tasks at sporadic intervals can effectively avert the performance decrement (Ariga and Lleras 2011).

With HRV declining as a function of period on watch, this trend would indicate that participants were working harder as the vigil progressed (and see Fig. 3). This finding coincides with the performance data and subjective reports of workload and stress provided by the participants. However, these results do not hold with a previous study that found an increase in correct detections with time on task (as this study did), that was instead accompanied by an increase in inter-beat interval (and therefore a rise in HRV) which was not the case in the present study (Pattyn et al. 2008). Further research is therefore needed to clarify potential sources of dissonance. While perceived stress and workload were not differentially affected based on the manipulation of cueing, results from the DSSQ and NASA-TLX did reveal that the experimental task imposed significant stress and workload on participants in a fashion commensurate with stress profiles observed in other vigilance tasks (Grier et al. 2003; Temple et al. 2000; Szalma et al. 2004; Warm et al. 2008).

The current results therefore cannot confirm unequivocal empirical support for the proposition that cueing based on an operator's physiological state, as opposed to environmental indicators of signal onset probability, would help to improve vigilance. Yet this proposition still remains a hopeful one. Future efforts should incorporate a canned vigil engineered so that all trials are 5 s in duration with a 1 s inter-stimulus interval between each trial, and then manipulate the placement of trials such that there is a minimum of 20–25 s between any two trials containing critical signals. The length of the vigil may also be extended in order to increase the event rate and target rate. Coupling these higher event rates and target rates with the successive-discrimination nature of the task should consequently engender the conditions to observe the decrement (Parasuraman 1979). A more powerful manipulation for neurofeedback cueing could also be adopted. In addition to the verbal instructions, a display that shows the participant's psychophysiological output could well be included.

The current work found no evidence that cues based on an individual's physiological state were effective in improving vigilance performance. Such results may be due to the fact that the vigilance decrement was not observed. Further empirical study adopting a task wherein the decrement has already been reliably observed, or field studies in relevant performance domains, is needed to address this on-going and practical research issue.

References

Aiken EG, Lau AW (1967) Response prompting and response confirmation: a review of recent literature. Psychol Bull 68:330–341

Annett J, Patterson L (1967) Training for auditory detection. Acta Physiol 27:420–426

Ariga A, Lleras A (2011) Brief and rare mental "breaks" keep you focused: deactivation and reactivation of task goals preempt vigilance decrements. Cognition 118(3):439–443

Arruda JE, Amoss RT, Coburn KL, McGee H (2007) A quantitative electroencephalographic correlate of sustained attention processing. Appl Psychophysiol Biofeedback 32(1):11–17

Ayaz H (2005) Analytical software and stimulus-presentation platform to utilize, visualize and analyze near-infrared spectroscopy measures. Master's thesis, Drexel University, Philadelphia

Ayaz H (2010) Functional near infrared spectroscopy based brain computer interface. PhD thesis, Drexel University, Philadelphia

Belyavin A, Wright NA (1987) Changes in electrical activity of the brain with vigilance. Electroencephalogr Clin Neurophysiol 66(2):137–144

Bergum BO, Lehr DJ (1963) End spurt in vigilance. J Exp Psychol 66(4):383–385

Davies D, Parasuraman R (1982) The psychology of vigilance. Academic, London

De Joux NR, Wilson K, Russell PN, Helton WS (2015) The effects of a transition between local and global processing on vigilance performance. J Clin Exp Neuropsychol 37(8):888–898

Gill GW (1996) Vigilance in cytoscreening: looking without seeing. Adv Med Lab Prof 8:14–15

Grajales L, Nicolaescu I (2006) Wearable multisensory heart rate monitor. In: Proceedings of the 3rd international workshop on wearable and implantable body sensor networks, BSN 2006. IEEE Computer Society, Washington, DC, pp 157–161

Grier RA, Warm JS, Dember WN, Matthews G, Galinsky TL, Szalma JL, Parasuraman R (2003) The vigilance decrement reflects limitations in effortful attention, not mindlessness. Hum Factors 45(3):349–359

Gunn DV, Warm JS, Nelson WT, Bolia RS, Schumsky DA, Corcoran KJ (2005) Target acquisition with UAVs: vigilance displays and advanced cuing interfaces. Hum Factors 47 (3):488–497

Hancock PA (2014) Finding vigilance through complex explanations for complex phenomena. Am Psychol 69(1):86–88

Hancock PA (2013) In search of vigilance: the problem of iatrogenically created psychological phenomena. Am Psychol 68(2):97–109

Harris DH (2002) How to really improve airport security. Ergon Des 10:17–22

Hart SG, Staveland LE (1988) Development of NASA-TLX (Task Load Index): results of empirical and theoretical research. In: Hancock PA, Meshkati N (eds) Human mental workload. Elsevier, Amsterdam, pp 139–183

Helton WS, Hollander TD, Warm JS, Tripp LD, Parsons K, Matthews G, Dember WN, Parasuraman R, Hancock PA (2007) The abbreviated vigilance task and cerebral hemodynamics. J Clin Exp Neuropsychol 29(5):545–552

Hitchcock EM, Dember WN, Warm JS, Maroney BW, See J (1999) Effects of cueing and knowledge of results on workload and boredom in sustained attention. Hum Factors 41:365–372

Hitchcock EM, Warm JS, Matthews G, Dember WN, Shear PK, Tripp LD, Mayleben DW, Parasuraman R (2003) Automation cueing modulates cerebral blood flow and vigilance in a simulated air traffic control task. Theor Issues Ergon Sci 4:89–112

Jerison HJ, Pickett RM (1964) Vigilance: the importance of the elicited observing rate. Science 143(3609):970–971

Jerison HJ (1963) On the decrement function in human vigilance. In: Buckner DN, McGrath JJ (eds) Vigilance: a symposium. McGraw-Hill, New York, pp 199–216

Jevon P (2010) An introduction to electrocardiogram monitoring. Nurs Crit Care 15(1):34–38

Johnson RF, Merullo DJ (1996) Effects of caffeine and gender on vigilance and marksmanship. In: Proceedings of the human factors and ergonomics society annual meeting, vol 40, no 23. SAGE Publications, pp 1217–1221

Kamzanova AT, Kustubayeva AM, Matthews G (2014) Use of EEG workload indices for diagnostic monitoring of vigilance decrement. Hum Factors 56(6):1136–1149

Mackworth N (1948) The breakdown of vigilance during prolonged visual search. Q J Exp Psychol 1(1):6–21

Matthews G, Campbell SE, Falconer S, Joyner LA, Huggins J, Gilliland K, Grier R, Warm JS (2002) Fundamental dimensions of subjective state in performance settings: task engagement, distress, and worry. Emotion 2(4):315

Matthews G, Joyner L, Gilliland K, Campbell SE, Huggins J, Falconer S (1999) Validation of a comprehensive stress state questionnaire: towards a state 'Big Three'? In: Mervielde I, Deary IJ, De Fruyt F, Ostendorf F (eds) Personality psychology in Europe, vol 7. Tilburg University Press, Tilburg

Matthews G, Szalma JL, Panganiban AR, Neubauer C, Warm JS (2013) Profiling task stress with the Dundee Stress State Questionnaire. In: Cavalcanti I, Azevedo S (eds) Psychology of stress: new research. Nova, Hauppauge, pp 49–90

McCormack PD (1959) Performance in a vigilance task with and without knowledge of results. Can J Psychol/Revue canadienne de psychologie 13(2):68–71

Metzger U, Parasuraman R (2001) The role of the air traffic controller in future air traffic management: an empirical study of active control versus passive monitoring. Hum Factors 43:519–528

Nuechterlein KH, Parasuraman R, Jiang Q (1983) Visual sustained attention: image degradation produces rapid sensitivity decrement over time. Science 220:327–329

O'Hanlon JF (1965) Adrenaline and noradrenaline: relation to performance in a visual vigilance task. Science 150(3695):507–509

Parasuraman R, Mouloua M (1987) Interaction of signal discriminability and task type in vigilance decrement. Percept Psychophys 41(1):17–22

Parasuraman R (1979) Memory load and event rate control sensitivity decrements in sustained attention. Science 205:924–927

Pattyn N, Neyt X, Henderickx D, Soetens E (2008) Psychophysiological investigation of 164 vigilance decrement: boredom or cognitive fatigue? Physiol Behav 93:369–378

Raymond J, Varney C, Parkinson LA, Gruzelier JH (2005) The alpha/theta neurofeedback on personality and mood. Cogn Brain Res 23(2–3):287–292

Szalma JL, Daly TN, Teo GL, Hancock GM, Hancock PA (2018) Training for vigilance on the move: a video game-based paradigm for sustained attention. Ergonomics 61(4):482–505

Szalma JL, Schmidt TN, Teo GWL, Hancock PA (2014) Vigilance on the move: video game-based measurement of sustained attention. Ergonomics 57(9):1315–1336

Szalma JL, Warm JS, Matthews G, Dember WN, Weiler EM, Meier A, Eggemeier FT (2004) Effects of sensory modality and task duration on performance, workload, and stress in sustained attention. Hum Factors 46(2):219–233

Szalma JL, Hancock PA, Dember WN, Warm JS (2006a) Training for vigilance: the effect of KR format and dispositional optimism and pessimism on performance and stress. Br J Psychol 97:115–135

Szalma JL, Hancock PA, Warm JS, Dember WN, Parsons KS (2006b) Training for vigilance: using predictive power to evaluate feedback effectiveness. Hum Factors 48:682–692

Teichner W (1974) The detection of a simple visual signal as a function of time on watch. Hum Factors 16:339–353

Temple JG, Warm JS, Dember WN, Jones KS, LaGrange CM, Matthews G (2000) The effects of signal salience and caffeine on performance, workload, and stress in an abbreviated vigilance task. Hum Factors J Hum Factors Ergon Soc 42(2):183–194

Valentino DA, Arruda JE, Gold SM (1993) Comparison of QEEG and response accuracy in good vs poorer performers during a vigilance task. Int J Psychophysiol 15(2):123–133

Warm JS, Parasuraman R (2007) Cerebral hemovelocity and vigilance. In: Neuroergonomics: the brain at work. MIT Press, Cambridge

Warm JS, Matthews G, Parasuraman R (2009a) Cerebral hemodynamics and vigilance performance. Mil Psychol 21:75–100

Warm JS, Matthews G, Parasuraman R (2009b) Cerebral hemodynamics and vigilance performance. Mil Psychol 21(S1):S75

Warm J (1984) Sustained attention in human performance. Wiley, Chicester

Warm JS, Matthews G, Finomore VS (2008) Workload, stress, and vigilance. In: Hancock PA, Szalma JL (eds) Performance under stress. Ashgate, Brookfield, pp 115–141

Wiener EL, Attwood DA (1968) Training for vigilance: combined cueing and knowledge of results. J Appl Psychol 52(6):474–479

Maximum Acceptable Work Time for the Upper Limbs Tasks and Lower Limbs Tasks. Workload Limits

Juan Carlos Velásquez V[1]([⊠]) [ID], Leonardo Briceño A[2] [ID],
Diana Marcela Velasquez B[3] [ID], and Silvio Juan Viña B[4]

[1] Universidad del Valle, Cali, Colombia
`juan.carlos.velasquez@correounivalle.edu.co`
[2] Universidad del Rosario, Bogotá, Colombia
`leonardo.briceno@urosario.edu`
[3] Universidad Libre, Cali, Colombia
`diana.counsel88@gmail.com`
[4] Instituto Superior Politécnico José Antonio Echeverria, CUJAE, Havana, Cuba
`silviovi@ind.cujae.edu.cu`

Abstract. Introduction: The physical workload is a major occupational risk factor for workers. Currently the used methods to assess physical dynamic workload evaluate working with the whole body and do not discriminate the load of the body segments. Objective: Determine the maximum acceptable dynamic work time when the work is involves the whole body, the upper limbs and the lower limbs. Methods: Oxygen consumption measurement by ergospirometry and heart rate were monitored in 30 workers exposed to various loads executed with the whole body, legs and upper limbs. Anaerobic threshold was determined by respiratory quotient. This was used to calculate the acceptable dynamic work time. Results: Statistically significant differences were found between acceptable dynamic work time for upper limbs and lower limbs. Negative exponential correlation was found between the workload time, oxygen consumption and heart rate, so we found that R > 0.9 in all cases. We propose six regression equations to determine the acceptable dynamic work time. Conclusions: The acceptable dynamic work time for lower limbs and whole body is similar. The acceptable dynamic work time for upper limbs was significantly lower than acceptable dynamic whole body work time. The relative heart rate seems to be the best indicator to measure acceptable dynamic work time.

Keywords: Oxygen uptake · Workload · Acceptable work time
Occupational health

1 Introduction

Physical work load can be measured considering oxygen consumption (VO_2) or heart rate (HR) [1, 2]. This one is a very accessible variable for monitoring and is frequently used to determine the energy consumption in different workplaces. That's why it is a

© Springer Nature Switzerland AG 2019
S. Bagnara et al. (Eds.): IEA 2018, AISC 827, pp. 282–290, 2019.
https://doi.org/10.1007/978-3-319-96059-3_31

broadly used method for the measure of physical activity at work, even if it is influenced by weather and mental stress and other factors that can introduce bias [3, 4].

The relationship between HR and VO_2 has been deeply studied [5]. It has been demonstrated the direct relationship between both of them [1, 5–10]. The relative cardiac cost index, (RCC index) lead to a more adjusted estimation of the response to a work load [9, 11–13].

The relative cardiac index cost is defined as:

$$RCC\ index = \frac{HRwork - HRrest}{HRmax - HRrest} \times 100 \tag{1}$$

The balance between the physical workload and the respiratory and cardiovascular capacity should be expected in a properly designed work [14]. This concept can be used to establish the maximum acceptable time in a dynamic job (TMAT), this is, the maximum time to which an individual must be exposed to a workload without presenting fatigue. The limit of accumulated energy expenditure (LAEE) is another indicator used to calculate the maximum acceptable working time. This one requires knowing the maximum oxygen consumption or the physical capacity of work (PCW) of the individual to perform the calculation. The limit of accumulated energy expenditure allows providing elements for the design of the work system and the regime of breaks [15].

The LAEE for works done with the whole body is defined as:

$$LAEE = PWC \times \{1, 1 \times (0.33 \times \log T)\}T \tag{2}$$

Time T: in minutes

When the total energy expenditure of work surpasses the limit of accumulated energy expenditure an energy expenditure barrier is established, from this point the risk of fatigue increases. If the energy expenditure of work is not corrected the fatigue will increase exponentially.

Studies carried out in Colombia populations demonstrate the usefulness of this method in the characterization and analysis of the physical workload and in the design-redesign of the jobs, especially the work-rest system [16–18]. However, these studies have been conducted observing the global physical load and not by body segments, an aspect that if not taken into account could underestimate the risk derived from exposure to physical load. This article presents the correction made for the calculation of the maximum acceptable working time and the accumulated energy expenditure limit when the work is done with upper limbs or with lower limbs.

2 Method

Thirty young and healthy workers participated in this study (17 women, 13 men).

A Cosmed K4B2 ergospyrometer was used for the assessment of oxygen consumption, CO_2 production, basal metabolic rate, respiratory frequency, heart rate and respiratory coefficient [19]. Before each evaluation a calibration of the equipment with gases mixture was done for the analysis of O_2 and CO_2.

For the evaluation of upper limbs a Monark handgrip cycloergometer was used and for lower limbs assessment we used a Monark pedal cycloergometer. Oxygen consumption and heart rate test.

Basal heart rate and oxygen consumption were measured in a noiseless room (<60 DbA) with 22 °C ($\pm1°$) air average temperature. The time of the day, elapsed time since the last food intake and the physical activity performed 24 h before the test were registered.

The criteria for the evaluation of the maximum consumption of oxygen were that each individual reached at least 90% of the theoretical maximum heart rate (220- age), a respiratory coefficient higher that 1,1 and presence of plateau at the peak of the slope with no apparent growth.

For lower limbs and total body assessment the tests were performed administering 50 W in two minutes duration loads [20]. The first load had a 60 rpm rhythm.

For upper limbs, the first load was 25 W with growing increases of 25 W two minutes duration and 60 rpm rhythm.

Maximum acceptable time assessment

(1) The load at which the maximum VO2 was obtained was taken in the protocol for upper limbs and lower limbs.
(2) The calculation of the 50%, 40%, 30% and 20% of the maximum load obtained for the test with lower limbs and upper limbs was performed.
(3) Each individual was evaluated during four consecutive days for upper limbs and for lower limbs with continuous workloads until reaching a risk level. This level was set considering the respiratory quotient and/ or the decision to suspend the test due to fatigue in the individual evaluated. Individuals had a break for two days between upper and lower limbs assessment.

Statistical analysis

Fort the analysis the SPSS predictive analytics software V 13.0 was used. Student's t-test and Shapiro-Wilk tests were performed. Heteroscedasticity was evaluated.

A statistical significance level at $\alpha = 0.05$ was established. A p value less than 0.05 was considered statistically significant difference.

3 Results

The average age of the population was 24 ± 3.7 years, ranging between 22 and 37 years, no statistical difference between the ages of men and women were found.

The mean body weight of the participants was 65.8 ± 7.7 kg. The gender difference was not statistically significant.

The maximum oxygen consumption was between 20.5–41.2 ml/ kg-min with a mean of 30.4 (±5.5 ml/ kg-min).

The maximum oxygen consumption in the upper limb test was between 13.6–27.9 ml/ kg-min with an average 21.03 (±4.2 ml/ kg-min) and the maximum oxygen consumption in the lower limb test was between 218.9–36.3 ml/ kg-min with an average of 26.3 (±4.6 ml/ kg-min).

We found statistically significant differences between the oxygen consumption of upper limbs and lower limbs tests (p = 0.000).

The relationship between the maximum acceptable work time and the relative cardiac cost index in tasks were evaluated with upper limbs.

The acceptable working times oscillate according to the administered charge. When the workload increases, the maximum acceptable time decreases exponentially.

The equations that explain this behavior showed a high correlation (r) between the variables. A significant difference between men and women was observed, both genders showed a similar pattern in the form of the correlation curve.

Greater engagement in heart rate was evident when the work was done with upper limbs compared to work done with legs and entire body (Table 1).

Table 1. Relationship between the Maximum acceptable work time and the relative cardiac cost index. (RCC index) in work with upper limbs.

MAWT min	RCC index			
	50%	40%	30%	20%
MAWT global	19	32	55	96
MAWT Hombres	30	53	93	164
MAWT Mujeres	18	31	53	93

The models obtained from the relationship between MAWT and RCC index with upper limbs were exponential with a high negative correlation and are explained by the following formulas:

Relative cardiac cost index

$$MAWT\ mmss = 278.4e^{-0.055ICCR} \qquad r = 0.92 \qquad (3)$$

$$MAWT\ hombres\ mmss = 501.2e^{-0.056ICCR} \qquad r = 0.96 \qquad (4)$$

$$MAWT\ mujeres\ mmss = 287.1e^{-0.055ICCR} \qquad r = 0.93 \qquad (5)$$

Limits for physical workload in relation to the exposure time to the load and the relative cardiac cost index were estimated Table 2.

Relationship between maximum acceptable work time and heart rate relative cost in tasks performed with legs.

The relationship between MAWT and index of cardiac relative cost was obtained by regression analysis. An exponential decrease in MAWT with increasing workload was observed.

In all cases when crossing the variable maximum acceptable work time (MAWT) with the relative cardiac cost index (RCC index) an exponential model with a negative

Table 2. Load limits of physical work suggested for the study population by exposure time based on the upper limbs.

Work time (hours)	RCC index
4	–
2	15
1	30
0.5	40
0.3	50

trend and a high determination coefficient was obtained with a statistically significant difference (p = 0.000).

It was observed also a decrease in the maximum acceptable work time as the RCC index increased as occurred with upper limbs Table 3.

Table 3. Relationship between MAWT and relative cardiac cost index. Lower limbs work.

MAWT en minutos	RCC index			
	50%	40%	30%	20%
MAWT global	100	1850	330	600
MAWT Hombres	150	250	430	730
MAWT Mujeres	90	170	320	615

The following formulas explain the prediction models of acceptable maximum working time for lower limbs.

MAWT and relative cardiac cost index.

$$MAWT\ mmii = 1980.8e^{-5.921ICCR} \qquad r = 0.93 \qquad (6)$$

$$MAWT\ hombres\ mmii = 2116.2e^{-5.34ICCR} \qquad r = 0.98 \qquad (7)$$

$$MAWT\ mujeres\ mmii = 2253.1e^{-6.484ICCR} \qquad r = 0.95 \qquad (8)$$

The above findings permit calculate the acceptable maximum working time in hours depending on the RCC. Table 4.

Table 4. Limits for physical workload suggested in the study population for work with lower limbs in relation to the exposure time.

Time of work (hours)	RCC index
12	10
8	24
4	40

Based on the proportion of oxygen consumption obtained for upper limbs and lower limbs in relation to the oxygen consumption obtained with the whole body, corrections were made to the limit of accumulated energy expenditure (LAEE) for limbs and lower limbs.

$$LAEE\ uper\ limbs = PWC \times 0.6 \times \{1.1 - (0.33 \times \log T)\}T \qquad (9)$$

$$LAEE\ legs\ limbs = PWC \times 0.8 \times \{1.1 - (0.33 \times \log T)\}T \qquad (10)$$

Time T: in minutes

4   Discussion

Statistically significant differences between men and women in the RCC index and MAWT were found. Similar results were documented by Wu et al. [9].

The maximum acceptable work times obtained in cycle ergometer for lower limbs and whole body load had similar behavior.

The maximum acceptable working time was much lower when work was performed with upper limbs. The findings that correlate the MAWT with RCC index, were similar to those proposed by Wu et al. [9] and Rodgers et al. [14] Table 5.

Table 5. Suggested limits of workload for working times of 12, 8 and 4 h. Whole body – lower limbs.

Working time - hours	Colombian population		Taiwanese population		European population
	Average age 24		Average age 26		Average age 26
	% VO$_2$ máx.	RCC index	% VO$_2$ máx.	RCC index	% VO$_2$ máx.
12	28,6	10,4	28.5	16	28
8	33	24,6	34	24.5	33
4	46.8	40,4	43.5	39	45

Load limits expressed in terms of VO2 max. and RCC index in the Colombian population were similar those in the study of Taiwanese and European population [9, 13].

The results showed that the MAWT was significantly correlated with % VO2max and RCC index. Results showed high correlation in each of the variables, with a correlation coefficient over 73%, especially when the RCC index was over 90%. The maximum acceptable work time decreased exponentially in all variables, including gender.

In the review there were no studies that showed the relationship between % VO2max, RCC index and MAWT in upper limbs.

This study shows that the maximum acceptable time job for work performed with upper limbs is lower than MAWT obtained for loads in lower limbs, as shown in Table 6 and Graphs 1 and 2.

Table 6. Relationship between maximum acceptable time and RCC index in upper and lower limbs.

	RCC index		
	50	40	30
MAWT Upper limbs, hours	0,25	0,5	1
MAWT Lower limbs, hours	2	4	6

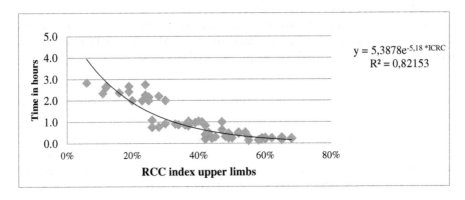

Graph 1. Relationship between maximum acceptable time and RCC index. Upper limbs.

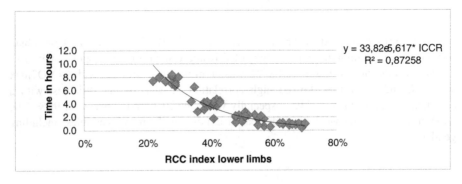

Graph 2. Relationship between maximum acceptable time and RCC index. Lower limbs.

5 Conclusions

Common features of all three models can be summarized:

(1) To the extent that physical workload increases a continuous decrease in MAWT is observed.
(2) When there is an increase in the physical workload, the MAWT decreases rapidly and exponentially, aspect to consider in the design of work.
(3) MAWT approached zero as the physical workload got heavier and the RCC index is equal to or greater than 75%.
(4) The MAWT in the upper limbs test is significantly lower than for similar workloads in the lower limbs test.
(5) The relative heart rate (RCC index) seems to be a very good estimator of MAWT.

The relative cardiac cost index seems to be the best indicator to assess the relationship between physical work capacity, workload and maximum acceptable working time with upper and lower limbs time due to its low cost, simplicity and sensitivity for measurement.

References

1. Astrand PO, Saltin B (1961) Maximal oxygen uptake and heart rate in various types of muscular activity. J Appl Physiol 16(6):977–981
2. Becker TJ, Morrill JM, Stamper EE (2008) Applications of work physiology science to capacity test prediction of full-time work eight hour work day. Rehab Profess 15(4):45–56
3. Malchaire J, Kampmann B, Havenith G, Mehnert P, Gebhardt HJ (2000) Criteria for estimating acceptable exposure times in hot working environments: a review. Int Arch Occup Environ Health 73:215–220
4. Vallejo NG, Lopez GD, De Paz Fernandez JA (2005) Diferentes modelos de regresión para describir la relación entre VO_2-FC para estimar el VO_2 a diferente intensidad de esfuerzo. Cultura ciencia y deporte 3:131–135
5. Hart K (1998) Work standard assessment using: heart rate monitoring. IIE Sol 30(9):36–43
6. Lehmann G (1960) Fisiología Practica del Trabajo, Aguilar, Madrid
7. Swain DP, Leutholtz BC (1997) Heart rate reserve is equivalent to % VO_2Reserve, not to% VO_2max. Med Sci Sports Exer 29(3):410–414
8. Smolander J, Aminoff T, Korhonen I, Tervo M, Shen N, Korhonen O, Louhevaara V (1998) Heart rate and blood pressure responses to isometric exercise in young and older men. Eur J Appl Physiol Occup Physiol 77(5):439–444
9. Wu H-C, Wang MJJ (2002) Relationship between maximum acceptable work time and physical workload. Ergonomics 45(4):280–289
10. Asmussen E, Hemmingsen I (1958) Determination of maximum working capacity at different ages in work with the legs or with the arms. Scand J Clin Lab Inves 10(1):67–71
11. Saha PN, Datta SR, Banerjee PK, Narayane GG (1979) An acceptable workload for Indian worker. Ergonomics 22(9):1059–1071
12. Je Graves, Ml Pollock, Jf Carroll (1994) Exercise, age, and skeletal muscle function. South Med J 87(5):17–22
13. Rodgers SH, Kenworth DA, Eggleton EM (1986) Ergonomic design for people at work. Van Nostrand Reinhold publisher, New York, p 2

14. Aminoff T, Smolander J, Korhonen O, Louhevaara V (1998) Prediction of acceptable physical workloads based on Responses to prolonged arm and leg exercise. Ergonomics 41:109–120

15. Viña SY, Gregori E (1987) Ergonomía. C y E, La Habana

16. Velásquez JC, Gutiérrez RC, Ospina N (2011) Cornejo. R. Natalia. Carga Física en Trabajadores de un cultivo florícola de Suesca, Cundinamarca. Momentos de Ciencia 8 (1):64–72

17. Velásquez JC, Guzman N (2013) Efecto de una intervención tecnológica sobre la carga física durante el proceso de coquización en una empresa de Colombia. Momentos de Ciencia 10 (2):117–123

18. Helena AL, Alvaro I (2005) Carga Física y tiempo máximo aceptable de trabajo en trabajadores de un supermercado en Cali Colombia. Rev Salud pública 2:145–156

19. Hausswirth C, Bigard AX, Le Chevalier JM (1997) The cosmed K4 telemetry system as an accurate device for oxygen uptake measurements during exercise. J Sports Med 18(6):449–453

20. Maidorn K, Mellerowics H Der (1962) Sportarzt 11:355

21. Fc Hagerman, Ra Lawrence, Mc Mansfield (1988) A comparison of energy expenditure during rowing and cycling ergometry. Med Sci Sports Exer 20(5):479–488

22. Hoffman MD, Kassay KM, Zeni AI, Clifford PS (1996) Does the amount of exercising muscle alter the aerobic demand of dynamic exercise? Eur J Appl Physiol 74:541–547

Mediating Role of Loneliness and Organizational Conflict Between Work Overload and Turnover Intention

Serpil Aytac[1(✉)] and Oguz Basol[2]

[1] Uludag University, 16056 Bursa, Turkey
saytac@uludag.edu.tr
[2] Kırklareli University, 39100 Kırklareli, Turkey
oguzbasol@klu.edu.tr

Abstract. The fact that organizations offer employees a positive organizational climate not only increases employees' happiness levels but also raises the level of creativity of employees. When it is thought that the only way for organizations to survive in a competitive environment is innovative and creative applications, the positive dynamics within the organization once again emerges. This research focuses on negative behaviors such as work overload, loneliness, organizational conflict and turnover intention which affect employees' creativity and it aims to discover the mediating role of loneliness and organizational conflict between work overload and turnover intention. In the present research, a survey was conducted including 145 service sector employees. To analyze the demographic characteristics of the participants, internal consistency and correlations of scales, SPSS 22.0; and to test the mediating role, SmartPLS 2.0 were used. According to the analysis results; it is determined that between work overload and turnover intention, loneliness and organizational conflict have a mediating role. When all the results are evaluated together, work overload in the organization forces employees to extreme behavior and it is found that employees feel loneliness or conflict, and as a result, their turnover intention increases. Related to this, it is possible to say that the overloaded employees get further away from creativity.

Keywords: Work overload · Loneliness · Conflict · Turnover intention

1 Introduction

The fact that the most important source of organizations is human value has turned into an undeniable truth. In this regard, organizations aim to affect the attitude of their employees positively and improve the organizational climate in order to boost their creativity and thus, turn into an organization with a highly competitive power at global scale.

The present study focuses on the dynamics that affect organizational creativity negatively. Thus, it takes an inverse perspective on the matter. Whereas a lot of studies in the literature focus on the question of "What practices are useful for increasing creativity?", the present study focuses on "How are the relationships between the

© Springer Nature Switzerland AG 2019
S. Bagnara et al. (Eds.): IEA 2018, AISC 827, pp. 291–301, 2019.
https://doi.org/10.1007/978-3-319-96059-3_32

dynamics that affect creativity negatively?". Within this context, the study handles the relationships among work overload, loneliness at work, interpersonal conflict and turnover intention and questions the mediating role of loneliness at work and interpersonal conflict in the relation between work overload and turnover intention.

2 Literature

The workload is defined as the amount of work given to an employee to do [1]. The load or the amount of the work that is performed is a subjective matter that changes considerably according to the perception of an individual. In every organization, there are regulations concerning the work to be done (tasks) within the working hours and organizations schedule the tasks considering other organizations, the areas they compete in and the capacities of their employees. Thus, employees are given workloads that they can fulfill at the optimum level. The situation in which they are responsible for a workload which is even higher than their capacities is called "work overload" [2]. When the workload is excessively higher or lower, this situation affects the job satisfaction of employees negatively [3]. Previous studies have shown that work overload isolates the employees [4–6], leads to conflicts [7, 8] and increases employees' turnover intention [9].

Loneliness in business life comes out when employees experience the loneliness that they suffer from in their daily lives at work. Wright (2005) states that loneliness at work occurs due to the effect of various environmental (culture, family support, social support etc.), organizational (workload, control area, organizational communication, organizational climate, managerial support, coworker support etc.) and personal (personality traits, shyness, self-confidence, pessimism and self-concentration etc.) factors on individuals [10]. In business life, loneliness at work occurs as a result of the exclusion of an individual out of an organizational social environment he/she is in and shows itself in the form of a problem arising from some deficiencies of good interpersonal skills [11, 12]. Studies have indicated that the loneliness of employees at work affects their well-being, job satisfaction decreases their productivity and increases their dissatisfaction and turnover intention [13–15].

As for interpersonal conflict, it is described as the "the conflict of aims, behavior, value judgements within groups or among the individuals in an organization" in general terms [16]. Thus, it is considered as a situation that comes out as a result of the disparity of aims and expectations among individuals or groups in an organization [17]. It represents negative situations that threaten the psychological well-being of employees [18].

Interpersonal conflict in organizations might harm mental and physical health; lead to a hostility among employees; cause a waste of time, money and energy as well as reducing the trust in the organization [19]. For this reason, the determination of interpersonal conflict cases taking place in organizations can both help to encourage a positive environment in the organization and ensure that the employees work with high motivation. Furthermore, it should be remembered that organization culture can be affected by interpersonal conflicts negatively [20]. The findings also indicate that interpersonal conflicts influence employees' turnover intention [21].

The concept of "turnover intention" is defined as an employee's giving up his/her job on his/her own accord and willingly [21]. According to another definition, it is described as "a negative behaviour demonstrated by employees when they are not satisfied with the working conditions" [22]. The turnover intention can be considered as the most severe reaction that an employee can give when his/her expectations from an organization are not satisfied. It is considered that the turnover intention depends on organizational factors (status, organization culture, and environment, the lack of organizational support etc.), work-related factors (work overload, stress, autonomy, income and shift work) and personal factors (age, experience, and sector) [21–25].

Based on the aforementioned studies, the following hypotheses were suggested:

H1: There is a positive relationship between work overload and turnover intention.
H2: There is a positive relationship between work overload and loneliness at work.
H3: There is a positive relationship between work overload and interpersonal conflict.
H4: There is a positive relationship between loneliness at work and turnover intention.
H5: There is a positive relationship between interpersonal conflict and turnover intention.

In the literature, it is possible to find opinions on the relationship between work overload and turnover intention. In the present study, some expectations are indicated regarding the mediating role of loneliness and interpersonal conflict, in addition to these relationships and the following hypotheses are suggested:

H6: Loneliness plays a mediating role between work overload and turnover intention.
H7: Interpersonal conflict plays a mediating role between work overload and turnover intention.

3 Material and Methods

The study was carried out with 145 public sector employees. Through random sampling, survey forms were distributed to 200 employees in closed envelopes and were collected within 2 days. In total, 152 survey forms were collected. As 7 out of these 152 forms included wrong or missing data, they were eliminated from the study; thus 145 survey forms were evaluated in total.

3.1 Instruments

The first part consisting of 5 questions was designed to find out the demographic characteristics (e.g. gender, marital status etc.) of the participants.

Also, *"Work Overload Scale"* consisting of 5 items developed by Spector and Jex [26] and adapted to Turkish by Keser et al. [27] was used in the study (An example of the items: To be asked to work too much). The average is taken into account to evaluate the scale. The higher the evaluation score got, the higher became the level of the workload.

Besides, *"Loneliness at Work Scale"* developed by Wright et al. [11] and adapted to Turkish by Doğan et al. [28] was used in the study. The scale consists of 16 items. 9 of them gather under the sub-factor of "emotional deprivation" while 7 are under "social friendship". For a general evaluation of the scale, the items relevant to the social friendship sub-factor were reverse-coded and the average of all items was calculated. The higher the evaluation score got, the higher became the level of loneliness felt by employees at the workplace (An example of the items: I feel that my colleagues stay away from me most of the time).

In addition, the *"Interpersonal Conflict Scale"* consisting of 4 items developed by Spector and Jex [26] was used in the study. The average was taken into account for the evaluation of the scale. The higher the evaluation score got, the higher became the interpersonal conflict level (An example of the items: People shout at you).

Lastly, the *"Turnover Intention Scale"* developed by Cook et al. [29] and adapted to Turkish by Teoman [30] was used in the study. In total, there were 4 items in the scale, 1 of them being reverse. For a general evaluation of the scale, the items were reverse-coded and the average of all items was calculated. The higher the evaluation score got, the higher became the level of turnover intention (An example of the items: Do you ever think of continuing your career at another organization, in the future?).

In order to find out the demographic characteristics of the participants and the internal consistency measures of the scales and to test the first 5 hypotheses concerning the relationships, SPSS 22.0 was used. Also, SmartPLS 2.0 was conducted to test H_6 and H_7 hypotheses involving the mediating effect test.

4   Results

This section assessed the survey results. The following Table 1 gives a summary of all demographic variables.

On Table 1, the demographic characteristics of the participants are given. The study was carried out with the participation of 145 public sector employees. 20 of the participants (13.8%) were single while 125 (86.2%) were married. A considerable share of the participants were officers (83 people – 85.5%) whereas technical staff (14 people, 9.7%), auxiliary staff (3 people, 2.1%), managers (3 people, 2.1%) and an expert (1 person, 0.7%) also took part in the study. A high share of the participants had a high-school education or equivalent (83 people, 57.2%) while there were also other employees with a higher education or university degree (51 people, 35.2%). On the other hand, few numbers of people had only primary school education (9 people, 6.2%) and lastly, 2 participants (1.4%) had received a post-graduate education. The age of the participants ranged from 24 to 58 and the average age was 41. The participants had an experience of 1 to 35 years in their career while the average year of experience was 13 in the institution.

Table 1. Demographic variables.

Variables	N	Percent (%)
Marital status		
Single	20	13.8
Married	125	86.2
Position		
Manager	3	2.1
Expert	1	0.7
Officer	124	85.5
Technical staff	14	9.7
Auxiliary staff	3	2.1
Educational Status		
Primary school	9	6.2
High school	83	57.2
University	51	35.2
Master degree	2	1.4
Total	**145**	**100**

On Table 2, information is provided about the scales used in the study. Accordingly, it was determined that the work overload, loneliness at work, interpersonal conflict and turnover intention scales had internal consistency (Cronbach's Alpha > 0,700). According to the results of the Kolmogorov-Smirnov test, it was found out that work overload, loneliness at work, interpersonal conflict and turnover intention scales were not normally distributed. Thus, non-parametric analysis methods were used in the analyses involving the aforementioned variables.

Table 2. Cronbach's Alpha reliability coefficients and Kolmogorov-Smirnov test results of measurement for each scale

Scales	Item no.	Cronbach's Alpha	Kolmogorov-Smirnov test	
			Statistics	p
Work overload	5	0,888	0,116	0,00
Loneliness at work	16	0,872	0,087	0,00
Interpersonal conflict	4	0,866	0,224	0,00
Turnover intention	4	0,792	0,160	0,00

On Table 3, the values relevant to the correlation among the variables are shown. These values involve the results of the first 5 hypotheses. In this regard, a positive relationship was found between work overload and turnover intention (r: 0.208; $p < 0.05$); between work overload and loneliness at work (r: 0.324; $p < 0.01$) and between work overload and interpersonal conflict (r: 0.535; $p < 0.01$). In the light of this, H1, H2, H3 hypotheses were accepted. On the other hand, there was a positive relationship between loneliness at work and turnover intention (r: 0.314; $p < 0.01$); and

Table 3. Correlation analysis between variables

Scales	Mean	SD	1	2	3
1. Work overload	2.31	1.1	–		
2. Loneliness at work	2.2	0.83	**.324****	–	
3. Interpersonal conflict	1.81	1.06	**.535****	**.523****	–
4. Turnover intention	2.43	1.19	.208*	**.314****	**.297****

*p < 0.05 **p < 0.01
In the analysis of the correlation between the variables,
Spearman's rho value was taken as a basis

between interpersonal conflict and turnover intention (r: 0.297; p < 0.01). Thus, H4 and H5 hypotheses were also accepted.

At this stage of the study, the results of the analyses of the hypotheses (H_6 and H_7) suggested in relation to the mediating effect are given place. For the analyses of the hypotheses regarding the mediating effect, SmartPLS 2.0 was used. In the program, the number of observations was 145 while the bootstrapped samples were 5000.

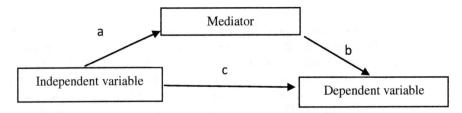

Fig. 1. Mediating effect conditions

Figure 1 shows the conditions in which the mediator effect can take place. Thus, it is necessary that

- the path between the independent variable and the mediator variable is significant;
- the path between the dependent variable and the mediator variable is significant;
- the path between the dependent variable and the independent variable is significant [31, 32].

To talk about the mediating effect, it is necessary that the "c" path between the independent and dependent variable remains insignificant after the mediator variable is included in the model. When the "c" path obtained after the mediator variable is included in the model is considered as "c_1", it is referred to as full mediation effect if "c_1" is completely insignificant whereas it is referred to as half mediating effect if the coefficient decreases while "c_1" remains significant.

Figure 2 shows the coefficient and "t" values with regards to the first model. As a result of the analysis, a significant relationship was found between work overload and turnover intention (0.239; t: 2.780). Accordingly, an increase in work overload leads to an increase of 23.9% in employees' turnover intention. Also, 5.7% of the change occurring in the turnover intention variable is associated with work overload.

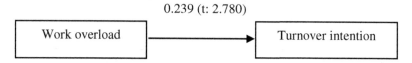

Fig. 2. Coefficient and 't' values with regards to the first model

Figure 3 shows the analysis results with regards to the mediating role of loneliness in the relationship between work overload and turnover intention. According to the results, it was determined that loneliness had a full mediation effect on the relationship between work overload and turnover intention. The work overload and turnover intention relationship that was significant (0.239; t: 2.780) before was found to be insignificant as a result of the analysis (0.146; t: 1.708). This result indicates that loneliness at work variable had a full mediating effect on the relationship between work overload and turnover intention. Additionally, it was determined that there was a significant relationship between work overload and loneliness at work (0.252; t: 2.694) and between loneliness at work and turnover intention (0.371; t: 4.035).

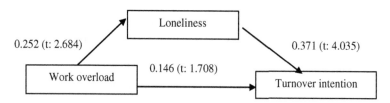

Fig. 3. The coefficients and 't' values with regards to the mediating effect of loneliness at work variable in the relationship between work overload and turnover intention.

In the light of these, it was discovered that an increase occurring in work overload did not increase the turnover intention directly because work overload increased the feeling of loneliness at work in the first place which led to an increase in the turnover intention of an employee feeling lonely at the workplace. Furthermore, the explanation rate of turnover intention increased from 5.7% to 18.6%. Thus, it can be said that the explanatory power of the model is higher in comparison to the first model (Fig. 2).

Figure 4 shows the analysis result with regards to the mediating role of interpersonal conflict in the relationship between work overload and turnover intention. According to the results, it was determined that conflict had a full mediating effect between work overload and turnover intention. Also, the relationship between work overload and turnover intention that was significant before (0.239; t: 2.780) was found to be insignificant as a result of the analysis (0.046; t: 0.395). This result demonstrates that the interpersonal conflict variable plays a full mediating role in the relationship between work overload and turnover intention. Also, significant relationships were found out between work overload and conflict (0,553; t: 6.015) and between conflict and turnover intention (0.349; t: 2.821).

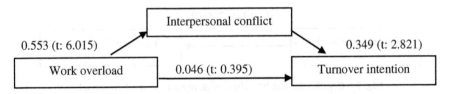

Fig. 4. The coefficients and 't' values with regards to the mediating effect of interpersonal conflict in the relationship between work overload and turnover intention

Accordingly, it was discovered that an increase occurring in work overload did not increase the turnover intention directly because work overload firstly increased the level of conflict at the workplace which led to an increase in the turnover intention of an employee having a conflict with other colleagues. Additionally, the explanation rate of turnover intention increased from 5.7% to 14.2%. Thus, it can be said that the explanatory power of the model is higher in comparison to the first model (Fig. 2).

5 Discussion

Focusing on the relationships among work overload, loneliness at work, interpersonal conflict and turnover intention and the mediating role of loneliness at work and interpersonal conflict in the relationship between work overload and turnover intention, the present study provides data about a group of 145 public sector employees. In the study carried out with the participation of officers, the majority of whom were married; among whom the average age was 41 and the average year of experience was 13, positive and significant relationships were found between work overload and turnover intention; work overload and loneliness at work; work overload and interpersonal conflict; loneliness at work and turnover intention; and between interpersonal conflict and turnover intention.

These results demonstrate how harmful work overload can be for organizational dynamics. Thus, an increase that occurs in work overload increases employees' turnover intention, isolates them and causes conflicts among the employees. Naturally, new organizational dynamics, such as downsizing and use of external resources, increase the workload of employees; however, a permanent increase of the work overload can lead to the isolation of employees, the rise of conflict and even their resignation, as it has been demonstrated.

The results obtained from the study are in parallel to the results of other national and international studies. The findings indicate that work overload isolated employees [4, 5], caused conflicts [7, 8] and increased their turnover intention [9]. It was also shown that interpersonal conflict increased the turnover intention among the employees. These results demonstrate that loneliness and interpersonal conflict have destructive effects on organizational balances. In the light of the results obtained, it can be said that the organizations that cannot control the state of relationships between the employees and the managers nor develop strategies to manage conflicts that arise in the workplace are likely to lose human resources which are considered as the most valuable element.

This result is in parallel to the results of other national and international studies. Thus, the positive relationship between the loneliness of employees and the turnover intention has also been proven in different studies [13, 14]. As for the relationship between interpersonal conflict and turnover intention, it is still investigated in several studies in national and foreign literature [20, 21].

The analysis results concerning the mediating effect provide data about how the potential effects of work overload on the intention to leave are shaped. Thus, it has been concluded that loneliness at work and interpersonal conflict variables played a full mediating role in the relationship between work overload and turnover intention. In the light of this, it was determined that an increase occurring in work overload did not increase the turnover intention directly because work overload increased the feeling of loneliness at work in the first place which led to an increase in the turnover intention of an employee feeling lonely at the workplace. Similarly, it was discovered that an increase occurring in work overload did not increase the turnover intention directly because work overload firstly increased the level of conflict at the workplace which led to an increase in the turnover intention of an employee having a conflict with other colleagues.

Even though previous studies in the literature point to a positive relationship between work overload and turnover intention, the present study has demonstrated that there are other variables that play a mediating role in this relationship. However, the results indicated that work overload was harmful to employees and the organization under any circumstances. This factor isolates the employees or leads to the rise of a conflict among them and thus forces the employee, the presence that contributes to organizational creativity, to leave the job.

In the light of this data, future studies should be designed so as to involve this mediating effect because it can enable a more realistic and multidimensional determination of organizational dynamics. Given that interpersonal conflicts and loneliness increase the turnover intention, the investigation of the mediating effect of different organizational dynamics can be useful to obtain a well-rounded scientific information. The results of this study can be considered of valuable guidance for further research on these issues.

Acknowledgements. The authors thank MBT Translation and Interpreting Services for them assistance in proof-reading and editing of this manuscript.

Funding. The authors received no financial support for the research, authorship, and/or publication of this article.

Declaration of Conflicting Interests. The authors declared no potential conflicts of interest with respect to the research, authorship, and/or publication of this article.

References

1. Qureshi MI, Iftikhar M, Abbas SG, Hassan U, Khan K, Zaman K (2013) Relationship between job stress, workload, environment and employees turnover intentions: what we know, what should we know. World Appl Sci J 23(6):764–770
2. Cam E (2004) Çalışma Yaşamında Stres ve Kamu Kesiminde Kadın Çalışanlar. Uluslararası İnsan Bilimleri Dergisi 1(1):1–10
3. Keser A, Kümbül Güler B (2016) Çalışma Psikolojisi. Umuttepe Yayınları, Kocaeli
4. Coşkuner S, Şener A (2013) Akademisyenlerin İş ve Aile Karakteristiklerinin Evlilik, Aile ve Yaşam Tatmini ile İlişkisi: İş ve Aile Çatışmasının Aracı Rolü. Hacettepe Üniversitesi Sosyolojik Araştırmalar E-Dergisi, 1–25 Kasım 2013
5. Dağdeviren N, Musaoğlu Z, Ömürlü İK, Öztora S (2011) Akademisyenlerde İş Doyumunu Etkileyen Faktörler. Balkan Med J 28(4):69–74
6. Demirbaş B, Haşit D (2016) İş Yerinde Yalnızlık ve İşten Ayrılma Niyetine Etkisi: Akademisyenler Üzerine Bir Uygulama. Anadolu Üniversitesi Sosyal Bilimler Dergisi 16 (1):137–158
7. Radzali FM, Ahmad A, Omar Z (2013) Workload, job stress, family-to-work conflict and deviant workplace behavior. Int J Acad Res Bus Soc Sci 3(12):109–115
8. Suozzo SH (2015) The effect of workload, job satisfaction, and role conflict on the timing of leaving of nursing faculty from their current faculty position. PhD Thesis, Seton Hall University
9. Xiaoming Y, Ma B-J, Chang CL, Shieh C-J (2014) Effects of workload on burnout and turnover intention of medical staff: a study. Stud Ethno-Med 8(3):229–237
10. Wright SL (2005) Loneliness in the Workplace. PhD Thesis in Psychology, University of Canterbury. Christchurch. New Zealand
11. Wright SL, Burt CDB, Strongman KT (2006) Loneliness in the workplace: construct definition and scale development. NZ J Psychol 35(2):59–68
12. Çetin A, Alacalar A (2016) İş Yaşamında Yalnızlığı Yordamada Kişilik Özellikleri ile Algılanan Sosyal ve Örgütsel Desteğin Rolü. Uluslararası Yönetim İktisat ve İşletme Dergisi 12(27):193–216
13. Erdil O, Ertosun ÖG (2011) The relationship between social climate and loneliness in the workplace and effects on employee well-being. Procedia Soc Behav Sci 24:505–525
14. Ertosun ÖG, Erdil O (2012) The effects of loneliness on employees' commitment and intention to leave. Procedia – Soc Behav Sci 41:469–476
15. Chana SH, Qiu HH (2011) Loneliness, job satisfaction, and organizational commitment of migrant workers: empirical evidence from China. Int J Hum Resour Manag 22(5):1109–1127
16. Rahim MA (2002) Toward a theory of managing organizational conflict. Int J Conflict Manag 3(13):206–235
17. Tozkoparan G (2013) Beş Faktör Kişilik Özelliklerinin Çatışma Yönetim Tarzlarına Etkisi: Yöneticiler Üzerine Bir Araştırma. Ekonomik ve Sosyal Araştırmalar Dergisi 9(9):189–231
18. Wickham RE, Williamson RE, Beard CL, Kobayashi CLB, Hirst TW (2016) Authenticity attenuates the negative effects of interpersonal conflict on daily well-being. J Res Pers 60:56–62
19. Leon-Perez JM, Medina FJ, Arenas A, Munduate L (2015) The relationship between interpersonal conflict and workplace bullying. J Manag Psychol 30(3):250–263
20. Karcıoğlu F, Kahya C, Buzkan K (2012) Çatışma Yönetim Stratejisinin Tahmin Edicileri Olarak Örgütsel Kültür Tipleri. Atatürk Üniversitesi İktisadi ve İdari Bilimler Dergisi 26(1):77–91

21. Takase M (2010) A Concept analysis of turnover intention: implications for nursing management. Collegian 17:3–12
22. Onay M, Kılcı S (2011) İş Stresi ve Tükenmişlik Duygusunun İşten Ayrılma Niyeti Üzerine Etkileri: Garsonlar ve Aşçıbaşılar. Organizasyon ve Yönetim Bilimler Dergisi 3(2):363–372
23. Galletta M, Portoghese I, Battistelli A, Leiter MP (2013) The roles of unit leadership and nurse-physician collaboration on nursing turnover intention. J Adv Nurs 69:1771–1784
24. Nelsey L, Brownie S (2012) Effective leadership, teamwork and mentoring essential elements in promoting generational cohesion in the nursing workforce and retaining nurses. Collegian 19:197–202
25. Kelly LA, McHugh MD, Aiken LH (2011) Nurse outcomes in magnet(r) and nonmagnet hospitals. J Nurs Adm 41:428–433
26. Spector PE, Jex SM (1998) Development of four self-report measures of job stressors and strain: interpersonal conflict at work scale, organizational constraint scale, quantitative workload inventory, and physical symptoms inventory. J Occup Health Psychol 3:356–367
27. Keser A, Öngen Bilir B, Aytaç S (2017) Niceliksel İş Yükü Envanterinin Geçerlik ve Güvenirlik Çalışması. İş-Güç Endüstri İlişkileri Dergisi 19(2):55–78
28. Doğan T, Çetin B, Sungur MZ (2009) İş Yaşamında Yalnızlık Ölçeği Türkçe Formunun Geçerlilik ve Güvenilirlik Çalışması. Anadolu Psikiyatri Dergisi 10:271–277
29. Cook JD, Hepworth SJ, Wall TD, Warr PB (1981) The experience of work a compendium and review of 249 measures and their use. Academic Press, London
30. Teoman DD (2007) Performans Değerlendirme Sürecinde Oluşan Adalet Algısı, Bu Algının İç, Dış ve Sosyal Ödüllerle Olan İlişkisinin İşten Ayrılma Niyetine Olan Etkisi, Master Thesis of Psychology, Istanbul University
31. Baron MR, Kenny D (1986) The moderator-mediator variable distinction in social psychological research: conceptual, strategic, and statistical considerations. J Pers Soc Psychol 51(6):1173–1182
32. Bayram N (2010) Yapısal Eşitlik Modellemesine Giriş: AMOS Uygulamaları. Ezgi Kitabevi, Bursa

Multiple Factors Mental Load Evaluation on Smartphone User Interface

Meng Li[1,2(✉)] ⓘ, Armagan Albayrak[1] ⓘ, Yu Zhang[2],
and Daan van Eijk[1]

[1] Delft University of Technology,
Landbergstraat 15, 2628CE Delft, Netherlands
m.li-4@tudelft.nl
[2] Xi'an Jiaotong University, Xianning Road 28, Xi'an 710029, China

Abstract. Smartphone is nowadays the most prevalent computer system, thus a lot of attention from academia and industries has been put to evaluate its quality of use. However, Smartphone has more complex interaction modes and usage scenarios than PC and laptop. And therefore assessing its quality using a conventional usability evaluation is not sufficient. Meanwhile, the mental load serves as an acknowledged index of effort that operators have put in human-machine interaction, especially under high-demanding context. Mental load contains a set of parameters in multiple dimensions, such as primitive task performance, biological measurement(s) and subjective mental load scale, which assesses the efforts of tasks under a particular environment and operating conditions. Thus, it is suitable for evaluating complex mental work, and may indicate the use of Smartphones.

The aim of this paper is to apply a multi-dimensional method to assess the mental load of users, and find out which measurement(s) is the most suitable one to evaluate the efforts for using a smartphone. During this study, the effort on conducting tasks with four difficulty levels were assessed using measurements in three dimensions, which were (1) user performance (task accomplishment and secondary task), (2) subjective rating (NASA-TLX scale) and (3) physiological function (EDA). The values of these measurements were compared across novice, average and skilled users. The results show that: task duration and number of usability error are significantly related with mental load and change with the difficulty level of tasks; in subjective rating, *Mental Demand*, *Effort* and *Frustration* were highly related with mental load.

Keywords: Mental load evaluation · Usability · Smartphone

1 Introduction

1.1 Quality of Use

Usability is an international standard for evaluating quality of use of computer systems, which is widely applied on vertical display terminals (VDTs). The first international standard mentioned usability is ISO/IEC 9126, which described "usability" as an index for assessing software quality from users' perspective, and it should include

© Springer Nature Switzerland AG 2019
S. Bagnara et al. (Eds.): IEA 2018, AISC 827, pp. 302–315, 2019.
https://doi.org/10.1007/978-3-319-96059-3_33

understandability, learnability, operability and attractiveness [1]. The acknowledged definition of "usability" is from ISO 9241, which defines usability as "The effectiveness, efficiency and satisfaction with which specified users achieve specified goals in particular environments" [2]. Though ISO 9241 did not regulate a uniform test method of these usability parameters, but it suggested the number of usability errors and task duration as variables for effectiveness and efficiency of a computer system. Li found that in real environment user's behaviours often not conform to the action phases of Robicon model from motivational psychology. Thus, he suggested a compound user model [3], which is composed of cognitive and action errors that users make when performing a task, to evaluate the quality of use. However, the prevalence of Smartphone gave computational terminals more mobility and flexibilities under variable scenarios, while the conventional user testing mainly focus on static and single task setting, thus makes the usability evaluation of them more difficult.

According to ISO 9241-11, the cognitive demand in human-computer dialogue, also known as Mental Load, influences the usability, an therefore suggests a mental load evaluation method [2]. However, mental load measurement is not widely adopted on user interface evaluation. It is probably because of the complexity of mental workload measurement [4]. Mental load (ML) originates from the 1960s to evaluate complicated operation system of aircraft in a high-speed environment. In the 1980s, an integrated system of methodology on mental load measurement started establishing [5]. ML measured the efforts of operators when they execute tasks in specific environments and operational affordances. The researches on ML mainly focus on operation of aviation tasks and vehicle driving, in order to assess usability of man-machine interaction in these systems [6, 7]. Since the Smartphone currently integrated more and more multi-task functions, and often used under dynamic environment, the mental effort of the usage of a Smartphone could be comparable with the dual-task diagram on driving or flying a plane.

1.2 Mental Load Measurements

Because of the complexity of ML, in last three decades many measurements for ML were developed, which could be classified into three dimensions [8]:

Behaviour Measurements. They assess the behavioural performance of the operators, to estimate operators' mental capacity objectively. Primitive task performance is often solely applied, or combined with secondary task(s) to measure the entire mental consumption of users [13, 14].

Subjective Measurements. They consist of structured or non- structured questions to probe ML perceived by the operators. Self-report, Cooper-Harper Questionnaire, NASA-TLX Questionnaire, SWAT Questionnaire, MRQ are widely applied methods for subjective measurement [9–12]. NASA-TLX proved to be reasonably easy to use and reliably sensitive in various experimental settings in last twenty years [11].

Physiological Function Measurements. They are based on the symptoms that mental effort influences on physiological processes, such as oxygen consumption of brain [15], eye-blink and pupil dilation [16], p-wave of heart beat [17] and muscle tension [18].

Electrodermal Activity (EDA) was proved being sensitive to measure stress level [19] and often used in medical settings like nursery tasks [20]. Currently a wearable sensor, Affectiva Q sensor 2.0, can measure and record the EDA without interrupting of daily activities and causing discomfort [21].

Each category of methods measures a specific aspect of ML with one or two dozen of different parameters [4], so ML research in the last ten years developed multi-dimensional models to integrate these different methods, and these models usually base on expert rating, neural network and Multiple Resource Theory [22].

1.3   Mental Load Evaluation with Computer Systems

Currently, ML evaluation on human-computer interaction mainly focuses on VDTs. For instance, ML of arithmetic task on visual display terminal was firstly measured in 2006 [23]. Li et al. analyzed interaction ML in internet search and dual-task diagram in 2009, and combined factor analysis, back propagation neural network and self-organized neural network to establish a synthesis assessment model [24]. A 20-task navigation usability test and post-test NASA-TLX were applied to compare the ML of enhanced sound menu and visual menu on mobile terminal [25]. The electrodermal activity, electrocardiogram, photoplethysmo-graphy electroencephalogram were used as ML indicators for web browsing task [26]. According to these research cases, it is common to test typical tasks of a computer system with about 30 student participants under lab environment for Smartphone ML assessment.

The purpose of this experiment is to explore a comprehensive method for Smartphone ML, and compare different measurements to find out easy-to-use and sensitive indexes.

2   Method

2.1   Participants

This study was conducted with 33 college students (8 females, 25 males; average age 20.1, SD = 1.3), who received course credits for their participation. Participants separated into three groups: Novice, Average and Skilled users, according to a pre-test questionnaire on their knowledge on Smartphone usage [27], as shown in Table 1.

Table 1. Participants in three groups according to smartphone using proficiency

User group	Male	Female	In-total
Novice	2	3	5
Average	14	4	18
Skilled	9	1	10

The conditions of the pre-test was setting functions into levels easy, medium, hard and top by the degree of difficulty and frequency of usage. Then participants were asked to fill in a questionnaire containing the questions on their experience in number of years of smartphone using and the different functions they know, in order to determine in which user group they belonged.

2.2   Research Design

The usage of Smartphone is a process of cognitive action, which mainly depends on user's perception and thinking abilities [5], so the ML supposed to including four main dimensions: (1) primitive performance measurement from usability test, (2) secondary tasks performance as environmental interference, (3) subjective ML scoring and (4) stress level from Electrodermal Activity (EDA).

The method from Donnell was applied on selecting secondary tasks and evaluated them with five indexes (sensitivity, diagnosis, interference, demands of manipulation and acceptance of operator) [28], see Table 2. The difficulty gradient between four main tasks was also checked in this pilot test. After analyzing validity and reliability of the pilot test, the experiment design was modified.

Table 2. Secondary task comparison in pilot test

Secondary tasks	Sensitive	Diagnosis	Interference	Demands of manipulation	Acceptance of operator
Beat rhythm	−	+	O	−	O
Time estimate	O	−	−	O	O
Words memory	O	−	−	O	O
Mental calculation	+	−	−	−	O
Random number memory	−	+	O	−	+

Note: "−" means unsuitable; "O" means neutral; "+" means suitable.

Error record of the observing researcher indicated the standard usage and alternative paths as criteria to record the usability problems.

2.3   Measurements

According to the Compound User Model [3], participants separated in novice, average and skilled user groups based on their experience. The novice users have less than a half year experience; the average users have around one and half year experience and know the basic functions of the Smartphone OS; the skilled users should have at least three years of experience and have knowledge on advance functions in Smartphone OS.

The ML of Smartphone was evaluated using the following categories of measurements [29]:

Performance of Use. Users make errors during using which may indicate that interface design challenges the cognitive and action capabilities of users. Besides, these errors also prolong the task duration. Thus, *the number of usability error* and the *task duration* are parameters of primitive task performance.

Secondary Tasks. The secondary task was *Random Number Memory* (RNM), which is asking the participants remember a set of random digits in less than one minute and recall them after task. Since chosen as secondary task, *Random Number Memory* (RNM) indicates the capability of short-term memory, which represents mental resource occupation of human brain. The fewer the memory of numbers after operation, it means the larger mental resource occupation of the just finished task. The less the digits remembered, the larger the mental load is.

Subjective Rating. NASA-TLX is a widely applied questionnaire to indicate general mental load with *mental demand, physical demand, temporal demand, performance, effort* and *frustration* indexes. It measures the ML perceived by operators. A higher score means a higher perceived subjective ML.

Physiological Function. *Electrodermal Activity* (EDA) closely relates to stress in mind, so EDA (in μ Siemens) represents the degree of nervous excitement and alertness levels. Thus, it could imply the degree of attention. Higher EDA means more concentrated mind status.

The basic assumption is that when the task difficulty increases, the user's mental load increases, causing an increase in the number of usability errors, a longer task duration, decline in short-term memory, increased EDA values, and higher score on mental demand, temporal demand, effort and frustration of NASA-TLX.

Moreover, users with different usage proficiency will show a different mental load distribution in four main tasks, that is: The novice users could have a higher ML than other user groups in all main tasks; the average users might have higher ML in and above medium tasks; skilled users may only experience high ML with high and top main tasks.

2.4   Tasks

In this study, the experiment conducted with fixed posture under a quiet indoor environment for easier operation and better experiment control [4]. At first, the participants attended a pre-test interview about their experience on Smartphone, mental and physical status, personality type, environmental distractions (e.g. ambient noise), and the inform consent alike. Then, they needed to wear Affectiva Q sensor 2.0 on his/her distal forearm, and relaxed in five minutes to get stable physiological signal (EDA at 32 Hz) as their baseline. Before the test started, the participants asked to remember seven random digits, and the number of correct answers was recorded as a benchmark of RNM. At last, they received an introduction on how to fill NASA-TLX Questionnaire and learn the basic configuration of the test device (Samsung Galaxy S III with Android 4.1 OS).

The experiment contained 17 fundamental functions of smartphone OS, which were divided into four main tasks from low, medium, high to top difficulty levels. Each main task included 3–5 sub-tasks, e.g. "save the missed call and name it as XX". Thus, the participants could finish each task chain in a similar duration, if they have no usability problem in their operation. Moreover, the design of task chains ensured that these features have no significant overlap in the operation path. The various interface elements of the smartphone operating system (e.g. functions, meta-interactions, icons, controls, interface structure) distributed relative evenly across each chain.

Each participant finished all main tasks in random sequence on the test device. When the participants were conducting the tasks, the researchers recorded their number of usability errors and task duration of each main task. Each main task was followed by a new set of RNM task. When participants completed the RNM task, they evaluated their subjective mental load of this main task using NASA-TLX questionnaire. There was a three-minute break between each main task. The total experiment lasted between 20 and 30 min.

3   Results of Experiment

3.1   User Performance

Task durations across main tasks with different difficulty level were compared and it was found out that task duration was increased at level of hard (shown in Table 3). The SD values of time duration were increasing on primitive performance across main tasks.

Table 3. Task duration in different task level

Task duration	Low	Medium	High	Top
Mean	156.37	233.37	256.00	322.11
SD	56.26	92.52	101.78	177.99
SE	10.83	17.81	19.59	34.26

The differences in duration between low and other main tasks were all significant ($p < 0.01$); the differences between top and all other main task were significant too ($p < 0.05$, paired samples T-test shown in Table 4).

Table 4. Task duration paired sample T-test in four chains

	Low	Medium	High	Top
Low	–	77.00**	99.63**	165.74**
Medium		–	22.63	88.74*
High			–	66.11*

Figure 1 shows that task duration across different users increased, when the main tasks became more difficult. The duration of the novices grew fastest. However, there are only significant difference between novice and skilled users on top main task (mean difference 255.35, at p < 0.05 in one-way ANOVA).

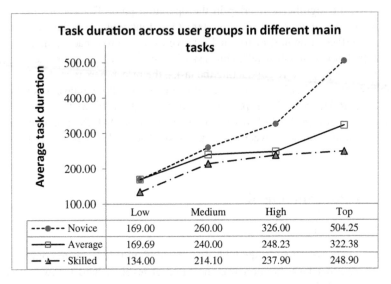

Fig. 1. Task durations of three users group across four chains.

As shown in Fig. 2, all users made more errors with the increasing difficulty of the main tasks (p < 0.01 in two-way t-test). The number of usability errors between novice and skilled users was significantly different, and the value between average and skilled users alike (p < 0.05 in paired t-test). For the skilled users, their error counts highly related to their task duration (Pearson $r_{Sut} = 0.99$). In general, the number of usability errors and task duration had medium correlation (Pearson $r_{ut} = 0.69$).

3.2   Secondary Task

The pre-test and post-test numbers of RNM in different user groups were compared across main task (see Table 5 Memory ability pre and post each main task). The memory of novice and average users increased in lower level tasks, while decreased slightly after main tasks that were more difficult. The memory of skilled users just decreased evenly after the main tasks.

With the increasing difficulty, the RNM value decreased slightly (see Table 6), as subjective ML score also shown below.

3.3   Subjective Mental Load

The scores of skilled users were significantly higher than the novice and average users as shown in Table 7 (p < 0.01). The values of the *physical demand* index were significantly lower than other indexes, while the *performance* values were significantly highest (p < 0.05 in t-test).

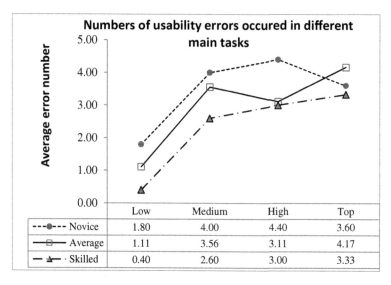

Fig. 2. The number of usability errors occurred in four main tasks.

Table 5. Memory ability pre and post each main task

User groups	Pre-task	Low	Medium	High	Top
Novice	4.80	5.20	4.60	3.80	4.40
Average	4.67	5.17	3.78	3.67	4.22
Skilled	5.40	4.10	4.20	4.20	4.30

Table 6. The proportion of users who has memory decline in three user groups

The proportion of users	Low	Medium	High	Top
Novice (%)	20.00	60.00	60.00	40.00
Average (%)	31.58	57.89	47.37	42.11
Skilled (%)	44.44	56.56	56.56	4.44

Table 7. NASA-TLX score of three user groups

	Mental demand	Physical demand	Effort	Frustration	Temporal demand	Performance
Novice	3.05	2.33	3.28	2.98	3.23	3.98
Average	3.50	2.72	3.40	2.90	3.61	4.32
Skilled	4.11	3.38	4.53	3.77	4.90	5.33
Average	3.56	2.81*	3.73	3.21	3.91	4.54*

As shown in Fig. 3, though the *physical demand* was low in Smartphone tasks, it still highly correlated *to mental demand* (Pearson $r_{mdpd} = 0.90$).

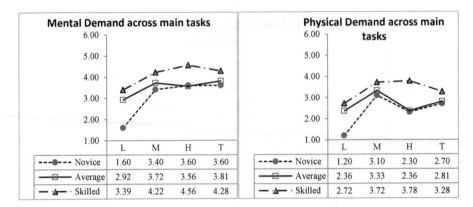

Fig. 3. The mental and physical demand of three user groups.

On *mental demand, frustration, effort* and *physical demand*, the skilled user's subjective mental load increased slightly, when the main tasks became more difficult, as shown in Figs. 3 and 4. Moreover, the skilled user's values on all these indexes highly related to the number of usability errors (Pearson r_{Smd} = 0.95, Pearson r_{Sf} = 0.95, Pearson r_{Se} = 0.85, Pearson r_{Spm} = 0.79).

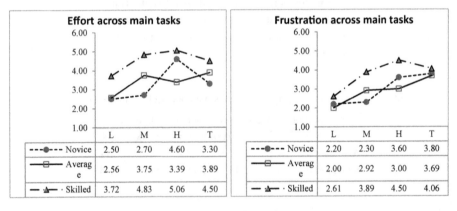

Fig. 4. The effort and frustration demand of three user groups.

The subjective mental load of both novice and skilled users fluctuated slightly across different main tasks. Similar to the skilled users, *mental demand* of average and novice users also highly related to their number of usability errors (Pearson r_{Amd} = 0.99, Pearson r_{Nmd} = 0.95). The novice user's *physical demand* and their number of usability errors was also highly correlated (Pearson r_{Npd} = 0.97).

In general, the *effort* and *frustration* were relative highly correlated among users, as shown in Fig. 4 (Pearson r_{ef} = 0.87). However, only for the skilled and average users, their *effort* and *frustration* had high correlation to their number of usability errors (Pearson r_{Ae} = 0.96, Pearson r_{Af} = 0.99).

Like *mental demand, temporal demand* had also high correlation with the number of usability errors across different users (Pearson r_{Std} = 0.94, Pearson r_{Atd} = 0.82, Pearson r_{Ntd} = 0.88). However, the users rated their *performance* relatively low for simpler task chains, which was negatively related to their number of usability errors (Pearson r_p = −0.81). Detailed data is shown in Fig. 5.

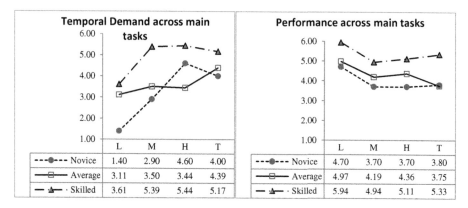

	L	M	H	T
---●--- Novice	1.40	2.90	4.60	4.00
—▱— Average	3.11	3.50	3.44	4.39
— ▲ · Skilled	3.61	5.39	5.44	5.17

	L	M	H	T
---●--- Novice	4.70	3.70	3.70	3.80
—▱— Average	4.97	4.19	4.36	3.75
— ▲ · Skilled	5.94	4.94	5.11	5.33

Fig. 5. The temporal demand and performance of three user groups.

3.4   Electrodermal Activity Values (EDA)

The value of the user's EDA varied from 0.04 to 24.91 µs, so the average EDA in different main tasks did not show significant difference across three user groups. Therefore, the minimal and maximal points of EDA were picked up in each main task. Although individual differences are large, but we can see a slow rise on EDA after difficulty increased (Fig. 6). Only at the minimal EDA level, the novice and skilled user show significant differences ($p < 0.05$ in t-test). Besides, there were also significant differences in low-high comparison and low-top comparison ($p < 0.05$ in t-test). The minimal EDA values had weak correlation (Pearson r_{eda} = 0.30) to the number of usability errors.

4   Discussion

Comparison of Secondary Task Methods. This study rated five secondary tasks in a pilot study, including beat rhythm, time estimate, words memory, mental calculation and random number memory, according to sensitivity, diagnosis, interference, demands of manipulation and acceptance of operators in a pilot test. The RNM had the highest acceptance value of operator. Therefore, it was chosen in order to control the mental load of the secondary task itself.

User Performance. In difficult main tasks, the novice users reduced their operating speed in order to avoid the errors. In the top task chain, because the novices consumed much more time, carefully learning concepts of advance functions at first, their task

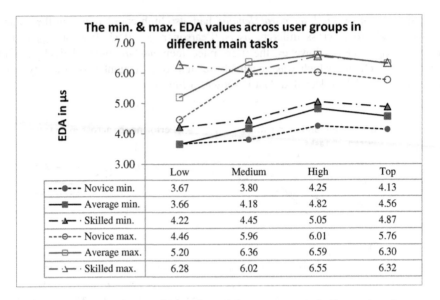

The min. & max. EDA values across user groups in different main tasks

EDA in μs

	Low	Medium	High	Top
---●--- Novice min.	3.67	3.80	4.25	4.13
---■--- Average min.	3.66	4.18	4.82	4.56
---▲--- Skilled min.	4.22	4.45	5.05	4.87
---⊖--- Novice max.	4.46	5.96	6.01	5.76
---☐--- Average max.	5.20	6.36	6.59	6.30
---⊿--- Skilled max.	6.28	6.02	6.55	6.32

Fig. 6. The min. & max. EDA values of three user groups in four main tasks

duration increased very fast whereas the number of usability errors dropped. Moreover, the average user made more errors in medium level than high level, but not consumed much more time. This indicates that they feel confident adopting the "trial and error" strategy in familiar operation and could find the correct path fast. The main cause of increased task duration was the usability errors in operation due to their correlation coefficient; especially for skilled user they showed a near linear correlation. When the main tasks become more difficult, the performance differences across the users are larger, as indicated by the standard deviation increase on task duration.

Therefore, task duration could better reflect the ML of learning for novice users, while the number of usability errors is suitable to evaluate ML in familiar tasks for average and skilled user. Moreover, the number of usability errors also shows high correlation to subjective mental load.

Secondary Task. The memory of novice users and average users increased in lower level tasks, which may due to a higher degree of brain excitability. Since these tasks had no time requirement, short-term memory was not influenced by temporal stress. The rising of difficulty level in tasks was the main cause of growing mental effort. The skilled user had more memory loss might due to their higher self-expectation on task accomplishment, which consists of the high mental load on performance.

Thus, RMN easily reflects user's self-expectation of their performance.

Subjective Scale Evaluation. *The mental demand, effort, frustration* and *temporal demand* indexes in NASA-TLX were all highly related to the number of usability errors. Especially for *mental demand* index, they are linear correlated across user groups.

Since the participants conducted all main tasks in a quiet lab with fixed sitting posture, the real physical demand was low. However, the correlation of *physical demand* for a slight physical task like operating a Smartphone, indicate that the perceived *physical demand* is highly influenced by *mental demand*.

Mental demand, frustration and *temporal demand* of the skilled users were all nearly linear correlated to their number of usability errors. For the average users, *mental demand* and *effort* showed the highest correlations to their number of usability errors. For the novice users, their *mental demand* strongly related to their *physical demand*.

On *performance* values, they had high negative correlation to the number of usability errors. Thus, it could be explained that the participants had higher self-expectation on task accomplishment in lab environment than in real using context, especially the skilled users. This experienced the psychological pressure could trigger a higher mental load.

EDA Measurement. In the pilot test, the influence of gender on the EDA was obvious, that female participants had higher EDA level than male, and the fluctuation was more drastic.

Actually, EDA indirectly related with mental load via user's emotional fluctuation. For the person who said s/he is extrovert in per-test interview, the relation between EDA and workload is not significant. For the person who is introvert, when pressure generated from tension and unconfident mood in more difficult tasks, the EDA fluctuation would become evident, thus the average EDA values would rise. Therefore, the EDA is more like a qualitative measurement for Smartphone ML.

However, there was an only a weak correlation between minimal EDA values and the number of usability errors. The small sample size of novice users may cause the high fluctuation on EDA values. Further research is needed with more participants to find out a specific relationship between EDA and Smartphone mental load.

5   Overall Conclusion

Though the mental load assessment for Smartphone is more complex than conventional usability tests, it could offer richer information on the correlation between different factors that may influence the quality of use.

Furthermore, different measurements could suit to different proficiency of users. For instance, task duration is a better proxy understanding the mental load of novice users, while the number of usability errors is better to evaluate the mental load of the average and skilled user.

Due to the high correlation between subjective scale and the number of usability errors, further researches could focus on simplify this test using less measurements with similar validity.

Besides, RNM and EDA are easily influenced by psychological pressure, which are more related to self-expectation and personality. Further researches could explore their relationships to Smartphone ML under time pressure.

References

1. ISO (2001) ISO/IEC 9126-1:2001 Software engineering – Product quality – Part 1: Quality model. International Standard. International Organization for Standardization, Switzerland
2. ISO: ISO9241-11(1998) 1998 Ergonomic requirements for office work with visual display terminals (VDT's) – Part 11: Guidance on usability. International Standard, International Organization for Standardization, Switzerland
3. Li LS (1999) Action theory and cognitive psychology in industrial design: User models and user interfaces. Dissertation, Art University of Braunschweig, Braunschweig
4. Kantowitz BH (1987) Mental Workload. In: Hancock PA. (ed) Advances in Psychology, vol 47, pp 81–121
5. Liao JQ (1995) Mental workload and its measurement. J Syst Eng 10(3):119–123
6. Kang WY, Yuan XG, Liu ZQ, Liu W (2008) Synthetic Evaluation method of mental workload on visual display interface in airplane cockpit. Space Med Med Eng 21(2):103–107
7. Li L, Yuan M (2011) Influential factors analysis of drivers' mental workload with the use of vehicle navigation system. J Saf Environ 11(6):202–204
8. Cui K, Sun LY, Feng TW, Xing X (2008) New developments in measurement methodologies of mental workload. Industr Eng J 11(5):1–5
9. Cooper GE, Harper RP, Jr (1969) The Use of Pilot Rating in the Evaluation of Aircraft Handling Qualities. Report No NASA TN-D-5153. Technical Report, Ames Research Center, National Aeronautics and Space Administration. Moffett Field
10. Hart SG (2006) NASA-task load index (NASA-TLX); 20 years later. In: Proceedings of the human factors and ergonomics society 50th annual meeting, vol 50. Sage Publications, Los Angeles, pp 904–908
11. Reid GB, Nygren TE (1988) The subjective workload assessment technique: a scaling procedure for measuring mental workload. Adv Psychol 52:185–218 Elsevier Science Publishers, North Holland
12. Boles DB, Bursk JH, Phillips JB, Perdelwitz JR (2007) Predicting dual-task performance with the multiple resources questionnaire (MRQ). Hum Factors 49(1):32–45
13. Galy E, Cariou M, Mélan C (2011) What is the relationship between mental workload factors and cognitive load types? Int J Psychophysiol 83(3):269–275
14. Shingledecker CA, Crabtree MS, Simons JC et al (1980) Subsidiary Radio Communications Tasks for Workload Assessment in R&D Simulations I. Task Development and Workload Scaling. Technical Report, Systems Research Labs Inc, Dayton Ohio
15. Horst RL, Johnson R, Donchin E (1980) Event-related brain potentials and subjective probability in a learning task. Mem Cognit 8(5):476–488
16. Ahlstrom U, Friedman-Berg FJ (2006) Using eye movement activity as a correlate of cognitive workload. Int J Industr Ergon 36(7):623–636
17. Gunn CG, Wolf S, Block RT et al (1972) Psychophysiology of the cardiovascular system. In: Greenfield NS, Sternbach RA (eds) Handbook of psychophysiology. Holt, Rinehart & Winston, New York, pp 457–483
18. Suzuki S, Kumano H, Sakano Y (2003) Effects of effort and distress coping processes on psychophysiological and psychological stress responses. Int J Psychophysiol 47(2):117–128
19. Reinhardt T, Schmahl C, Wüst S, Bohus M (2012) Salivary cortisol, heart rate, electrodermal activity and subjective stress responses to the mannheim multicomponent stress test (MMST). Psychiatry Res 198(1):106–111

20. Moya-Albiol L, Sanchis-Calatayud MV, Sariñana-González P, De Andrés-García S, Romero-Martínez Á, González-Bono E (2012) P03-425 - Electrodermal activity in response to a set of mental tasks in caregivers of persons with autism spectrum disorders. Eur Psychiatry 26(1):1595

21. Affectiva (2012) Liberate yourself from the lab: Q Sensor measures EDA in the wild. Affectiva QTM Solutions White Paper

22. Wang J, Fang WN, Li GY (2010) Mental workload evaluation method based on multi-resource theory model. J. Beijing Jiaotong Univ. 34(6):107–110

23. Peng XW, He QC, Ji T, Wang ZL, Yang L (2006) Mental workload for mental arithmetic on visual display terminal. Chin J Industr Hyg Occup Dis 24(12):726–729

24. Li JB, Xu BH (2009) synthetic assessment of cognitive load in human-machine interaction process. Acta Psychologica Sinica 41(1):35–43

25. Yu YH, Li ZJ (2011) Study of sonically enhanced menu interaction for mobile terminals. Appl Res Comput 28(10):3742–3745

26. Jimenez-Molina A, Retamal C, Lira H (2018) Using psychophysiological sensors to assess mental workload during web browsing. Sensors 18(2):458

27. Li M (2008) Comparison of Usability Evaluation Method Based-on Needs of Software Development. Master Thesis, Xi'an Jiaotong University, Xi'an

28. O'Donnell RD, Eggemeier FT (1986) Workload assessment methodology. In: Boff KR, Kaufman L, Thomas JP (eds) Handbook of perception and human performance, vol II. Wiley, New York, pp 42–43

29. Galy E, Cariou M, Mélan C (2012) What is the relationship between mental workload factors and cognitive load types? Int J Psychophysiol 83(3):269–275

Effect of Age on Heart Rate Responses and Subjective Mental Workload During Mental Tasks

Hiroyuki Kuraoka[1,2(✉)], Chie Kurosaka[2], Chikamune Wada[1], and Shinji Miyake[2]

[1] Kyushu Institute of Technology, Kitakyushu, Fukuoka 8080196, Japan
[2] University of Occupational and Environmental Health, Kitakyushu, Fukuoka 8078555, Japan
`h-kuraoka@health.uoeh-u.ac.jp`

Abstract. We investigated the effect of age on heart rate (HR) responses and subjective mental workload during mental tasks. In this study, 55 male participants performed a mental arithmetic (MA) task and a mirror tracing (MT) task for 5 min each. Electrocardiogram (ECG) was recorded during these mental tasks and resting periods. Low frequency component (LF), high frequency component (HF), and LF/HF ratio of heart rate variability (HRV) were derived from ECG signals. NASA Task Load Index (NASA-TLX) was used to evaluate the subjective mental workload during each task. In the results, HR was significantly larger in the MA task than the resting period. LF was significantly smaller in the MA than the resting period before the task. On the other hand, no HR changes from the baseline in the MT task were found, suggesting that Pattern 2 response might appear. In the results of HRV indices, decrease in LF suggests the inhibition of the parasympathetic nervous system activity by the task. These results are consistent with those of our previous studies. No significant main interaction with age and blocks (MA, MT and resting periods) were found in physiological responses except for LF/HF and subjective mental workload score. Therefore, the physiological responses induced by mental tasks might be more susceptible to task characteristic than to aging. This study partly supported the previous study.

Keywords: Heart rate · Heart rate variability · Mirror tracing
Pattern 2 response

1 Introduction

Psychophysiological researches have shown that physiological responses induced by mental tasks vary according to task characteristics. Lacey's intake-rejection hypothesis is one of the representative examples to explain the effects of differences in task characteristics on cardiovascular system (Lacey 1959). For example, sensory rejection task such as mental arithmetic (MA) increases heart rate (HR), blood pressure, and peripheral vasodilation. In contrast, sensory intake tasks such as the mirror tracing (MT) task reduce HR. Subsequently, the former is expressed as Pattern 1 response, and

© Springer Nature Switzerland AG 2019
S. Bagnara et al. (Eds.): IEA 2018, AISC 827, pp. 316–321, 2019.
https://doi.org/10.1007/978-3-319-96059-3_34

the latter as Pattern 2 response (Schneiderman and McCabe 1989). However, these previous studies have mainly targeted healthy and young participants. It is well known that aging is associated with the decline in cardiovascular systems (e.g. Liao et al. 1995). Cardiovascular responses to mental tasks in older individuals have been inconsistent with those of young participants. One previous study found that increased HR induced by MA task was significantly greater in young than old males (Barnes et al. 1989). Another study concluded that the older group demonstrated significantly higher HR, systolic blood pressure, total peripheral resistance levels, and HRV during Stroop task than the younger group (Boutcher and Stocker 1996). Therefore, cardiovascular responses induced by mental tasks in relation to Pattern 1 response might decline with age. However, little has been reported on the relationship between aging and cardiovascular responses during sensory intake tasks such as MT task that evokes Pattern 2 response. Therefore, this study investigated the effect of aging on HR, HRV, and subjective mental workload during mental tasks including MA task as a sensory rejection task and MT task as a sensory intake task.

2 Method

2.1 Participants

Fifty-five males (mean age: 44.4 ± 1.8 years) participated in this study. Three participants were excluded from the analysis due to failure of ECG recording and frequent arrhythmia. Therefore, data of a total of 52 participants were analyzed. This study was approved by the Ethics Committee of University of Occupational and Environmental Health.

2.2 Procedures

Upon the arrival of the participants at the laboratory, after obtaining informed consent and providing instructions for the experiment, ECG electrodes were attached to participants; they then took rest for 10 min to adapt to the experimental environment. Following a 5-min pre-test rest (PRE) in a sitting position, the participants were asked to perform the MA and MT tasks for 5 min each. The order of the tasks was counterbalanced. Lastly, they took a 5-min post-test rest (POST) again. Subjective mental workload assessment using NASA Task Load Index (NASA-TLX) was conducted after each task.

2.3 Task

The MA task is based on the MATH algorithm proposed by Turner et al. (1986). In this task, a problem appears on the PC screen for 2 s, followed by the word "EQUALS" for 1.5 s. An answer then appears for 1.5 s. Therefore, a problem appears every 5 s. Participants were required to press the left button of the mouse if the presented answer is correct and the right button if it is incorrect. Participants must respond quickly within 1.5 s. The MA task contains five levels of difficulty: level one, level two, level three, level four, and level five, comprising 2-digit + 1-digit, 2-digit − 1-digit, 2-digit ± 2-digit, 3-digit + 2-digit problems, and 3-digit − 2-digit problems, respectively. All subtractions yield positive answers. The first problem presented is always at level 3. Thereafter, the levels of the subsequent problems depend on the participant's responses. When the participant's response is correct, the level of the next problem increases by one. If an incorrect response or no response is given within the time limit, the level goes down. For correct responses to level five problems and incorrect responses to level one problems, the level of the next problem remains the same. Participants responded to total 60 problems for 5 min. In the MT task, a complex zig-zag pathway is presented on a PC screen. Participants were asked to trace this pathway with a mouse as precisely as they could. The horizontal and vertical axes of controls of the mouse were interchanged.

2.4 Physiological Measurement

ECG (Nihon Kohden Pocket ECG Monitor Pocket-ECG WEC-7101) from lead II was recorded at 500 Hz sampling rate during tasks (MA and MT) and resting periods (PRE and POST). In the HRV analysis, low frequency components (LF), high frequency components (HF), and LF/HF ratio were derived from ECG signals.

2.5 Subjective Assessment

NASA-TLX was used for the evaluation of subjective mental workload during each task. This index consists of six subscales: Mental Demand, Physical Demand, Temporal Demand, Own performance, Effort, and Frustration. The weighted average score (Adaptive Weighted Workload: AWWL) of these six subscales was calculated using the weighting coefficients defined by the rank order of the raw scores without the paired-comparisons (Miyake and Kumashiro 1993).

2.6 Statistical Analysis

All physiological indices were standardized across four blocks (PRE, MA, MT, and POST) for each participant. The participants were divided into two age groups on the basis of their mean age: young (n = 27, mean age: 32.4 ± 5.4 years) and old (n = 25, mean age: 55.9 ± 7.6 years) group. All physiological indices were analyzed by age groups (2) × blocks (4) two-way repeated measures ANOVAs to assess the difference between factors and interactions with age groups and blocks. NASA-TLX scores were assessed with age groups (2) × task (2) repeated measures ANOVAs. Greenhouse-Geisser correction of the degrees of freedom was applied. A post-hoc analysis using Bonferroni corrections was conducted.

3   Results

The changes in HR and HRV indices for young and old males are shown in Fig. 1. HR in the MA task was significantly greater than those in PRE, MT, and POST (p < 0.01, Fig. 1a). LF during MA task were significantly smaller than during PRE (p < 0.01, Fig. 1b). HF were significantly lower in the MA than the PRE and POST (p < 0.01, Fig. 1c). No significant main effects were found in LF/HF (Fig. 1d). These results are identical in both age groups. Although there was no significant interaction between age and task for HR, LF, and HF, a significant interaction was found for LF/HF (p < 0.05). Figure 2 demonstrates the changes in AWWL scores. No significant main effects and interaction between age and block were observed.

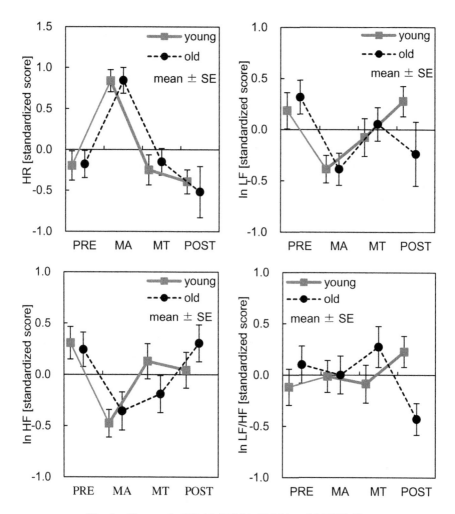

Fig. 1. Changes in HR (a), LF (b), HF (c) and LF/HF (d).

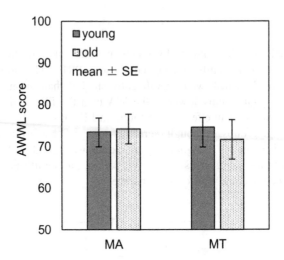

Fig. 2. Changes in AWWL scores

4   Discussion

The results of HR indicated that the physiological responses in MA task clearly showed Pattern 1, in which HR increases. LF and HF decrease in the MA task suggested the inhibition of the parasympathetic nervous system activity by the task. These results are consistent with those of our previous studies (e.g. Miyake et al. 2014). However, HR acceleration occurred equally regardless of age difference. Thus, the machine-paced MA task may induce Pattern 1 response more clearly. On the other hand, there was no HR change from the base line in MT task, suggesting that Pattern 2 response might be induced. However, our previous studies indicated that HR in the MT task for young females decreased significantly (Kuraoka et al. 2015). As for Pattern 2 responses to MT task, they might appear more clearly in females without aging effects. Namely, the participant's coping behavior to the task may be more affected by gender than aging. Consequently, HR and HRV are likely to be influenced by several factors including aging and gender. Some studies have addressed that the HR responses to psychological stress should be observed at the individual level. Blood pressure may be a more sensitive overall index of aging effect on psychological stress (Bert et al. 2010). Therefore, multi-dimensional evaluation using several indices such as blood pressure, which is a robust task characteristic, and total peripheral resistance may be more feasible for the age-related mental workload assessment.

5   Conclusion

This study has demonstrated that the age-related physiological responses induced by mental tasks were not observed clearly. These responses may be less susceptible to aging. However, the physiological responses vary according to the task characteristic regardless of age difference, even though subjective mental workload scores are identical. This discrepancy should be fully considered in the mental workload assessment using physiological indices.

Acknowledgement. The authors thank Takehiko Hiei, Kazuhisa Shigemori, Chiaki Yasumoto, and Sayo Toramoto in DAIKIN INDUSTRIES, Ltd. for their collaborative support in this experiment.

References

Barnes RF, Raskind M, Gumbrecht G, Halter JB (1989) The effects of age on the plasma catecholamine response to mental stress in man. J Clin Endocrinol Metab 54:64–69

Berntson GG, Cacioppo JT, Fieldstone A (1996) Illusions, arithmetic, and the bidirectional modulation of vagal control of the heart. Psychophysiology 44(1):1–17

Uchino BN, Birmingham W, Berg CA (2010) Are older adults less of more physiologically reactive? A meta-analysis of age-related differences in cardiovascular reactivity to laboratory tasks. J Gerontol Ser B Psychol Sci Soc Sci 65:154–162. https://doi.org/10.1093/geronb/gbp127 Epub 6 January 2010

Boutcher SH, Stocker D (1996) Cardiovascular response of young and older males to mental challenge. J Gerontol B Psychol Sci Soc Sci 51:261–267

Kasprowicz AL, Manuak SB, Malkoff SB, Krantz DS (1990) Individual differences in behaviorally evoked cardiovascular response: temporal stability and hemodynamic patterning. Psychophysiology 27(6):605–619

Kuraoka H, Kazuki T, Wada C, Miyake S (2015) Effects of a sensory intake task on heart rate and heart rate variability. In: Proceedings 19th triennial congress of the IEA (IEA 2018), Paper 2004

Lacey JI (1959) Psychophysiological approaches to the evaluation of psychotherapeutic process and outcome. In: Rubinstein EA, Parloff MB (eds) Research in psychotherapy. American Psychological Association, Washington, DC, pp 160–208

Liao D, Barnes RW, Chambless LE, Simpson RJ, Sorlie P, Heiss G (1995) Age, race, and sex differences in autonomic cardiac function measured by spectral analysis of heart rate variability—the ARIC study. Am J Cardiol 76:906–912

McCanne RT, Hathway MK (1979) Autonomic and somatic responses associated with performance of the embedded figures test. Psychophysiology 16(1):8–14

Miyake S, Kumashiro M (1993) Subjective mental workload assessment technique - an introduction to NASA-TLX and SWAT and a proposal of simple scoring methods. Jpn J Ergon 29:399–408

Miyake S, Kuraoka H, Wada C (2014) Heart rate variability as a mental workload index. In: Proceedings of the international conference on applied human factors and ergonomics (AHFE 2014), pp 19–23

Sato N, Miyake S (2004) Cardiovascular reactivity to mental stress: relationship with menstrual cycle and gender. J Physiol Anthropol Appl Hum Sci 23(6):215–223

Schneiderman N, McCabe M (1989) Psychophysiologic strategies in laboratory research. In: Schneiderman N, Weiss SM, Kaufmann PG (eds) Handbook of research method in cardiovascular behavioral medicine. Plenum Press, pp 349–364

Turner JR, Heiwitt JK, Morgan RK, Sims J, Carrol D, Kelly KA (1986) Graded mental arithmetic as an active psychological challenge. Int J Psychophysiol 3(4):307–309

Evaluation of Muscle Load of Hand and Forearm During Operation of Cross-shaped Switch by Thumb

Motonori Ishibashi[✉] and Risa Hashimoto

Nihon University, 1-2-1 Izumi-cho, Narashino, Chiba 275-8575, Japan
ishibashi.motonori@nihon-u.ac.jp

Abstract. The purpose of this study is to try an evaluation of muscle load of hand and forearm during the cross-shaped switch operation. Five young and right-handed college students participated in the experiment in which the portable game device was used. They performed four kinds of operation task with metronome. EMG was derived from 4 muscles and transformed to the ratio to maximal voluntary contraction (%MVC). Uncomfortable feeling was also measured after the each task. From a correlation analysis between the %MVC and the uncomfortable feeling, higher correlation was found in the muscles which have direct connection with the push motion. Through a regression analysis which adopted the %MVC as an explanatory variable and the uncomfortable feeling as an objective variable, the regression coefficients in the flexor pollicis brevis muscle were higher than those of the extensor pollicis brevis muscle. In conclusion, possibility of the evaluation of the muscle load during the operation of cross-shaped switch by thumb was shown.

Keywords: Electromyogram · Operability · Uncomfortable feeling

1 Introduction

There are many products for daily living which require users to operate cross-shaped switch by thumb such as portable game device or automotive steering switch. This type of device is held two-handed, and the cross-shaped switch is pushed, rotated and slid by users when it is used.

In order to keep operability of these devices, it is necessary to consider user characteristics on the hand and finger operation broadly. Database of anthropometric dimension has been applied to the design of products [1]. In the case of the switch operation by thumb, it is desirable to consider length and breadth of thumb in the design.

However, of course, reach area of finger defined by the length of thumb is different from easy work area. The users with short thumb are forced to stretch their fingers during the operation even if their thumbs reach the cross-shaped switch, and it leads to generate larger muscle load. The users with narrow range of joint motion may feel large muscle load when they are forced to move their thumbs widely. Furthermore, since holding the device also requires the muscular strength, the users with weak grip or

© Springer Nature Switzerland AG 2019
S. Bagnara et al. (Eds.): IEA 2018, AISC 827, pp. 322–328, 2019.
https://doi.org/10.1007/978-3-319-96059-3_35

forearm feel larger muscle load. Thus, it is necessary to consider the muscle load in addition to the anthropometric dimension. In the case of isometric contraction, it is known that muscular fatigue may appear even if the ratio of the muscle contraction to maximal voluntary contraction (MVC) is equivalent to 10% which is regarded as low %MVC [2]. It can be a reference value to inhibit the muscular fatigue. In the same way, practical finding on the relationship between muscle load and feeling is also required in the case of dynamic situation involving the motion of thumb.

The aim of this study was to try to evaluate muscle load of hand and forearm which has relationship with the cross-shaped switch operation by thumb.

2 Methods

2.1 Participants

Five college students who were 20s and were familiar with portable game participated in the experiments as volunteers. They were all right-handed. Two of them were male whose thumb lengths were 60 mm. Three of them were female, and their thumb length was each 50 mm, 53 mm and 54 mm. From the reference [1], the thumb lengths of the participants were shorter than the 50 percentile of the Japanese young adults (N = 421).

This study was approved by Research Ethics Committee concerning Research with Human Subjects in College of Technology, Nihon University.

2.2 Operation Task

Portable game device, New NINTENDO 3DS LL (3DS), was used for operation task. The participant was required to hold the 3DS by both hands and operate buttons (X, Y, A, B in Fig. 1) by pushing or sliding. Speed of task was 148 beats per minute which is equal to that of existing Japanese game application (Taiko-no-tatsujin). It was quite fast, and the participant pushed or slid each button with metronome. The way to operate the button is shown in Fig. 1.

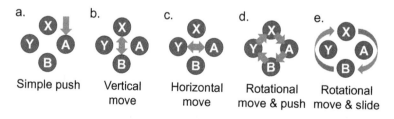

Fig. 1. Button operation in the task

- Simple push task. The participant pushed only the A button. (a)
- Vertical move task. The participant pushed the X and B button alternately. (b)
- Horizontal move task. The participant pushed the Y and A button alternately. (c)

- Rotational move task with push operation. The participant pushed the X, A, B, Y button in this order. (d)
- Rotational move task with slide operation. The participant pushed the X, A, B, Y button in this order with sliding his/her thumb along the button. (e)

The way of operating the button was set based on the game. The participant performed each task for 60 s in that order (from a. to e.). As for the dimension of 3DS, vertical size was 92 mm, horizontal size was 160 mm, and its thickness was 12 mm. Diameter of each button was 7 mm.

2.3 Measurement of Electromyogram

The authors thought that muscle activity of cross-shaped switch operation by thumb consisted of two elements. One was the muscle activity of thumb which had relationship with operation directly (#1). The other was the muscle activity of the forearm which had relationship with hold of 3DS by four fingers except thumb (#2). Through a preliminary test to observe the electromyogram (EMG) during 3DS operation, as test muscle corresponding to flexion and extension of thumb (#1), extensor pollicis brevis muscle and flexor pollicis brevis muscle in the right hand were selected. In a similar way, corresponding to flexion and extension of four fingers to hold 3DS (#2), flexor digitorum superficialis muscle and extensor digitorum muscle in the right forearm were selected.

For the measurement of EMG, Polymate Mini AP108 (Miyuki Giken Co., Ltd.) which is a telemetry system was used with a time constant of 0.01 s, hi-cut filter of 100 Hz, and notch filter. EMG was digitally recorded with PC at the sampling frequency of 500 Hz.

2.4 Subjective Evaluation

Item of subjective evaluation was the "uncomfortable feeling" by the switch operation. From the author's preliminary investigation (not published), it was found that "unnatural", "clumsy", "uncomfortable" and "uneasy" feeling were summarized to this item. After each task, uncomfortable feeling elicited by the task was rated with 6-point scale (1: absolutely not, 2: little, 3: slightly, 4: moderately, 5: quite, 6: very).

2.5 Procedure

Two types of posture were adopted in the experiment. One was "bending arms" and the other was "stretching arms" (see Fig. 2).

After instruction, electrodes were attached to the participant. At the start of the experiment, maximal voluntary contraction (MVC) was measured in each posture by clenching the right hand with all his/her strength. Then, the participant performed the operation task (from a. to e.) in the "bending arms" condition, followed by the "stretching arms" condition. The participant was asked to take a natural posture to hold 3DS as usual, and he/she supported the wrist on the work table so that the hand and forearm did not move.

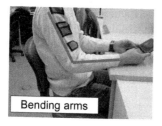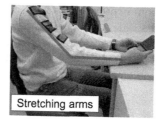

Bending arms

Stretching arms

Fig. 2. Posture during the operation task

2.6   EMG Analysis

Figure 3 shows the signal processing of EMG. In each position, after EMG was filtered by 100 Hz low-pass digital filter, it was rectified and was integrated with respect to time. The interval of integration for the EMG during the task was 1 min which was the time for each task, and the integrated value was converted to the value per second (divided by 60 s). It was defined as MC_{task} (abbreviation for muscle contraction). MVC was also rectified and integrated, and it was converted to the value per second. Then, %MVC which was an index for muscle activity was calculated by the ratio of MC_{task} to MVC multiplied by 100.

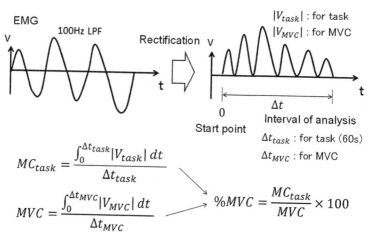

$$MC_{task} = \frac{\int_0^{\Delta t_{task}} |V_{task}| \, dt}{\Delta t_{task}}$$

$$MVC = \frac{\int_0^{\Delta t_{MVC}} |V_{MVC}| \, dt}{\Delta t_{MVC}}$$

$$\%MVC = \frac{MC_{task}}{MVC} \times 100$$

Fig. 3. Signal processing of EMG

3   Results

3.1   Overall Muscle Load

In each operation task and posture, mean %MVC among the test muscles and participants, and mean score of uncomfortable feeling are shown in Fig. 4. From this, %MVC was larger in the "bending arms" condition in whole than in the "stretching arms" condition. Especially it was found in the rotational move task with slide operation (rotation&slide, task etc.). Uncomfortable feeling was low in whole, however, it was a little higher in the "stretching arms" condition. Correlation coefficient between mean %MVC and mean score of uncomfortable feeling was 0.661 in the "bending arms" condition, and was 0.751 in the "stretching arms" condition.

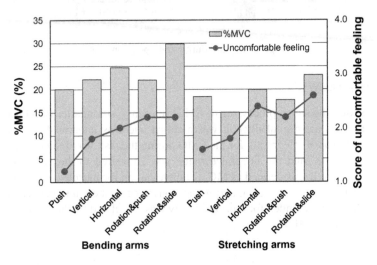

Fig. 4. %MVC (mean among the test muscles and participants) and uncomfortable feeling

3.2   Relationship Between %MVC and Uncomfortable Feeling

After mean %MVC among participants was calculated in each operation task, test muscle, and posture, correlation between mean %MVC and mean score of uncomfortable feeling was analyzed. Correlation coefficients in each posture and muscle are shown in Table 1. From this, the muscles activity of forearm (flexor digitorum superficialis muscle, extensor digitorum muscle) which had relationship with holding 3DS had little correlation with uncomfortable feeling except extensor digitorum muscle in "bending arms" condition. On the other hand, the muscles activity of thumb (flexor pollicis brevis muscle, extensor pollicis brevis muscle) which had relationship directly with switch operation had high correlation with uncomfortable feeling, and they were clearly higher than the forearm. In addition, the correlation coefficients were higher than those between overall four muscles activity and the feeling.

Next, focusing on the extensor pollicis brevis muscle and flexor pollicis brevis muscle which were more highly correlated muscles with the uncomfortable feeling, a regression analysis which adopted the %MVC as an explanatory variable and the uncomfortable feeling as an objective variable was performed (see Fig. 5). The flexor pollicis brevis muscle corresponds to push operation, while the extensor pollicis brevis muscle corresponds to pull operation. As a result, the regression coefficients in the flexor pollicis brevis muscle were higher than those of the extensor pollicis brevis muscle. Particularly it was the highest in the "stretching arms" condition.

Table 1. Correlation coefficients between %MVC and uncomfortable feeling in each muscle

	Forearm (holding 3DS)		Thumb (switch operation)		Overall four muscles
	Flexor digitorum superficialis muscle	Extensor digitorum muscle	Flexor pollicis brevis muscle	Extensor pollicis brevis muscle	
Bending	−0.112	0.523	0.778	0.731	0.661
Stretching	−0.155	−0.284	0.685	0.913	0.751

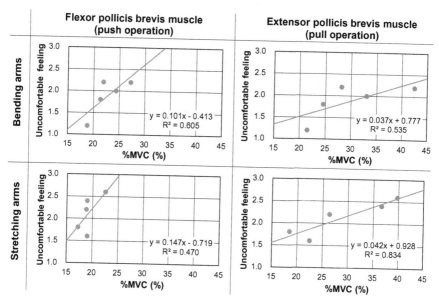

Fig. 5. Uncomfortable feeling as a function of %MVC in muscle activity of thumb

3.3 Relationship Between Thumb Length and %MVC

In each operation task and posture, correlation coefficient between mean %MVC among the test muscles and thumb length of the participant was calculated. As shown in Table 2, high correlation was found in whole, especially in the "bending arms" condition the correlation coefficients were more than 0.90.

Table 2. Correlation coefficient between mean %MVC and thumb length

	Push	Vertical	Horizontal	Rotation&push	Rotation&slide	All (mean)
Bending	−0.984	−0.901	−0.951	−0.948	−0.979	−0.966
Stretching	−0.902	−0.520	−0.836	−0.708	−0.735	−0.735

4   Discussion and Conclusion

The overall muscle load was higher in the "bending arms" condition than in the "stretching arms" condition, while the result of subjective score of uncomfortable feeling, which was low level in whole, was inverted. However, the correlation between the %MVC and the subjective score was quite high in each posture. Basically there was some relationship between them. As for the posture, the "bending arms" posture was more natural for the switch operation than the "stretching arms" posture, so generation of muscle power might be easier. Therefore, it was thought that the %MVC reflected not only uncomfortable feeling but also easiness of generation of muscle power, and its generation was easier in the "bending arms" condition.

Whereas a problem mentioned above was remained, this study made the following contribution. First, there was a possibility that the uncomfortable feeling for the switch operation had more relationship with %MVC of thumb than overall four muscles on hand and forearm, which means that the uncomfortable feeling of the participant may be caused by the muscle load of thumb. Next, it was thought that the muscle load for push operation had higher sensitivity than pull operation to the uncomfortable feeling. Furthermore, it was suggested that the participant with shorter thumb was forced more muscle load by the operation even if the thumb reached the switch.

This study showed a possibility of the evaluation of the muscle load during the operation of cross-shaped switch by thumb. However, for instance, the other study [3] indicated that the complex operation of consumer products could be evaluated by the amplitude probability distribution function (APDF) [4]. Like this, there is a possibility that the other method and index may be more suitable than %MVC. Moreover, physical attribute, test muscles and posture should be widely examined. More systematic experiments will be required in future.

References

1. National Institute of Bioscience and Human-Technology (1994) Reference manual of anthropometry in ergonomic design, 1st edn. Japan Publication Service, Tokyo. (in Japanese)
2. Kato Z, Okubo T (2006) Shogakusha-no-tameno seitai-kinou-no hakarikata, 2nd edn. Japan Publication Service, Tokyo (in Japanese)
3. Tomioka K (2001) Study on myoelectric APDF for consumer product development and evaluation. Toshiba Rev 56(6):58–61 (in Japanese)
4. Hagberg M (1979) The amplitude distribution of surface EMG in static and intermittent static muscular performance. Eur J Appl Physiol Occup Physiol 40(4):265–272

Affective Evaluation of a VR Animation by Physiological Indexes Calculated from ECGs

Kodai Ito[1(✉)], Naoki Miura[2], and Michiko Ohkura[2]

[1] National Institute of Advanced Industrial Science and Technology,
2-3-26 Aomi, Koto-ku, Tokyo 135-0064, Japan
[2] Shibaura Institute of Technology,
3-7-5, Toyosu, Koto-ku, Tokyo 135-8548, Japan
nb15501@shibaura-it.ac.jp

Abstract. Virtual reality (VR) continues to make remarkable advances. Many devices and contents for VR systems have been developed. In our previous research, we evaluated the feeling of excitement experienced through VR systems using biosignals and proposed the standard deviation of NN intervals (SDNN) and R-R interval variability (RRV) from electrocardiograms (ECGs) as useful physiological indexes to estimate the degree of feeling. In this paper, we evaluated a commercially available VR animation content developed for PlayStation VR (PSVR). Based on our evaluation results, we constructed models to express the feelings evoked by the events in the content by regression analysis using various physiological indexes of ECG as variables.

Keywords: Affective evaluation · Virtual reality · ECG

1 Introduction

Virtual reality (VR) has recently made remarkable progress. It had a banner year in 2016 and has attracted substantial media attention because many companies have launched head-mounted displays (HMDs) and related products [1]. In October 2016, Sony Interactive Entertainment released PlayStation VR (PSVR) as a home-use HMD [2]. Many kinds of game software for PSVR are being sold.

Evaluating the affective values of an interactive system just using such subjective evaluation methods as questionnaires is almost impossible because they suffer from the following drawbacks [3]:

- linguistic ambiguity;
- interfusion of experimenter and/or participant intentions with the results;
- interruption of the system's stream of information input/output.

Solving these problems is crucial to evaluate the degree of interest in and/or the excitement of a constructed interactive system and identifying the moment of excitement. We began our research into objectively evaluating interactive systems by

© Springer Nature Switzerland AG 2019
S. Bagnara et al. (Eds.): IEA 2018, AISC 827, pp. 329–339, 2019.
https://doi.org/10.1007/978-3-319-96059-3_36

quantifying sensations using biological signals that offer the following advantages and overcome the above questionnaire drawbacks:

- can be measured using physical quantities;
- can avoid influence from the intentions of experimenters and participants;
- can be continuously measured.

Affective experiences are described in two dimensions: valence and arousal [4]. Figure 1 shows an example of classifying emotional words. Even though there are many studies on affective evaluation using physiological indexes, few have evaluated high valence and high arousal feelings. Therefore, we evaluated the feelings of excitement by physiological indexes and proposed useful indexes [5–7]. For example, when participants feel excited, SDNN and RRV from electrocardiogram (ECG) averages are lower than when they do not feel excited. SDNN denotes the standard deviation of the intervals between the R-waves (RRIs) of the ECGs. RRV denotes the SDNN and RRI ratio [8].

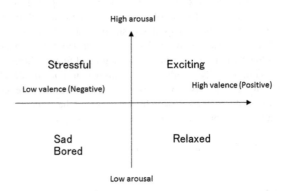

Fig. 1. Example of classifying expressional words by valence and arousal axes.

In this paper, we evaluated a retailed VR animation content developed for PSVR. Based on the evaluation results, we constructed models to express the feelings caused by the events in the content by regression analysis using various physiological indexes of ECG as their variables.

2 Preliminary Experiment

2.1 Content

The participants watched the following six scenes of VR animation content called *Allumette*, which is based on the Hans Christian Andersen story, *The Little Match Girl:*

1. Opening (104 s)
   Lights appear in darkened windows one after another, and silhouette people inside.
2. Night scene 1 (125 s)

As a girl walks in the city at night, the wind blows and a match falls to the ground. When she picks it up and strikes it, she is engulfed in light.

3. Previous scene 1 (270 s)

A flying boat's engine breaks down, but the girl's mother repairs it. After that, a mother and her daughter arrive at a town in the sky. The girl leaves the ship to sell matches. After a blind man drops his cane, the mother hands him a match as a substitute for the cane.

4. Night scene 2 (48 s)

The match goes out, and the scene returns to a city at night. The girl takes out another match and lights it, and she is again engulfed in light.

5. Previous scene 2 (257 s)

The girl notices the ship's accident and boards it alone. The ship becomes engulfed in flames, and the mother rescues her daughter. However, the mother continues to pilot the ship. The ship doesn't fall into the city and explodes in the sky.

6. Final scene (175 s)

The match goes out, and the scene returns to the night city. Finally, the girl is surrounded by light and is hugged by her mother.

The story is not changed by the participants' operation.

2.2   Experimental System

Figure 2 shows our experimental system. Participants wore HMDs and played the VR game. We measured their ECGs during the experiments.

2.3   Experimental Method

The following are our experiment's procedures:

1. We explained our experiment to the participants.
2. They put on the measuring instruments.
3. We started recording their ECGs and set the first 30 s as the resting state sections.
4. They experienced the content.
5. We stopped recording their ECGs.
6. We removed the instruments.
7. They answered questionnaires.

For our evaluation, we used the results of the questionnaires and the ECGs. Participants filled out questionnaires after they played the content. We defined specific six scenes in the content. Participants answered how they felt for each scene on a 10-step scale with the following emotional words:

- interesting
- exciting (*wakuwaku* in Japanese)
- pounding (*dokidoki* in Japanese)
- relaxing

- thrilling
- sad
- impressive

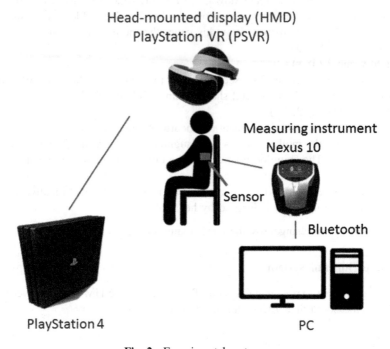

Head-mounted display (HMD)
PlayStation VR (PSVR)

Measuring instrument
Nexus 10

Sensor

Bluetooth

PlayStation 4

PC

Fig. 2. Experimental system.

3   Preliminary Experimental Results and Discussion

3.1   Outline

The preliminary experiment was performed with five male and five female students in their twenties. Figure 3 shows the experimental scene.

Fig. 3. Experimental scene.

3.2   Questionnaire Results

Figure 4 shows the questionnaire results. The vertical axis denotes the 10-step scale points. The horizontal axis denotes the following emotional words, which were selected by the greatest number of participants:

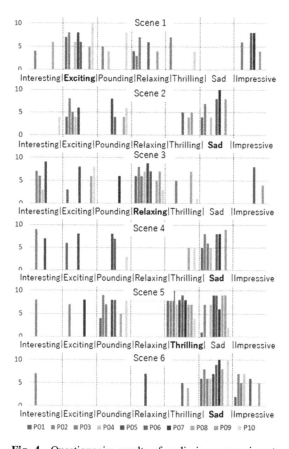

Fig. 4. Questionnaire results of preliminary experiment.

- Scene 1: exciting (7 people)
- Scene 2: sad (6 people)
- Scene 3: relaxing (9 people)
- Scene 4: sad (7 people)
- Scene 5: thrilling (10 people)
- Scene 6: sad (10 people)

From t-test results, we found no significant gender differences.

3.3   ECG Results

From the questionnaire results, we found that the feelings evoked in the participants in each scene were different. Therefore, in the ECG analysis, we separately scrutinized the six scenes and performed stepwise multiple regression analysis for the questionnaire results and the physiological indexes of ECG. We created the following regression formula using each adjective from the questionnaire as a dependent variable and employed various physiological indexes as independent variables. We calculated each physiological index by dividing the various time periods of the scenes:

$$Y = a_0 + \sum_l^2 \sum_k^n \sum_j^3 \sum_i^4 a_{ijkl} x_{ijkl},$$

(N depends on number of seconds of each scene)

where Y = score of 10-step scale evaluation for each participant. We employed the emotional words selected by the greatest number of participants. For example, we employed exciting for scene 1:

- i = type of physiological indexes.
  1: HR (heart rate), 2: SDNN, 3: RRV, 4: RMSSD
- j = length of target section for calculating physiological indexes.
  1: 10 (seconds), 2: 20, 3: 30
- k = starting position of target section. Starting time of each scene was set to 0.
  For example, the length of scene 1 is 104 s. If j = 20, the value of k is one of the followings: 1: 0, 2: 20, 3: 40, ..., 5: 80.
- l = Is the variable different from the resting state?
  1: Yes, 2: No

As an example of the result of multiple regression analysis, below we show the multiple regression equation obtained in scene 1. By the stepwise method, unnecessary variables were excluded. We replaced the subscript of x to each meaning of the variable for readability:

exciting $= 1.140 x_{RMSSD,10,40,N} - 0.594 x_{HR,10,80,Y} + 0.514 x_{RMSSD,10,70,N} + 0.485$
$x_{RMSSD,10,80,N} - 0.252 x_{RMSSD,10,10,N} - 0.095 x_{RRV,20,0,Y} + 0.063 x_{RMSSD,20,40,Y} + 0.047 x_{SDNN,20,0,Y}$ ($R^2 = 1.000$).

As a result of our multiple regression analysis, since RMSSD (length of target section was ten seconds) was obtained most frequently as indexes, it might be a physiological index of various emotions. However, it is difficult to determine meaning from this model because RMSSD has both positive and negative coefficients. The results of other scenes also resembled this result. The cause of these results is over-fitting. The model obtained R2 = 1.000; however it was too high since there were too few participants for the number of variables. Therefore, we increased the number of participants and reduced the number of variables in our next experiment.

We also focused on the starting position of the variable's target section. This suggests that our model can detect the important events in each scene. For example, in scene 1, the RMSSD (calculated from 40 to 50 s) has the highest standardization coefficient. The following are the details of the event from 40 to 50 s: "Several windows floated in the air. Next, the number of windows increased and the world spread across the screen." This event seems exciting, and employing 40 s as the starting position of this variable is appropriate.

In our previous study [4], SDNN and RRV were useful indexes for feelings of excitement. However, they were not obtained as useful indexes in these results. Since SDNN and RRV must be analyzed with sections of ten seconds or longer, we set the analysis section in advance. After that, our analysis used the differences of physiological indexes between the sections before and after the event. However, since the content is telling a story during this time, specifying the analysis target from many events was difficult. Therefore, we separately set the analysis section 10, 20, and 30 s from the scene's start.

In this analysis, we didn't address the SDNN and RRV changes before and after the event. The absolute value of SDNN or RRV in a target section or the differences of these indexes from the resting state were used as a variable. These differences of calculating indexes might have caused our results in this study. Perhaps we can obtain models with SDNN or RRV as a variable if we employ the differences before and after the event as variables.

4 Experiment

Our experimental method was identical as the preliminary experiment.

5 Experimental Results and Discussion

5.1 Outline

The experiment was done with nineteen male and four female students in their twenties. We analyzed the experimental results of 33 people with the participants of our preliminary experiment.

5.2 Questionnaire Results

Figure 5 shows the questionnaire results. The vertical axis denotes the 10-step scale points. The horizontal axis denotes the emotion candidates. The following are the emotional words selected by the greatest number of participants:

- Scene 1: exciting (28 people)
- Scene 2: sad (22 people)
- Scene 3: relaxing (30 people)
- Scene 4: sad (28 people)
- Scene 5: thrilling (30 people)
- Scene 6: sad (29 people)

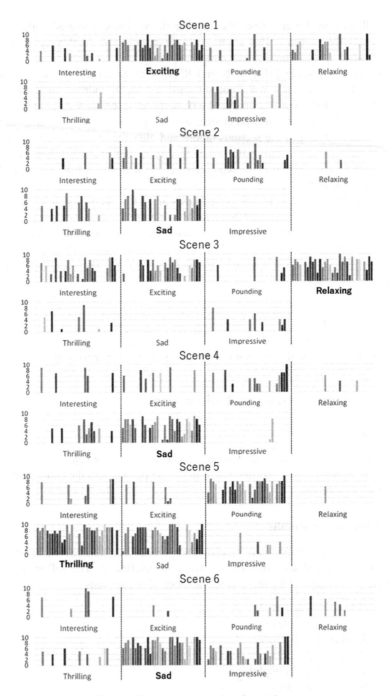

Fig. 5. Questionnaire results of experiment.

5.3   Questionnaire Results

We performed stepwise multiple regression analysis for the questionnaire results and the physiological indexes of ECG. We created the following regression formula using each adjective of the questionnaire as a dependent variable and employed various physiological indexes as independent variables. We calculated each physiological index by dividing various time periods of scenes:

$$Y = a_0 + \sum_{k}^{3} \sum_{j}^{3} \sum_{i}^{4} a_{ijk} x_{ijk},$$

(N depends on the number of seconds of each scene)

where Y = score of 10-step scale evaluation of emotional words representing each scene for each participant. For example, we employ exciting for scene 1.

i = type of physiological indexes.
    1: HR (heart rate), 2: SDNN, 3: RRV, 4: RMSSD
j = length of target section for calculating physiological indexes.
    1: 10 (seconds), 2: 20, 3: 30
k = starting position of target section. Starting time of each scene was set to 0.
We decided starting positions based on the preliminary experiment results. The following are the t values:
    Scene 1 ... 1: 40 (second), 2: 70, 3: 80
    Scene 2 ... 1: 30, 2: 70, 3: 100
    Scene 3 ... 1: 30, 2: 240, 3: 200
    Scene 4 ... 1: 10, 2: 20, 3: 30
    Scene 5 ... 1: 90, 2: 100, 3: 40
    Scene 6 ... 1: 20, 2: 30, 3:110

In our previous research [4], we obtained useful SDNN and RRV results by analyzing the differences 10 s before and 10 s after the event. Figure 6 shows an image of the analysis sections. We calculated physiological indexes at sections A and B and subtracted the value of B from A as variables. However, in the case of $j > k$, we did not calculate the variables:

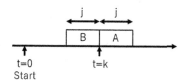

Fig. 6. Image of sections for analysis.

As an example of the result of multiple regression analysis, the multiple regression equation obtained from scene 1 is shown below. By the stepwise method, unnecessary variables were excluded. We replaced the subscript of x to each meaning of variable for readability:

$$exciting = 0.650x_{HR,10,80} - 0.514x_{SDNN,30,80} - 0.329x_{RRV,10,40} \ (R2 = 0.541).$$

We employed the differences of the physiological indexes before and after the target section as variables in our analysis. The standardization coefficient of $x_{HR, 10, 80}$ was 0.650, suggesting that if the HR of 80 to 90 s is higher than the HR of 70 to 80 s, the exciting score will also be higher. In the same way, the second and third variables mean that the exciting score will be higher if SDNN and RRV are lower.

The averages of HR, SDNNs and RRVs were consistent with the results of a previous study [4]. In addition, since the model's explanatory power was 0.541, it is appropriate.

We also solved the overfitting problem in these results. On the other hand, we failed to obtain a model from scene 4. By changing the selection method of the analysis section or adding another physiological index, we may be able to obtain models with higher explanatory power. However, when increasing the number of variables, we have to increase the number of participants to prevent overfitting.

6   Conclusion

We focused on the emotions evoked during a VR animation and we constructed models to express the feelings using various physiological indexes of ECG as variables. The VR animation's story was composed of six scenes. Based on the questionnaire and ECG results, we obtained useful models of emotions. Since some of our models have the same tendency as the results of previous research, our models are appropriate. By analyzing using our proposal, the emotional features of contents can be estimated by the physiological indexes of ECG. Future work will improve our method by setting the starting positions of the analysis sections and employing more variations of the length of the target sections for calculating physiological indexes. We also have to pay greater attention to overfitting.

References

1. (2016) The Year of VR?-Virtual Reality. http://www.vrs.org.uk/news/2016-the-year-of-vr
2. Sony Interactive Entertainment, "PlayStation VR,". https://www.playstation.com/en-us/explore/playstation-vr/
3. Ohkura M, Hamano M, Watanabe H, Aoto T (2011) Measurement of wakuwaku feeling of interactive systems using biological signals. Emotional Engineering. Springer, London, pp 327–343
4. Kensinger EA (2004) Remembering emotional experiences: The contribution of valence and arousal. Rev Neurosci 15(4):241–251
5. Ito K, Harada Y, Tani T, Hasegawa Y, Nakatsuji H, Tate Y, Seto H, Aikawa T, Nakayama N, Ohkura M (2016) Evaluation of feelings of excitement caused by auditory stimulus in driving simulator using biosignals. In: Advances in Affective and Pleasurable Design: Proceedings of the AHFE 2016, vol 483, pp 231–240

6. Ito K, Usuda S, Yasunaga K, Ohkura M (2017) Evaluation of 'feelings of excitement' caused by a VR interactive system with unknown experience using ECG. In: Advances in Affective and Pleasurable Design: Proceedings of the AHFE 2017, vol 585, pp. 292–302
7. Ito K, Ohkura M (2017) Evaluation of 'Dokidoki feelings' for a VR system using ECGs with Comparison between Genders. In: Proceedings of ICBAKE2017, pp 110–114
8. Acharya UR, Joseph P, Kannathal N, Lim CM, Suri J (2007) Heart rate variability: A review. Med Biol Eng Comput 44(12):1031–1051

Dynamic Adjustment of Interpersonal Distance in Cooperative Task

Yosuke Kinoe[(⊠)]

Hosei University, 2-17-1, Fujimi, Chiyoda-Ku, Tokyo 102-8160, Japan
kinoe@hosei.ac.jp

Abstract. This paper presented an empirical study which investigated a dynamic adjustment of interpersonal distance in a cooperative situation. In the experiment, three factors of the "task", "orientation", and "gender combination" were emphasized. Ninety-six data of interpersonal distances were obtained under four different conditions from twenty-four participants.

First, ANOVA and multiple comparisons revealed that the factors of the "task", "orientation", and "gender combination" had statistically significant influences on the adjustment of interpersonal distance. In particular, the results indicated that (1) preferred interpersonal distances were shortened when a dyad initiated a "cooperative task", (2) preferred interpersonal distance of "side-by-side" was shorter than "face-to-face", and (3) interpersonal distance of "female-female" pair was shorter than "male-male" pair and it was also shorter than "heterogeneous gender" pair.

Furthermore, correlation analyses, under "face-to-face" condition, revealed that, (1) evaluators who had longer interpersonal distances of "no task" shortened larger amount of their interpersonal distances when a dyad initiated a cooperation, and more interestingly, (2) a larger divergence of preferred interpersonal distances of "no task" condition among each dyad derived a larger amount of shortening of interpersonal distance when a dyad initiated a cooperation. The result also suggested that interpersonal distance of an evaluator and an approacher had a significant correlation under "cooperative task" condition.

Finally, beyond a self-centered view of personal space, this paper suggested an interesting view of phenomenon of adjusting interpersonal distance, from a dynamic interactive process of an evaluator and a confederate. An implication to the design of spatial behavior of social assistive robots was also discussed.

Keywords: Personal space · Spatial behavior · Cooperation

1 Introduction

1.1 Personal Space

Personal space concept is a useful tool to investigate human spatial behavior in our everyday social interpersonal transactions [1]. It can be defined as "an area individuals actively maintain around themselves into which others cannot intrude without arousing some sort of discomfort" [14]. People utilize personal space in everyday situations,

© Springer Nature Switzerland AG 2019
S. Bagnara et al. (Eds.): IEA 2018, AISC 827, pp. 340–349, 2019.
https://doi.org/10.1007/978-3-319-96059-3_37

however, most of time, people are unaware of its sophisticated functioning, while they are comfortable in interpersonal transactions with others.

The consideration of personal space issue gained the increased importance in the design of humane services and comfortable environment, for instance, in the domains including medical-care and nursing-care. Recently it has been emphasized as an interesting emerging issue in the design of proxemics in human-robot interaction [4, 10].

1.2 Dynamic Adjustment of Interpersonal Distance

Personal space is sometimes referred to an invisible "body buffer zone" [2, 7] we carry around with us. This description can be helpful, but we need to be aware that the dimensions of personal space are not fixed. Personal space can transform according to a situation [8]. We construct and adjust the amount of personal space that is appropriate between ourselves and other people.

Research findings suggested that interpersonal distance is a function of various factors, which can be classified into, at least, four broader categories: personal (*e.g.* [5]), social (*e.g.* [11]), physical (*e.g.* [3]), and cultural (*e.g.* [1]). There are more than thousand studies on the determinants of personal space to describe them. However, few studies were concerned with interpersonal distancing in a cooperative situation (*e.g.* [9, 12, 15]).

The present study aimed to investigate how people dynamically adjust interpersonal distance in a cooperative situation. The term, *interpersonal distance*, was used to describe human spatial behavior by using the measurement of the space between interacting individuals [13].

2 Experimental Study

There were three factors in the present study. The within-subject factors were "task" (2 levels: no particular task vs. cooperative task) and "orientation" of the body (2 levels: face-to-face vs. side-by-side). The between-subject factor was "gender combination" (3 levels: male-male vs. female-female vs. heterogeneous genders).

2.1 Method and Settings

Participants. Twenty four healthy, young adults (12 males, 12 females; Japanese; aged 18–23 years) recruited at a university in Tokyo, participated in the study. The participants were informed that the study dealt with their spatial preferences. They gave their informed consent before participation.

Procedure. All the participants were divided into twelve pairs, who were not acquaintances each other. The data collection was performed by each pair of participants (A and B). The distribution of gender combination was: male-male (4), female-female (4), and heterogeneous genders (4). The height differences among a dyad were less than 17 cm.

The stop-distance method [6] was employed to obtain interpersonal distances. At first, one of the participants (A) was assigned as an evaluator, and the other (B) took a role of an approacher. In "no task" condition, B approached A slowly from a distance of 3 M. The participant-A was asked to say stop when she/he felt uncomfortable about the closeness (Fig. 1). The remaining distance between the centers of their bodies was measured. On the other hand, in "cooperative task" condition, an approacher was asked to approach and an evaluator was asked to perform jigsaw-puzzles with an evaluator using a Macbook.

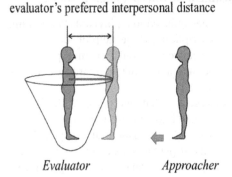

evaluator's preferred interpersonal distance

Evaluator   *Approacher*

Fig. 1. Evaluator's preferred interpersonal distance: an *evaluator-centered* view.

After four interpersonal distances under different conditions (2 tasks × 2 orientations) were obtained from each participant, then the participants exchanged their roles.

Settings. The data collection was carried out during daytime, in a quiet class room of a university, with a ceiling height of 3 M. The facility located in Tokyo. The brightness was appropriately maintained with an indoor lighting.

Measurements and Apparatus. In order to obtain interpersonal distances in a cooperative task, we applied the "center-center" model [8] which employed the distance between the vertexes of each dyad of participants (Fig. 1). The distances were measured by using a laser range finder (accuracy: $+/-1.5$ mm). In a cooperative task condition, participants were asked to play jigsaw puzzles application by using a notebook PC (Macbook Pro, 15 inch, 2.04 kg).

2.2 Data Analysis

Three factors ANOVA (analysis of variance), multiple comparisons with Bonferroni adjustment, and correlation analyses were carried out with SPSS statistics version 22.

3 Results

Ninety six data of interpersonal distances were obtained under four different conditions. Mean of observed distances was 68.49 cm (SD = 24.41). At first, three factors ANOVA and multiple comparisons with Bonferroni adjustment were carried out.

3.1   Effects on a Dynamic Adjustment of Interpersonal Distance

Task. The analysis revealed that there were statistically significant simple main effects of the "task" (F = 12.52, p < 0.01). In particular, the result indicated that preferred interpersonal distance of "cooperative task" was shorter than that of "no particular task" (difference = 16.8 cm, standard error (SE) = 4.73, p < 0.01) under either condition of the "orientation of the body" (Fig. 2).

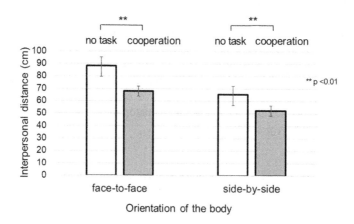

Fig. 2. Mean of interpersonal distance: task & orientation of the body.

Orientation. The analysis revealed that there were statistically significant simple main effects of the "orientation of the body" (F = 95.60, p < 0.001). In particular, the result indicated that preferred interpersonal distance of "side-by-side" was shorter than "face-to-face" (difference = 19.3 cm, SE = 2.50, p < 0.001).

Gender Combination. The analysis revealed that there were statistically significant simple main effects of the "gender combination" (F = 6.60, p < 0.01). In particular, the result indicated that interpersonal distance of "female-female" pair was shorter than "male-male" pair (difference = 20.7 cm, SE = 6.29, p < 0.05), and it was also shorter than "heterogeneous gender" pair (difference = 18.74 cm, SE = 6.29, p < 0.05).

3.2   Exploring Potential Effects on the Shortening of Interpersonal Distance

The above results of multiple comparison indicated that preferred interpersonal distance of the "cooperative task" was shorter than that of "no particular task", particularly in the face-to-face condition (Fig. 2). This subsection aimed to explore other potential determinant elements than the "task" factor on the shortening of interpersonal distance in a cooperation. Our focus was narrowed down to a cooperative situation standing face to face. Hereafter, preferred interpersonal distance of, for instance, an evaluator's "cooperative task" condition, can be abbreviated to "D_{ev_coop}".

Shortening of Interpersonal Distance When Initiating a Cooperative Task (Fig. 3). We carried out correlation analyses of the amount of the shortening of interpersonal distance when initiating a cooperative task with several other aspects of variables. Table 1 summarizes the results of correlation analyses related to this topic.

B-1. Correlation between the Amount of Shortening when Initiating a Cooperation and an Evaluator's Interpersonal Distance of "No Task" Condition. The analysis revealed that the amount of shortening of interpersonal distance between the condition of "no task" and "cooperative task" (D_{ev}_notask - D_{ev}_coop) had a strong positive correlation with interpersonal distances of "no task" condition (D_{ev}_notask) ($r = 0.803$, $p < 0.01$).

B-2. Correlation between the Amount of Shortening when Initiating a Cooperation and the Difference of Interpersonal Distance of "No Task" among a Dyad. On the other hand, the analysis revealed that the amount of shortening of interpersonal distance between "no task" condition and "cooperative task" condition (D_{ev}_notask - D_{ev}_coop) had a statistically significant positive correlation with the difference of interpersonal distances among a dyad of an evaluator and an approacher (D_{ev}_notask - D_{ap}_notask) ($r = 0.537$, $p < 0.01$).

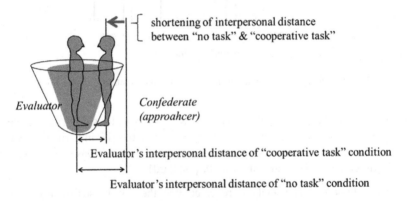

Fig. 3. Shortening of interpersonal distance between "no task" & "cooperative task".

The above results indicated that evaluators who had longer interpersonal distances of "no task" shortened larger amount of their interpersonal distances when a dyad initiated a cooperation. More interestingly, a larger divergence of preferred interpersonal distances of "no task" condition among each dyad (Fig. 4) derived a larger amount of shortening of interpersonal distance when a dyad initiated a cooperation. The results suggested that this shortening was influenced by not only an evaluator's own attribute, but also a certain factor related to an approacher's attribute.

Examining the Flexibility of Interpersonal Distance in a Cooperation. The "task" can be considered a complex factor involving at least "type of device". The experimental setting of the "cooperative task" condition should be precisely described by decomposing in more detail. It can be described as "playing together jigsaw puzzles" "face to face" using a "15 inch Macbook".

Table 1. Correlations between the amount of shortening of interpersonal distance when initiating a cooperation, and other variables.

Correlation analysis		Correlation coefficient	Frequency
Variable 1	Variable 2		
Amount of shortening of interpersonal distance between "no task" condition and "cooperative task" condition (D_{ev}_notask - D_{ev}_coop)	Evaluator's interpersonal distance of "no task" condition (D_{ev}_notask)	0.803(**)	24
	Difference of interpersonal distance of "no task" condition among evaluator & approacher (D_{ev}_notask - D_{ap}_notask)	0.537(**)	24

Note. Pearson's correlation coefficient, ** $p < 0.01$, * $p < 0.05$, n.s. (not significant).

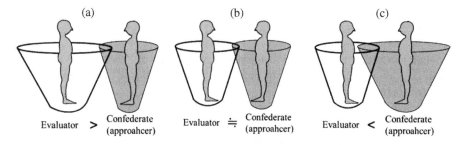

Fig. 4. Divergence of preferred interpersonal distances among an evaluator & a confederate.

The question was whether evaluators' interpersonal distances of "cooperative task" were converged into the single value, which was optimal distance simply determined by the task type (jigsaw puzzles) and the device type (15 inch Macbook). If it was fixed or converged, a longer interpersonal distance of "no task" (D_{ev}_notask) could simply cause a larger amount of its shortening (D_{ev}_notask - D_{ev}_coop).

B-3. Flexibility of Interpersonal Distance of "Cooperative Task". The Levene test was carried out to make sure that an evaluator's interpersonal distance in a cooperative task condition was flexible and that we could assume the equality among the distributions of interpersonal distances (face to face) of "no task" condition (mean = 88.37, SD = 28.78) and that of "cooperative task" condition (mean = 67.97, SD = 19.27). The result of the Levene statistics was 0.361 (n.s., p = 0.55). We could assume its equality of the distributions of interpersonal distances among them.

Examining Direct Influences of a Confederate's Interpersonal Distance. Then, next question was whether an evaluator's interpersonal distance of "cooperative task" condition was directly influenced by an approacher's interpersonal distance of "no task" condition. Table 2 summarizes the results of correlation analyses related to this topic.

B-4. No Correlation between Interpersonal Distance of "Cooperative Task" and that of "No Task". At first, the correlation analysis revealed no statistically significant

Table 2. Correlations between interpersonal distance of "no task" condition, "cooperative task" condition, and a confederate's interpersonal distance of "no task" condition.

Correlation analysis		Correlation coefficient	Frequency
Variable 1	Variable 2		
Interpersonal distance of "cooperative task" condition (D_{ev_coop})	Interpersonal distance of "no task" condition (D_{ev_notask})	0.167(n.s.)	24
Interpersonal distance of "cooperative task" condition (D_{ev_coop})	Approahcer's interpersonal distance of "no task" condition (D_{ap_notask})	0.121(n.s.)	24
Amount of shortening of interpersonal distance between "no task" condition and "cooperative task" condition (D_{ev_notask} - D_{ev_coop})	Approacher's interpersonal distance of "no task" condition (D_{ap_notask})	0.139(n.s.)	24

Note. Pearson's correlation coefficient, ** $p < 0.01$, * $p < 0.05$, n.s.: not significant.

correlation between an evaluator's interpersonal distance of "cooperative task" condition (D_{ev_coop}) and her/his interpersonal distance of "no task" condition (D_{ev_notask}) ($r = 0.167$, $p = 0.44$, n.s.).

B-5. No Correlation between Interpersonal Distance of "Cooperative Task" and an Approacher's Interpersonal Distance of "No Task". The analysis revealed there was no statistically significant correlation between interpersonal distance of "cooperative task" condition (D_{ev_coop}) and an approacher's interpersonal distance of "no task" condition (D_{ap_notask}) ($r = 0.121$, $p = 0.57$, n.s.).

B-6. No Correlation between the Amount of Shortening when Initiating a Cooperation and Approacher's Interpersonal Distance of "No Task" Condition. The analysis revealed there was no significant correlation between the amount of shortening of interpersonal distance between "no task" condition and "cooperative task" condition (D_{ev_notask} - D_{ev_coop}) and approacher's interpersonal distance of "no task" condition (D_{ap_notask}) ($r = 0.139$, $p = 0.52$, n.s.).

The above results suggested that interpersonal distances of "no task" condition of neither an approacher nor an evaluator had direct influences on an evaluator's interpersonal distance of "cooperative task" condition. The results also suggested that an approacher's interpersonal distance of "no task" condition didn't have a direct influence on the shortening of interpersonal distance when initiating a cooperation.

Possibility of an Influence of an Interaction Among an Evaluator and a Confederate. Then, we attempted further correlation analyses to shed light on another hidden dimension, for instance, an interactive process between an evaluator and an approacher. The Table 3 summarizes the results of correlation analyses related to this topic.

B-7. Correlation of Interpersonal Distance ("cooperative task" condition) among an Evaluator and an Approacher". Prior to the analysis, we simply assumed that evaluator's interpersonal distance was independent of that of an approacher's. However, in "cooperative task" condition, the analysis revealed a statistically significant positive correlation of interpersonal distance among a dyad of an evaluator (D_{ev_coop}) and an approacher (D_{ap_coop}) (r = 0.673, p < 0.001).

B-8. Weak Correlation of Interpersonal Distance ("no task" condition) among an Evaluator and an Approacher". Prior to the analysis, we simply assumed that evaluator's interpersonal distance of "no task" condition was independent of that of an approacher's. The analysis revealed there was no statistically significant but a very weak correlation between them, even under "no task" condition (r = 0.235, p = 0.27, n.s.).

Table 3. Correlations between an evaluators' interpersonal distance of "cooperative task" condition and a confederate's.

Correlation analysis		Correlation coefficient	Frequency
variable 1	variable 2		
Interpersonal distance of "cooperative task" condition (D_{ev_coop})	Approacher's interpersonal distance of "cooperative task" condition (D_{ap_coop})	0.673(**)	24
Interpersonal distance of "no task" condition (D_{ev_notask})	Approacher's interpersonal distance of "no task" condition (D_{ap_notask})	0.235(n.s.)	24

Note. Pearson's correlation coefficient, ** p < 0.01, * p < 0.05, n.s.: not significant.

The above results indicated that interpersonal distance of an evaluator and an approacher had a significant correlation under "cooperative task" condition, and also a weak correlation was revealed even under "no task" condition. This results need to be carefully re-examined in further statistical analyses of empirical studies, however, it suggested a possibility of an influence on interpersonal distance of dynamic interactive process of an evaluator and a confederate.

4 Discussion

Implication to the Design of Spatial Behavior of Socially Assistive Robots. Robot-human proxemics has recently been emphasized as one of the important component in the design of spatial behavior of socially assistive robots including a nursing-care robot. To enable situated human-robot interaction, an autonomous robot is desired to both understand and control the social use of space to employ natural communication mechanisms analogous to those used by humans [10]. Garzotto, et al. developed a mobile inflatable interactive robot for children with Neurodevelopmental Disorder (NDD), which employed their framework of "Interactional Spatial Relationship Model" [4]. It aimed to enable a robot's thoughtful spatial and emotional interactions for avoiding and mitigating undesirable situations with children.

On the other hand, the above results suggested, *beyond "self-centered" view* of interpersonal distance, the possibility of interactive process in which an evaluator and a confederate dynamically adjust *their* interpersonal distance in a cooperative situation. This study suggested the importance of not only the design of a robot's spatial behavior but also a robot-human collaboration of constructing an appropriate personal space for a comfortable cooperative environment.

5   Concluding Remarks

The present study investigated a dynamic adjustment of interpersonal distance in a cooperative situation. In this empirical study, three factors involving task, orientation of the body, gender combination, were emphasized. Ninety six data of interpersonal distances were obtained under four different conditions from twenty-four participants.

The results of three factors ANOVA and multiple comparisons revealed that the factors of the "task", "orientation of the body", and "gender combination" had statistically significant influences on dynamic adjustment of interpersonal distance. In particular, the results indicated that (1) preferred interpersonal distances were shortened when a dyad initiated a "cooperative task", under either condition of the "orientation of the body", (2) preferred interpersonal distance of "side-by-side" was shorter than "face-to-face", and (3) interpersonal distance of "female-female" pair was shorter than "male-male" pair and it was also shorter than "heterogeneous gender" pair.

Next, correlation analyses were conducted to explore potential factors related to the shortening of interpersonal distance when a dyad initiated a cooperation. The results, under a condition of standing face-to-face, revealed that, (1) evaluators who had longer interpersonal distances of "no task" shortened larger amount of their interpersonal distances when a dyad initiated a cooperation, and that, more interestingly, (2) a larger divergence of preferred interpersonal distances of "no task" condition among each dyad derived a larger amount of shortening of interpersonal distance when initiating a cooperation. Furthermore, the result also suggested that interpersonal distance of an evaluator and an approacher had a significant correlation under "cooperative task" condition.

The above results will need to be carefully re-examined in further empirical studies. However, importantly, beyond a self-centered view of personal space, this study suggested the possibility of an interesting view of phenomenon of adjusting interpersonal distance, from a dynamic interaction between an evaluator and a confederate.

Acknowledgement. We thank all the study participants, and Namiko M., who devotedly supported for conducting experiments.

References

1. Aiello JR (1987) Human spatial behavior. In: Stokels D, Altman I (eds) Handbook of environmental psychology. Wiley, New York, pp 385–504
2. Dorsey M, Meisels M (1969) Personal space and self-protection. J Pers Soc Psychol 11: 93–97

3. Evans GW, Lepore SJ, Schroeder A (1996) The role of interior design elements in human responses to crowding. J Pers Soc Psychol 70:41–46
4. Garzotto F, Gelsomini M, Kinoe Y (2017) Puffy: a mobile inflatable interactive companion for children with neurodevelopmental disorder. In: Bernhaupt R (eds) Human-computer interaction - INTERACT 2017, Lecture notes in computer science, vol 10514. Springer, pp 467–492
5. Gifford R (1982) Projected interpersonal distances and orientation choices: personality, sex, and social situation. Soc Psychol Q 45:145–152
6. Hayduk LA (1983) Personal space: where we now stand. Psychol Bull 94(2):293–335
7. Horowitz MJ, Duff DF, Stratton LO (1964) Body-buffer zone: exploration of personal space. Arch Gen Psychiatry 11:651–656
8. Kinoe Y, Mizuno N (2015) Situational transformation of personal space. In: Yamamoto S (eds) Human interface and the management of information, Information and knowledge in context, Lecture notes in computer science, vol 9173. Springer, Heidelberg, pp 15–24
9. Kinoe Y (2018) Interpersonal distancing in cooperation. In: Proceedings of the 20th international conference on human-computer interaction (in press)
10. Mead R, Matarić MJ (2017) Autonomous human–robot proxemics: socially aware navigation based on interaction potential. Auton Robot 41(5):1189–1201
11. Rosenfeld P, Giacalone R, Kennedy J (1987) Of status and suits: personal space invasions in an administrative setting. Soc Behav Pers 15:97–100
12. Sinha SP, Mukherjee N (1995) The effect of perceived cooperation on personal space requirements. J Soc Psychol 136(5):655–657
13. Sommer R (2002) Personal space in a digital age. In: Bechtel RB, Churchman A (eds) Handbook of environmental psychology. Wiley, New York, pp 385–504
14. Sommer R: Personal space: the behavioral basis of design. Bosko Books (2008). Updated
15. Tedesco JF, Fromme DK (1974) Cooperation, competition, and personal space. Sociometry 37:116–121

Evaluation of the Index of Cognitive Activity (ICA) as an Instrument to Measure Cognitive Workload Under Differing Light Conditions

Lisa Rerhaye[✉], Talke Blaser, and Thomas Alexander

Fraunhofer Institute for Communication, Information Processing
and Ergonomics (FKIE), Zanderstraße 5, 53177 Bonn, Germany
lisa.fromm@fkie.fraunhofer.de

Abstract. A straightforward and valid instrument for measuring cognitive workload would be heavily appreciated in many research areas, such as human-machine-interaction, driver behavior (e.g. automation and fatigue), usability and UI design (e.g. adaptive displays), training and education, or other areas, that are interested in the assessment of the cognitive state of a person. The Index of Cognitive Activity (ICA) is a promising but also controversially discussed instrument that could be of high relevance if it keeps its promises. The ICA is a patent from the year 2000, which claims to be an effective, light-independent recording method of mental workload.

On the basis of a literature research, we carried out a lab experiment to evaluate the ICA. Participants were equipped with an Eyetracking device and worked on a mental rotation task and a Stroop task under varying light conditions. The NASA-TLX was to be answered after each test condition to evaluate the subjective workload of the participants in each condition. If the ICA is truly light-independent, the ICA should show the same mental workload for each light condition. Results show expected ICA values for the Spatial Task, but inconclusive ICA values for the Stroop Task. Possible explanations and future work is discussed.

Keywords: Mental workload · Pupillometry · Index of Cognitive Activity

1 Theoretical Background of the Study

The Index of Cognitive Activity (ICA) has been patented in 2000. It claims to be an effective, light-independent recording method for mental workload [1]. Its validity has been controversially discussed: Neither the definition of the neuronal basics [2] nor the adequate measurement of mental workload succeeded in over 40 years of research. On the basis of a literature research, we carried out an experiment to evaluate the Index of Cognitive Activity in a laboratory experimental setting under varying light conditions to test its independence from light adaptations of the pupil.

1.1 The Concept of Mental Workload

Mental workload refers to the information processing power that the brain uses to perform one or more tasks within a certain period of time. What exactly is understood

© Springer Nature Switzerland AG 2019
S. Bagnara et al. (Eds.): IEA 2018, AISC 827, pp. 350–359, 2019.
https://doi.org/10.1007/978-3-319-96059-3_38

by mental workload, however, varies greatly [3]. Damos has defined mental workload as the resources or capacities of information processing that are needed to meet specific requirements [4]. The mental workload basically consists of two components: The input load and the effort. The input load refers to the tasks to be processed and is made up of the environment, the situation and the processes. It is basically the same for all persons. The effort is an individual size. It refers to subjective information processing and is influenced by a wide range of variables. Such variables range from the activation level and previous experience to personality traits or current moods. Measurements of mental workload largely refer to the measurement of effort [3]. Therefore, no distinction is made in the following between an assessment of the input load and the effort.

Even if the interpretation of the composition of mental workload is still an urgent research topic, a parameter that is sensible to workload differences would be very useful, e.g. to examine peaks of workload. Our findings emphasize the importance of a clear definition of what is really measured by the respective measurement method of mental workload before a study is conducted.

1.2   The Operationalization of Mental Workload

The operationalization of mental workload, meaning the process of defining mental workload into measurable factors, has not been successful so far. An important factor that makes the measurement of mental workload problematic is the great variety of influence variables. Thus, it is necessary to identify a measurable correlate of mental workload that was not triggered by one of its many side effects. With increasing mental workload, for instance, the stress level can also increase if the subjective assessment of the feasibility of the task turns out negative [2].

The neuronal processes associated with mental workload have not yet been clearly identified [2]. Therefore, the application of electrophysiological methods to assess mental workload is subject to great interpretation difficulties. Among others, this implies questions like: Are the measured neuronal signals triggered by the perception or processing of certain stimuli or do they even symbolize the emotional component? Is mental workload the interaction of all these processes or does it only refer to targeted information processing? Furthermore, procedures that visualize mental workload at the neuronal level are often not practicable due to the high technical effort involved at the application level.

Questionnaires are often considered the most suitable method to measure cognitive workload [5]. However, surveys also have considerable disadvantages. The selection of the scaling, the formulation of the question or the time of the survey can strongly influence the respondents' response behavior. It can be biased to such an extent that different results are achieved for the same underlying task only on the basis of the type of survey [6]. Due to the inconsistency of the definitions of mental workload and the great difficulties in recording it, it is still being discussed whether an adequate measurement of mental workload is possible.

Pupillometry for Measuring Mental Workload

Pupillometry is a frequently discussed method for recording mental workload. It uses the already replicated finding that the pupil widens under high mental effort [7, 8].

Although this dilation effect has been reliably replicated, it has not yet been adequately exploited [2, 7, 8]. This is partly because the pupil size is influenced by many other parameters. Loose [9] reports among other things fear, fatigue, intelligence, sympathy and reward as influencing variables. On the other hand, pupillometry can only be used in laboratory experiments, as the influence of changing light conditions on pupil size would otherwise make measurement impossible. Advantages are its objectivity or the online availability of the signal [2]. The most decisive advantage of pupillometry is the exact temporal resolution. The pupil reacts with a latency of less than 100 ms to an increase in mental workload [10] and reaches the peak of dilation about 1200 ms after stimulus onset [11]. The rapid and sensitive reaction of the pupil has led to the fact that pupillometry has already frequently been used to measure mental workload. It was, e.g., successfully used in a Stroop Task experiment, where the pupillometric recording of mental workload was validated [12].

Marshall presented a method based on pupillometry which claims to be able to record mental workload with high temporal resolution, low technical effort and application-oriented: The Index of Cognitive Activity (ICA). The ICA is a patented algorithm from the year 2000 [13]. However, since neither the identification of the neuronal basis [2] nor an adequate measurement of mental workload in over 40 years of research have been successful, doubts remain as to the effectiveness of the ICA. For this reason, the ICA is discussed below on the basis of extensive research and evaluated in a laboratory test.

1.3 Background and Function of the ICA

The ICA is based on the observation that increased mental workload is accompanied by dilation of the pupils [7]. This is the pupil dilation reflex [1, 9]. It has to be distinguished from the pupil light reflex, which causes a change in pupil size depending on the incident light [1]. Two muscle groups are involved in the change in pupil size: the annular sphincter and the radial dilator. The sphincter is essentially responsible for the reduction of the pupil under increased light incidence. According to Marshall, the dilator is only active when the cognitive workload increases and causes the typical pupil dilation through its contraction. Due to the different arrangement (annular and radial), the respective contraction takes place at different frequencies. More specifically, the rejuvenation process of the sphincter is described as slower and longer, whereas the dilation reflex is described as short-term and pulsating [13]. Hampson et al. also agree with these assumptions and found a connection between rapid, short-term movements of the pupils and mental workload [10]. In the case of the ICA, the different frequencies of dilation and light reflex are to be separated from each other using wavelet analysis, so that the ICA should be a light-independent measure of mental workload. The ICA can be recorded over any time interval, depending on the update rate of the recording device. In the case of an interval of one second and an update rate of 250 Hz, the ICA can therefore assume a value between 0 and 250. For easier handling, however, it is usually standardized, so that the ICA is reported in a range of 0–1. The ICA is collected with the help of an image acquisition device, for example the Dikablis system by Ergoneers GmbH. Through the combination of eye tracking and the analysis of mental workload, conclusions can be drawn about the causes of increased cognitive workload

[13]. For example, it could be investigated how much attention driver assistance systems require, or how much cognitive resources a firefighter has to spend for a decision in a critical situation.

Overall, it can be stated that the patent of the ICA describes the neuronal basis of the dilation reflex in low detail [13]. It names one comprehensive book as the only source for the underlying biological processes [14]. The biological processes adopted by Marshall were compared with the well documented data from Loose [9]. It has been shown that some assumptions critical for the evidence of the ICA are not mentioned in Loose [9] and that precisely this information is not clearly documented in Marshall. Nevertheless, there is a great deal of agreement.

Based on the literature, the following hypotheses are put forward: (1) The ICA distinguishes between high and low workload conditions, and (2) the ICA does not distinguish between the light variations.

2   Laboratory Test to Evaluate the ICA

For the application of the ICA in the field, it is necessary that the ICA measures validly and reliably even under strongly changing light conditions. Under the patent of Marshall, the ICA was tested under only two light conditions (dark vs. light). To the best of our knowledge, the ICA was not tested under permanently changing light conditions [1, 13], which is necessary for the application of the ICA in field research. For testing under these special conditions, a first evaluation experiment was carried out in laboratory of Fraunhofer FKIE.

2.1   Experimental Design

A laboratory test was conducted to evaluate the ICA. The question was whether the ICA can measure mental workload independently of the light influences. This required the selection of a task that can clearly induce mental workload in different gradations. The Criterion Task Set is a well evaluated tool for inducing mental workload [15]. The Criterion Task Set comprises various cognitive tasks that measure mental workload, which are: Continuous Recall, Grammatical Reasoning, Interval Production, Linguistic Processing, Mathematical Processing, Memory Search, Probability Monitoring, Spatial Processing and Unstable Tracking. Since the effort for an initial evaluation attempt should be kept as low as possible, a test procedure was used that is included in the test library of Millisecond's existing Inquisit Web Lab software. We decided for the spatial processing, as the load is particularly easy to vary. In addition, the Stroop Task was selected as a further experimental task, as it also varies the effort, but addresses different cognitive processes. Furthermore, the workload variation of the Stroop Task could already successfully be measured by pupil diameters [12].

The first independent variable is the variation of the stress level. As the second independent variable, the lighting conditions were varied in three gradations: 1. constant light (grey screen); 2. moderate variation (screen changes between white, gray and black with intervals of 4–6 s); and 3. strong variation (screen changes between white, grey and black with intervals of 1–3 s).

For the Spatial Processing and the Stroop Task the workload level was varied in blocks. Since the task of spatial processing has not yet been used in our laboratory experiments, the NASA Task Load Index (NASA-TLX) served as an additional dependent variable after each block. The NASA-TLX is a measure of subjective workload and is used here as an indicator of whether the intended manipulation of the workload level has worked [16].

The dependent variables in these tests are the ICA, the NASA-TLX and the performance in terms of reaction times. The ICA is output as a value between 0 and 1, while 1 stands for maximum workload. Both NASA-TLX values and performance are mainly used to check the manipulation of the stress level.

A sample size of 14 participants is proposed for a 2x3-factorial repeat measurement design with an alpha error of 5% and an average effect size. If a small effect is also to be excluded in an equivalence test, the sample should include 82 participants [17]. However, this high number of test persons cannot be realized, which is why the findings obtained can only serve as a rough assessment of the validity and reliability of the ICA.

2.2   Test Setup

The participants sat at a desk in front of a large screen (55 in.). The large screen was used to vary the lighting conditions. The rest of the room remained dark, so that the overall lighting conditions in the room could be varied by varying the background of the screen. The following three levels were used for the wallpaper: White (293.9 Lv), grey (69.02 Lv) and black (0.25 Lv). The three lighting conditions were chosen as following: fast changes (L1: change between the three light levels every 1 to 3 s), slow changes (L2: change between the three light levels every 4 to 6 s), and no change (L3: only grey light without changes). On a laptop in front of the screen, the participants worked on a mental rotation task and a Stroop task, each of the two varied in terms of their degree of difficulty. To avoid sequence effects, the light variations were permuted. The test was carried out with a 2 (workload level) × 3 (light variation) design.

Before the test phase a training block was completed. For the mental rotation task, the workload level was varied in terms of the complexity of bar diagrams that needed to be rotated. In the Stroop Task participants needed to name the font color of a color word (blue, red, green, yellow) shown on the screen. In the low workload condition the font color and the color word correspond (e.g., the word "BLUE" written in blue letters). In the high workload condition font color and color word are contradictory (e.g., the word "BLUE" written in red letters). The NASA-TLX was to be answered after each test condition to evaluate the subjective workload of the participants in each condition. If the ICA is truly light-independent, the ICA should show the same mental workload for each light condition.

2.3   Data Evaluation

A 2x3-factorial ANOVA with measurement repetition was applied. Additionally, a Pearson correlation between all dependent variables (ICA, NASA-TLX, and reaction time) was calculated. The sample size comprised X = 21, of which only X = 17 complete data were available. The missing data was caused by system errors during the test run.

2.4   Results

The descriptive statistics of the Spatial Task are summarized in Table 1. Means (M) and standard deviations (SD) are reported for the ICA (values between 0 and 1 with 0 = low workload and 1 = high workload), the NASA-TLX (values between 0 and 10 with 0 = low workload and 10 = high workload), and reaction times in milliseconds (RT in ms). In the left column the respective experimental condition is indicated. L stands for the lighting condition (L1: fast changes between the three light levels; L2: slow changes between the three light levels, and L3: only grey light without changes). The second part of the term stands for the distinction between low workload (L) and high workload (H) conditions.

Table 1. Descriptive data of the Spatial Task under varying light conditions.

	ICA left eye		ICA right eye		ICA both eyes		NASA-TLX		RT (in ms)	
	M	SD	M	SD	M	SD	M	SD	M	SD
L1_L	0,22	0,12	0,21	0,13	0,21	0,12	5,93	2,43	892,90	186,84
L1_H	0,24	0,11	0,22	0,13	0,23	0,11	7,77	2,38	1382,28	178,91
L2_L	0,18	0,09	0,19	0,10	0,19	0,09	5,98	3,13	874,92	233,44
L2_H	0,21	0,13	0,22	0,12	0,19	0,07	8,54	2,72	1399,34	289,03
L3_L	0,18	0,09	0,19	0,10	0,18	0,09	5,74	2,53	898,00	205,75
L3_H	0,20	0,12	0,21	0,13	0,21	0,12	7,78	2,24	1422,08	224,94

The main effect for light was not significant, only for the ICA of the left eye $F_{(2,15)} = 14{,}90$, $p < .001$. The main effect for load was significant for all dependent variables. The results of the ANOVA of the Spatial Task are presented in Table 2. The p-values and the partial eta squared are reported for the ICA, NASA-TLX, and reaction times. In the left column the respective factor (light, load and the interaction light * load) is indicated.

Table 2. Results of the ANOVA in the Spatial Task under varying light conditions.

	ICA left eye		ICA right eye		ICA both eyes		NASA-TLX		RT	
	p	η^2	p	η^2	p	η^2	p	η^2	p	η^2
Light	.000**	.550	.106	.258	.061	.192	.372	.060	.657	.026
Load	.000**	.548	.003*	.441	.003*	.429	.000**	.613	.000**	.881
Li * Lo	.675	.024	.427	.052	.504	.034	.403	.055	.786	.015

The significance level is *p < .05 and **p < .001.

Correlations were calculated between all dependent variables in the Spatial Task. Results are depicted in Table 3.

Table 3. Correlations of all dependent variables in the Spatial Task.

	ICA left eye	ICA right eye	ICA both eyes	NASA-TLX	RT
ICA left eye	1	.48**	.42**	.13	.33**
ICA right eye	.48**	1	.49**	.38**	.38**
ICA both eyes	.42**	.49**	1	.13	.17
NASA-TLX	.13	.38**	.13	1	.56**
RT	.33**	.38**	.17	.56**	1

** The correlation is significant at the level of 0.01 (2-sided).

The descriptive statistics of the Stroop Task are summarized in Table 4. For the Stroop Task N = 18 data sets were complete and could be used for analysis. Means (M) and standard deviations (SD) are reported for the ICA (values between 0 and 1 with 0 = low workload and 1 = high workload), the NASA-TLX (values between 0 and 10 with 0 = low workload and 10 = high workload), and reaction times in milliseconds (RT in ms). In the left column the respective experimental condition is indicated. The designations are as above.

Table 4. Descriptive data of the Stroop Task under varying light conditions.

	ICA left eye		ICA right eye		ICA both eyes		RT (in ms)	
	M	SD	M	SD	M	SD	M	SD
L1_L	0,20	0,09	0,19	0,10	0,20	0,09	771,93	142,41
L1_H	0,21	0,10	0,19	0,08	0,20	0,08	941,95	163,99
L2_L	0,18	0,08	0,18	0,08	0,18	0,08	782,13	163,75
L2_H	0,19	0,09	0,19	0,11	0,19	0,09	950,77	177,83
L3_L	0,15	0,09	0,16	0,10	0,16	0,09	769,50	124,77
L3_H	0,15	0,09	0,18	0,11	0,17	0,09	950,02	183,98

The main effect for light was significant regarding the ICA values, but not significant regarding reaction times. The main effect for load was not significant in terms of the ICA values, but significant in terms of the reaction times $F(2,16) = 104,60$, $p < .001$. The results of the ANOVA of the Stroop Task are shown in Table 5. The p-values and the partial eta squared are reported for the ICA, NASA-TLX, and reaction times. In the left column the respective factor (light, load and the interaction light * load) is indicated.

Correlations were calculated between all dependent variables in the Stroop Task. Results are depicted in Table 6.

Table 5. Results of the ANOVA in the Stroop Task under varying light conditions.

	ICA left eye		ICA right eye		ICA both eyes		RT	
	p	η^2	p	η^2	p	η^2	p	η^2
Light	.004*	.502	.254	.157	.012*	.426	.853	.020
Load	.448	.034	.104	.148	.136	.126	.000**	.860
Light * Load	.731	.038	.52	.078	.811	.026	.91	.012

The significance level is *p < .05 and **p < .001.

Table 6. Correlations of all dependent variables in the Stroop Task.

	ICA left eye	ICA right eye	ICA both eyes	RT
ICA left eye	1	.37**	.87**	.06
ICA right eye	.37**	1	.78**	.27**
ICA both eyes	.86**	.78**	1	.20**
RT	.06	.27**	.20*	1

** The correlation is significant at the level of 0.01 (2-sided).

3 Discussion

NASA-TLX results of the Spatial Task support that the manipulation of the induced workload worked out well. The data of the left eye are excluded from interpretations, because the infrared camera of the left eye showed many disturbances during the experiment. Regarding the Spatial Task, the first hypothesis "the ICA distinguishes between high and low workload conditions" is supported. The second hypothesis "the ICA does not distinguish between the light variations" is supported. The values of the NASA-TLX correlate with the ICA of the right eye and the reaction times, all different measurements of mental workload. This could be a hint for a similar construct underlying these measurements. Nevertheless, the results for application in field research should be treated with caution, as the differences in mean values are very small.

The results for the Stroop Task do not support any of the hypotheses. However, the results of the reaction times support both hypotheses. It seems that the ICA is not a proper tool to depict the specific mental workload induced by the Stroop Task. The Stroop Task is used to measure self-control as one of the important executive functions of the brain. Referring to the Criterion Task Set as an evaluated tool for inducing mental workload [15], self-control or inhibition of spontaneous reactions is not mentioned, but spatial processing is part of the Criterion Task Set. A possible explanation is that the specific mental workload induced by the Stroop Task cannot be depicted using the ICA. Findings suggest that ICA is a good measure of working memory activity [10, 18]. Moreover, the ICA has proven itself as an indicator for measuring cognitive workload in specific tasks, such as driving simulations [19] or language processing [20], which could also be related to working memory activity. This could explain why the ICA works for spatial processing, but not for inhibition like in the Stroop Task.

The use of the ICA to measure mental workload remains questionable. It seems that the validity of the ICA is highly task-specific. Before using the ICA, it must therefore be ensured that the ICA is applicable to the specific task. The underlying construct and definition of mental workload is still an issue. Furthermore, the ICA software is an impervious "black box": The intransperancy of the data processing to get the ICA values is a problem especially in research contexts that needs to be considered before using it.

References

1. Marshall SP (2002) The index of cognitive activity: measuring cognitive workload. In: Proceedings of the 2002 IEEE 7th conference on human factors and power plants in new century, new trends, Scottsdale, AZ, USA, p 7.5
2. Schwalm M (2009) Pupillometrie als Methode zur Erfassung mentaler Beanspruchungen im automotiven Kontext, Dissertation
3. Johannsen G (1979) Workload and workload measurement. In: Moray N (ed) Mental workload. Springer, Boston, pp 3–11
4. Damos DL (ed) (1991) Multiple-task performance. Taylor & Francis, London, Washington, DC
5. Korbach A, Brünken R, Park B (2017) Differentiating different types of cognitive load: a comparison of different measures. Educ Psychol Rev 16(6582):389
6. van Gog T, Kirschner F, Kester L, Paas F (2012) Timing and frequency of mental effort measurement: evidence in favour of repeated measures. Appl Cognit Psychol 26(6):833–839
7. Kahneman D (1973) Attention and effort. Prentice-Hall, Englewood Cliffs, N.J
8. Beatty J (1982) Task-evoked pupillary responses, processing load, and the structure of processing resources. Psychol Bull 91(2):276–292
9. Loose C (2004) Psychosensorische Pupillendilatation bei bewusster und unbewusster visueller Informationsverarbeitung: Untersuchungen an normalsichtigen Probanden und Hemianopikern
10. Hampson RE, Opris I, Deadwyler SA (2010) Neural correlates of fast pupil dilation in nonhuman primates: relation to behavioral performance and cognitive workload, (eng). Behav Brain Res 212(1):1–11
11. Janisse MP (1977) Pupillometry: The psychology of the pupillary response. Hemisphere Publishing Corporation; Distributed solely by Halsted Press, Washington, New York
12. Laeng B, Orbo M, Holmlund T, Miozzo M (2011) Pupillary Stroop effects, (eng). Cognit Process 12(1):13–21
13. Marshall SP (2000) Method and Apparatus for Eye Tracking and Monitoring Pupil Dilation to Evaluate Cognitive Activity, 6,090,051, US006090051A, USA, 18 July 2000
14. Loewenfeld IE, Lowenstein O (1993) The pupil: Anatomy, physiology, and clinical applications. Iowa State Univ. Press, Ames
15. Schlegel RE, Gilliland K, Schlegel B (1987) Factor structure of the criterion task set. Proc Hum Factors Soc Ann Meet 31(4):389–393
16. Hart SG, Staveland L (1988) Development of NASA-TLX (Task Load Index): Results of empirical and theoretical research. In: Human mental workload, pp 139–183
17. Bortz J, Döring N (2006) Forschungsmethoden und Evaluation für Human- und Sozialwissenschaftler, 4th edn. Springer Medizin Verlag, Heidelberg
18. Funahashi S (2006) Prefrontal cortex and working memory processes, (eng). Neuroscience 139(1):251–261

19. Schwalm M, Keinath A, Zimmer HD (2008) Pupillometry as a method for measuring mental workload within a simulated driving task. In: Waard D, Flemisch FO, Lorenz B, Oberheid H, Bookhuis KA (eds) Human factors for assistance and automation. Shaker Publishing, Maastricht, The Netherlands, pp 1–13
20. Demberg V, Sayeed A (2016) The frequency of rapid pupil dilations as a measure of linguistic processing difficulty, (eng). PLoS ONE 11(1):e0146194

Quantifying the Impact of Submersion in Water and Breathing Type on Cognitive Resource Utilization

Luke Goodenough[✉] and Swantje Zschernack

Human Kinetics & Ergonomics Department, Rhodes University, Grahamstown,
South Africa
l.goodenough@ru.ac.za

Abstract. The underwater environment has been seen as a potentially dangerous working environment, requiring high levels of situation awareness to function in safely. As a working environment, it is inherently complicated with multiple effects on the operator, including limited sensory input. Research into the effect of submersion and breathing type on cognitive functioning and the severity of any impacts would allow for adaptations in training and operation methods underwater, leading to reduced risk and increased efficiency in task performance. Literature surrounding the topic of cognitive function in the underwater environment is limited and was limited to assisted breathing only, no apnea based studies looked into cognitive functioning but rather focused on the human dive response and related physiological effects on being underwater. A pilot study focused on the impact of submersion in water and breathing modality (assisted breathing and apnea) on different stages of the information processing chain. This showed that only more complex tasks are affected, with no uniform reason as to why. Memory was impacted in terms of speed of recall in the apnea condition only. Visual detection was affected in terms of speed and accuracy in both underwater conditions, leading to the conclusion that submersion caused performance decrease. The recognition task was only affected in the assisted breathing condition, in terms of both speed and accuracy, indicating that the assisted breathing was the factor responsible for the decrease in performance.

Keywords: Underwater · Cognitive function · Apnea

1 Background

Literature surrounding the topic of cognitive function in the underwater environment is limited and was limited to assisted breathing only; no apnea based studies looked into cognitive functioning but rather focused on the human dive response and related physiological effects on being underwater [1–7]. Submersion in water impacts both the sympathetic and parasympathetic nervous systems, which impact heart rate and heart rate variability [8]. Apnea increases the activation of the parasympathetic nervous system, inducing bradycardia in the body [1, 6, 8]. Breath hold triggers an increase in muscle sympathetic nerve activation [2]. Assisted breathing can increase the

© Springer Nature Switzerland AG 2019
S. Bagnara et al. (Eds.): IEA 2018, AISC 827, pp. 360–365, 2019.
https://doi.org/10.1007/978-3-319-96059-3_39

parasympathetic response through the noise or feedback generated by the breathing equipment [8]. All these have an effect on the body and can influence a person's ability to perform cognitively.

2 Pilot Study

A pilot study focused on the impact of submersion in water and breathing modality (assisted breathing and apnea) on different stages of the information processing chain.

2.1 Aim

The pilot study aimed to determine what impact different breathing modalities underwater had on the cognitive performance of divers and which aspects of the conditions; breathing modality (apnea and assisted breathing) or submersion, caused any performance differences.

2.2 Method

The pilot study tested five different cognitive tasks, using standard ergonomics tests, in three conditions. Four of the five tasks were computer based with the fifth being a paper-based task. The three conditions tested were on land, underwater with apnea and underwater with assisted breathing; using a scuba diving tank and regulator to breathe while underwater. The testing underwater was conducted using a hydrostatic weighing tank with a glass window. A computer screen was mounted against the glass allowing the participant inside the tank to view the screen and a modified waterproof keyboard was used to allow for responses to the four computer based tests. The fifth task was done on laminated paper using a china marker (grease pencil). The dimensions of the underwater test setup were replicated on land for the land-based condition. The level of light and water clarity inside the hydrostatic weighing tank were maintained to the best possible level to avoid any confounding variables to the study. All tests were conducted in thirty second sessions to facilitate the apnea condition, allowing participants to complete each session of the tests without becoming distressed by running out of oxygen. Of the five cognitive tests used, four had two levels of difficulty. Eighteen participants were tested over the course of the pilot study. The data from the various tests were analyzed to test for significant differences between the different conditions and levels of difficulty.

2.3 Results

The results showed that only more complex tasks are affected, with no uniform reason as to why. Memory was impacted in terms of the speed of recall in the apnea condition only, leading to the determination that the apnea aspect of the condition was the cause of the performance decrement. Visual detection was affected in terms of response speed and accuracy in both underwater conditions, leading to the conclusion that submersion caused performance decrease. The recognition task was only affected in the assisted breathing condition, in terms of both speed and accuracy, indicating that the assisted breathing was the factor responsible for the decrease in performance.

2.4 Discussion

These findings run contrary to Dalecki [9] who found that simpler tasks were influenced by submersion in shallower water, tested at a 5 m depth, but the performance in the more complex tasks remained similar to that on land. Whereas, the pilot study found the inverse, with simpler tasks remaining unaffected in the underwater conditions. It must be acknowledged that the pilot study was conducted at a much shallower depth than the Dalecki [9] study and as a result the impact of the submersion at this reduced depth may affect the operator differently, accounting for the differences in the finding between the pilot study and Dalecki [9].

3 Future Directions of Research

3.1 Limitations of Pilot Study

The pilot study was limited considerably in terms of the study design. The tests used were not representative of real world tasks that divers would be conducting on a normal work basis. The environmental conditions for the testing were made as comfortable and as clear as possible and as a result do not represent the typical working conditions of divers in real world work, as the water clarity and temperature would vary greatly from one area of work to another. The study design was unbalanced, having two underwater conditions and only one land based condition, there was no normal breathing condition underwater and there was no assisted breathing or apnea condition on land, therefore making it difficult to distinguish whether the effect resulted from the submersion or the breathing type (assisted breathing or apnea). An improved approach would include a 'normal breathing' condition in water and test both breathing types (assisted breathing and apnea) both underwater and on land, with an 'unassisted' breathing condition on land as control condition to allow for comparison. This approach, to be used in the current study, would both balance the research design and allow for a refinement of the findings in determining the factor that is detracting from task performance.

The pilot study approached the understanding of cognitive functioning using Wickens [10] limited resource model, viewing the cognitive process as an information processing chain, having multiple processes chained together in an order of use. The proposed study will look into cognitive functioning in terms of multiple resource utilization [11], a set of functions occurring concurrently and sequentially using the same resource pool.

3.2 Theoretical Underpinning of Future Research

Measurement of cognitive workload will be done before, during and after the testing for the current study to assess the cognitive strain being experienced by the participants. Cognitive workload will be measured using both objective and subjective measures. The objective measures that will be used will be heart rate (HR) and heart rate variability (HRV). Other objective measures of cognitive effort are more difficult to use in an underwater setting as either the presence of the water or the pressure inhibit their use. Additional measures will still be looked into for their viability for use and if a solution for their use during the testing is found they will be included.

In addition to the assessments of cognitive workload, measurements of mood and emotion will be taken. Mood will be assessed in the form of a Multidimensional Mood Questionnaire (MDMQ) [12], an English language equivalent of the Der Mehrdimensionale Befindlichkeitsfragebogen (MDBF) [13]. This questionnaire will aid in evaluating participants emotions as well as levels of anxiety during the different testing conditions. Previous studies have utilized the MDMQ as a method of eliminating anxiety as an influencer on the outcomes of the study [12, 14]. When an emotion is being experienced, such as joy or fear, a person has a tendency to focus on the aspects of a situation most in line with their current emotional state [15]. As a result, a negative emotion may result in the memory of a task or event being more focused on the negative aspects involved or perceived threats in the environment around them, resulting in a poorer task performance [12]. With emotions playing a part in memory [15] and the type of emotion being experienced can impacting on which aspect of a memory is recalled, having the measures of mood and emotion will aid in understanding the results and eliminating confounding variables.

The limitations of the pilot study will be taken into account for the research project that is currently ongoing and will be used to improve the research design; the tests and measures used and will try to approach the topic using a more real world setup for the testing process.

3.3 Development of Research Concept

As a working environment, it is inherently complicated with multiple effects on the operator, including limited sensory input. In the underwater environment, the nature of tasks carried out varies greatly between different job types and recreational activities. Shallow water diving is used as the working depth for commercial divers and is a training environment for astronauts [9], it is also the operational area for recreational divers and spear fisherman. Knowledge of how operation in this environment is conducted is important as the different types of dives all expose the divers to the same risks. Commercial divers can conduct inspection tasks and repairs on structures off shore requiring the diver to operate in close proximity to the structure [16], exposing them to other risk factors posed by the structure itself. In all areas of work, especially in underwater work [17], situation awareness is of high importance for both task completion and operator safety. Situation awareness is "the perception of the elements in the environment within a volume of time and space, the comprehension of their meaning, and the projection of their status in the near future" [18]. Having high levels of situation awareness allows for focus to remain on the task at hand and avoids fixation on incorrect or irrelevant stimuli [17]. Achieving and maintaining a high level of situation awareness requires the operator to be aware of their surroundings and to be able to effectively process and understand what is happening around them [18]. This awareness would take into account information received by all of the senses of the body and the combination of all of the information would give the operator the fullest possible 'picture' of their environment of operation. This process is somewhat complicated in the underwater environment as reliable sensory input is limited to visual input alone as the other senses are limited or negated while submerged [19].

The underwater environment has been seen as a potentially dangerous working environment [20]. As a result, an operator may be expected to change tasks, for example, from an inspection task on a structure to a surveying task for threats that have been detected. This process of switching tasks may cause stress to the operator, which could reduce task performance. An operators' ability to switch between tasks while underwater may be reduced as task switching is linked to cognitive flexibility which is in turn controlled by the prefrontal cortex which is impaired by stress [21]. Alternatively, rather than switching between tasks the operators' may attempt to manage the different tasks concurrently, taking a multitasking approach to the their work where their attention becomes divided between task demands and factors in the environment around the operator. The processing of different factors concurrently may lead to a decreased ability to complete tasks effectively as the cognitive resources required by each aspect of the different tasks may overlap, resulting in a high demand on one resource [11]. The increased demand on a single cognitive resource would result in only one of the tasks, or aspects of a task, receiving the required amount of attention at the expense of the others as the resource is limited [11].

Operating in an environment that limits sensory input as well as containing dangers in multiple forms requires a high situation awareness but will also require attentional resources to be focused on multiple' different aspects of the environment as well as the task. Any loss of focus or attention on the task or the environment can result in an incomplete or incorrectly completed task. This may be because either the attention focused on the task was limited and there has been a performance deficit, or an aspect of the environment has interfered leading to an interruption of the task or a physical impact on the operator. Dangers in the environment would need to be monitored continuously throughout the task to compensate for them or avoid them completely.

4 Conclusion

Research into the effect of submersion and breathing type on cognitive functioning and the severity of any impacts would allow for adaptations in training and operation methods underwater, leading to reduced risk and increased efficiency in task performance. The findings may also be used to help improve current, and develop new, training techniques for underwater operators. This would offer an insight into areas where there may not have been enough focus before or changing the way in which certain types of operations or tasks are viewed or approached. Research of this nature could lead to or aid in blending academic and laboratory based research and teaching with the health, safety and training aspects of the industry.

References

1. Gooden BA (1994) Mechanism of the human diving response. Integr Physiol Behav Sci 29:6–16
2. Heusser K, Dzamonja G, Tank J et al (2009) Cardiovascular regulation during apnea in elite divers. Hypertension 53:719–724
3. Lemaître F, Polin D, Joulia F et al (2007) Physiological responses to repeated apneas in underwater hockey players and controls. Undersea Hyperb Med J Undersea Hyperb Med Soc Inc 34:407–414
4. Walterspacher S, Scholz T, Tetzlaff K, Sorichter S (2011) Breath-hold diving: respiratory function on the longer term. Med Sci Sports Exerc 43:1214–1219
5. Feiner JR, Bickler PE, Severinghaus JW (1995) Hypoxic ventilatory response predicts the extent of maximal breath-holds in man. Respir Physiol 100:213–222
6. Landsberg PG (1975) Bradycardia during human diving. S Afr Med J 49:626–630
7. Baddeley AD (1966) Influence of depth on the manual dexterity of free divers: a comparison between open sea and pressure chamber testing. J Appl Psychol 50:81–85
8. Schipke JD, Pelzer M (2001) Effect of immersion, submersion, and scuba diving on heart rate variability. Br J Sports Med 35:174–180
9. Dalecki M, Dern S, Steinberg F (2013) Mental rotation of a letter, hand and complex scene in microgravity. Neurosci Lett 533:55–59
10. Wickens CD (1984) Engineering psychology and human performance. Charles E. Merrill Publishing Company, Columbus
11. Wickens CD (2008) Multiple resources and mental workload. Hum Factors 50:449–455
12. Dalecki M, Bock O, Hoffmann U (2013) Inverse relationship between task complexity and performance deficit in 5 m water immersion. Exp Brain Res 227:243–248
13. Steyer R, Schwenkmezger P, Notz P, Eid M (1997) Der Mehrdimensionale Befindlichkeitsfragebogen (MDBF). Hogrefe, Gottingen
14. Dalecki M, Bock O (2014) Isometric force exaggeration in simulated weightlessness by water immersion: Role of visual feedback. Aviat Space Environ Med 85:605–611
15. Gooden DR, Baddeley AD (1975) Context-dependent memory in two natural environments: on land and underwater. Br J Psychol 66:325–331
16. Moan T (2005) Reliability-based management of inspection, maintenance and repair of offshore structures. Struct Infrastruct Eng 1:33–62
17. Heywood J (2012) Situation Awareness: a training paradox. In: Diver. http://divermag.com/situation-awareness-a-training-paradox/. Accessed 22 Sept 2016
18. Endsley MR (1995) Toward a theory of situation awareness in dynamic systems. Hum Factors 37:32–64
19. Hollien H, Rothman H (1971) Underwater sound localisation in humans, Gainesville
20. Baddeley AD (2000) Selective attention and performance in dangerous environments. Br J Psychol 63:537–546
21. Renner KH, Beversdorf DQ (2010) Effects of naturalistic stressors on cognitive flexibility and working memory task performance. Neurocase 16:293–300

Use of Presentation of Thermal Stimulus for Enhancing Excitement During Video Viewing

Ryota Tsuruno[1(✉)], Kentaro Kotani[2], Satoshi Suzuki[2],
and Takafumi Asao[2]

[1] Graduate School of Science and Engineering, Kansai University,
3-3-35 Yamate-cho, Suita, Osaka 564-8680, Japan
k322731@kansai-u.ac.jp
[2] Faculty of Engineering Science, Kansai University, 3-3-35 Yamate-cho,
Suita, Osaka 564-8680, Japan
{kotani,ssuzuki,asao}@kansai-u.ac.jp

Abstract. In this research, we focused on biofeedback technology to enhance perception of excitement during video viewing. This technology realizes viewers their emotion by presenting thermal stimuli at the moment they changed their emotion. Thermal stimulus was used as biofeedback. Biofeedback device generated thermal stimuli at the moment that SCR signals were observed. The thermal stimulus was designed to raise the skin temperature on the left forearm of the viewer from 33 °C to 36 °C at the rate of 0.2 °C/sec.

In the experiment, eight college students viewed three kinds of videos. After viewing of the videos, a questionnaire survey was applied to investigate whether or not perception of excitement was enhanced when stimuli were presented compared with normal condition. As a result of subjective responses, the system raised their stress levels, however failed to raise viewers excitement promptly.

Keywords: Biofeedback · SCR · Thermal stimulus · Video · Excitement

1 Introduction

In recent years, systems that improve the entertainment experience using biofeedback receive attention [1]. Biofeedback is a technology to detect biological signals which is hardly perceivable and presenting the information regarding the interpretation for such signals by using engineering methodology [2].

E3-player is one of the system enhancing perception of excitement using biofeedback during video viewing [1]. E3-player altered the video's volume level as biofeedback. The mechanism of E3-player was to alter the video's volume level accordingly whenever skin conductance response (SCR) signals appeared to cause the emotional changes. The E3-player has, however, a drawback, that is, participants may feel uncomfortable by altering the video's volume level. Therefore, they thought that E3-player cannot enhance participants' perception of excitement promptly during video viewing [1].

© Springer Nature Switzerland AG 2019
S. Bagnara et al. (Eds.): IEA 2018, AISC 827, pp. 366–370, 2019.
https://doi.org/10.1007/978-3-319-96059-3_40

This study focused on thermal stimulus, which has been reported to influence emotion strongly [3]. In this study, the system presenting thermal stimuli was introduced and SCR signals were observed during the video viewing with the system.

2 Method

2.1 Participants

Participants included eight college students (age of 21–22 years). All participants viewed videos through TV or Internet on daily. Each participant gave a written informed consent prior to the beginning of the study (Fig. 1).

Fig. 1. Experimental landscape (Appearance of presenting thermal stimuli during viewing video)

2.2 Experimental Apparatus

In this study, biofeedback device was used as experimental apparatus. This was made from a thermal stimulator and a measuring device for skin conductance changes (SCC). Peltier device (TES 1-12705), motor driver (TA7291P), thermistor (103JT-025) and Arduino UNO were used to construct the thermal stimulator. Arduino UNO was also used for the measurement of SCC with electrodes for measuring electrodermal activity (EDA) (PPS-EDA, Sekisui Plastics Tenri Co., Ltd.). A measuring device for SCC recorded the SCR signals when they were observed. A thermal stimulator presented thermal stimuli on the left forearm. Adapting skin temperature at the forearm was set to 33 °C and the room temperature was set to 24 °C.

2.3   Experimental Procedure

As a preparation of the experiment, participants sat on a chair with a comfortable posture, and a probe for the thermal stimulator was attached to their left forearms. Electrodes for SCRs were attached at the ventral part of the second and third fingers of their right hand. Participants put earphones (sealed inner ear receiver, MDR - EX 255). Videos were selected under the condition that professionals created them and participants have never seen them before. Videos were shown by a 10-in. monitor.

Participants viewed three videos under two conditions, with and without presenting thermal stimuli. After viewing videos, participants answered questionnaires regarding the level of affection during viewing videos. After waiting for participants to relax again, the trial was repeated. In order to eliminate the order effect, participants were divided into two groups; Group A watched videos with presenting thermal stimuli after a video without presenting thermal stimuli, and Group B did with videos without thermal stimuli first.

2.4   Data Analysis

In this experiment, participants answered questionnaires for investigating perception of excitement during video viewing. Questionnaires were made with reference to Emotion and Arousal Checklist (EACL) [3]. EACL was a checklist to evaluate temporary or persistent affection. EACL consisted of five subscales to assess emotions (fear, anger, sadness, disgust, and happiness) and four subscales to assess arousal (energetic arousal +, energetic arousal −, tense arousal +, and tense arousal −). In this experiment, energetic arousal + and tense arousal + were used. The average value of each scale was used for the analysis regarding excitement and tense during video viewing.

3   Results

Figure 2 shows the result for energetic arousal + scores by different videos and by the presence of thermal stimuli. There were no significant differences in scale scores of energetic arousal + between the two conditions, i.e., with and without presenting thermal stimuli, ($t = 2.36$, $df = 7$, n.s.). Figure 3 shows tense arousal + scores by different videos and by the presence of thermal stimuli. There were significant differences in scale scores of tense arousal + between the two stimulus conditions at the video B, ($t = 2.36$, $df = 7$, $p < .05$).

4   Discussion

As a result, thermal stimuli raising skin temperature on the viewers' left forearms from 33 °C to 36 °C at the rate of 0.2 °C/sec failed to enhance perception of excitement during video viewing. ThermOn enhanced perception of pleasure when listening to the music by using thermal stimuli with feedforward control [5]. The rate of temperature changes by ThermOn was set higher than biofeedback device in this study (1.5 °C/sec).

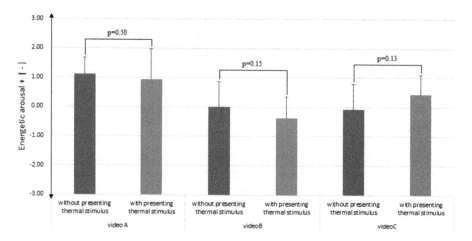

Fig. 2. Result of questionnaire on energetic arousal +. This bar chart was based on the summary of questionnaire during viewing video for each video program. Blue bars indicate that participants watched videos without presenting thermal stimuli. Orange bars indicate that participants watched videos with thermal stimulus presentations. (Color figure online)

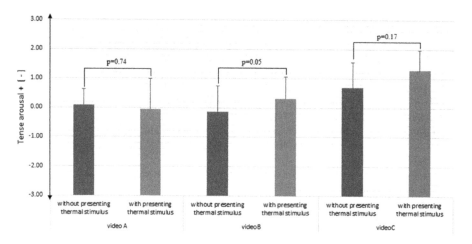

Fig. 3. Result of questionnaire on tense arousal +. This bar chart was based on the summary of questionnaire during viewing video for each video program. Blue bars indicate that participants watched videos without presenting thermal stimuli. Orange bars indicate that participants watched videos with thermal stimulus presentations. (Color figure online)

It was considered that biofeedback device in this study showed the same results as the performance by using ThermOn. Thus, enhanced perception of excitement during video viewing may be observed more clearly if the rate of temperature change would be raised.

The results implied that it was necessary to evaluate and examine the perception of excitement during video viewing by using other physiological indices for evaluating emotion. EDA can evaluate levels of emotional arousal, however, our results indicated that it seemed difficult to distinguish, in current settings, whether changes in the levels of emotional arousal were due to excitement or tension. It was necessary to reexamine the perception of excitement by using physiological indices for showing the effectiveness of the biofeedback device.

5 Conclusions

In this study, whether or not perception of excitement can be enhanced by using thermal stimuli presentation was investigated. The device resulted in amplifying perception of stress, instead of increasing perception of excitement.

It was necessary to review further for the physiological indices and redesign biofeedback device by changing the settings regarding the rate of temperature changes.

Acknowledgment. This study was supported by KUMP, 2016 MEXT Program "Branding Projects in the Academic Research of Private Universities", in part.

References

1. Shirokura T, Munekata N, Ono T (2013) E3-Player: video player enhance emotional excitement of viewer. In: 2013 information processing society of Japan, pp 272–277
2. Nishimura C (1988) Biofeedback. Biomed Eng 2(9):618–625
3. Doi K, Nishimura T, Seo A, Kushiyama K, Baba T (2012) Sensory characteristics of temperature and its discrimination in the human palm. Int J Affect Eng 11(3):419–425
4. Oda Y, Takano R, Abe T, Kikuchi K (2015) Development of the emotion and arousal checklist (EACL). Psychol Res 85(6):79–589
5. Akiyama S, Sato K, Makino Y, Maeno T (2013) ThermOn – Thermo-musical interface for an enhanced emotional experience. In: Proceedings of the 2013 international symposium on wearable computers, pp 45–52
6. Lynette J, Hsin-Ni H (2008) Warm or cool, large or small? The challenge of thermal displays. IEEE Trans Haptics 1(1):53–70
7. Yoshimura H, Fukuda M (1952) Studies on seasonal changes of thermal regulation of the Japanese in their normal living. Med Biol 25(1):24–28

Measuring Experience and Competence by Means of Electrodermal Activity – The Model of Experience-Dependent Somatic Activation

Tobias Heine[✉] and Barbara Deml

Institute of Human and Industrial Engineering, Karlsruhe Institute
of Technology, Engler-Bunte-Ring 4, 76131 Karlsruhe, Germany
tobias.heine@kit.edu

Abstract. When designing automated systems where users are still expected to take over control (e.g. automated driving up to level 3), it is important to prevent a critical loss of experience-based competence. Based on the somatic marker hypothesis, a model of experience-dependent somatic activation is proposed. It offers a theoretical linkage between the level of experience of a user and the mean somatic activation that can be measured by means of electrodermal activity. The model is consistent with both empirical findings and the theoretical conception of somatic markers.

Keywords: Electrodermal activity · Somatic marker hypothesis
Psychophysiology

1 Introduction

From an ergonomic point of view, human and machine form a work system. The goal of a work system is to complete a work task [1]. The quality of the work result depends on both the system elements (human and machine) and the interaction between these two elements. The interaction is thereby substantially determined by the division of functions between man and machine. For many decades, ergonomists have dealt with the question of how functions and responsibilities should be organized between human and machine. Finding adequate answers to this question is one of the most important tasks of human factors researchers [2].

The ongoing megatrend of automation leads to major changes of the division of tasks between human and machine. One area in which automation is currently strongly promoted is the automotive sector. The path to fully autonomous cars can be divided into 6 levels (level 0 – level 5), whereby humans are still integrated into the control loop up to level 3 [3]. In level 3 the car is already able to perform all driving tasks in many situations, nevertheless the driver is conceptualized as fallback option when the car reaches a system boundary or undergoes a system failure.

If a person no longer carries out routine activities, as they are taken over by the machine, this results in a loss of skills and competence. However, in the case of a system failure, a human driver must be able to solve the problem and/or take over

© Springer Nature Switzerland AG 2019
S. Bagnara et al. (Eds.): IEA 2018, AISC 827, pp. 371–376, 2019.
https://doi.org/10.1007/978-3-319-96059-3_41

control - sometimes within a short period of time. This phenomenon is known as irony of automation [4]. When designing automated systems, it is therefore essential to prevent a critical loss of competence. The prerequisite for this is the valid measurement of the current level of competence and, even more important, the detection of a critical change of competence. In this article, we present a model which postulates a relationship between the electrodermal activity and the driving experience. Therefore, it provides the opportunity to assess the level of experience and associated skills and competencies by means of electrodermal activity. The model is based on data from two experiments in the driver-vehicle domain. The theoretical background for the measurement and analysis of electrodermal data is the somatic marker hypothesis.

The somatic marker hypothesis is a neuroscientific theory which postulates that unconscious bodily signals play a central role in the decision-making process [5]. These bodily signals are called somatic markers (soma [Greek]: body) and represent emotional previous experience which is later reactivated in a similar situation. Somatic markers help to decide advantageously by indicating possible hazards: "[A somatic marker] [...] forces attention on the negative outcome to which a given action may lead, and functions as an automated alarm signal which says: Beware of danger ahead if you choose the option which leads to this outcome "[5, p. 173].

On a physiological level, somatic markers can be measured for example by changes of the electrodermal activity [6]. Since somatic markers are based on previous experiences, it can be derived that there is a connection between the amount of experience-based competence and the strength of somatic markers. Additionally, this experience-based competence is thus measurable by means of electrodermal activity.

2   Empirical Data: Somatic Markers and Driving Experience

In a previous experiment, Kinnear et al. applied the somatic marker hypothesis to the driver-vehicle-domain [7]. Their findings suggest a linear relationship between the experience level of a driver and the amount of somatic markers. The more experience a driver has, the more somatic markers are generated.

While in Kinnear et al.'s experiment the participants watched videos without having to make decisions, we wanted to further investigate somatic markers in more realistic situations where active decision-making is required (see [8] for more details).

In a first experiment, participants were asked to watch videos of potentially hazardous traffic situations. Their task was to decide, when they would abort or no longer initiate a given manoeuvre (e.g. overtaking another car). Somatic markers were measured by means of electrodermal activity and operationalized as anticipatory somatic activation in a time window of three seconds prior to a decision. For novice drivers the mean anticipatory somatic activation of each participant correlated positively with driving experience ($r = .56$, $p < .05$; novice drivers were defined as persons being currently in driving school or having completed it a maximum of three months ago). In contrast, mean anticipatory somatic activation and driving experience correlated negatively for expert drivers ($r = -.50$, $p < .05$; expert drivers were defined as persons who have travelled at least 20,000 km within the last 12 months; note that the mean experience of these drivers was therefore twice as high as that of the most experienced drivers in the experiment of Kinnear et al.). Interestingly, the comparison of the mean somatic activation of novice and expert drivers showed no significant difference.

In a second experiment, participants were confronted with different potentially hazardous traffic situations in a driving simulator (e.g. turn left at a T-junction while another car is approaching). In this experiment no distinction was made between novice and expert drivers. Overall, the level of experience corresponded more to that of the expert drivers from the first experiment. In addition to that, a comparable negative correlation was found between the mean anticipatory somatic activation of each participant and the level of driving experience ($r = -.50$, $p < .01$).

3 The Model of Experience-Dependent Somatic Activation

Our experimental results suggest that the mean strength of somatic markers increases up to a certain level of experience-based competence after which their strength again decreases. A closer examination of the somatic marker hypothesis reveals that our findings are in good agreement with theoretical expectations, since somatic markers can influence decision making in *two* ways; a fact which is rarely taken into account.

3.1 Body Loop and As-if-Loop

Decision making does not always require the activation of the entire body. According to [9] there are two decision paths: the body loop and the as-if-loop: "In the body loop mechanism, an appropriate emotional (somatic) state is actually re-enacted, and signals from its activation are then relayed back to subcortical and cortical somatosensory processing structures, especially in the somatosensory (SI and SII) and insular cortices" [9, pp. 37–38]. The as-if-loop forms a faster reaction chain: "after emotions are learnt, one possible chain of physiologic events is to by-pass the body altogether, activate the insular/somatosensory cortices directly, and create a fainter image of an emotional body state than if the emotion were actually expressed in the body" [9, p. 38]. The more uncertain the result of a decision, the more likely is the activation of the body-loop [9]. This has direct consequences for the measurability of somatic markers.

3.2 Somatic Marker Type 1 and 2

Only body-loop activated somatic markers produce signals that can be measured by means of electrodermal activity. Somatic markers that are generated via the as-if-loop do not result in changes of the electrodermal activity. In order to account for this important difference, we suggest distinguishing between two different types of somatic markers: Type I markers are the body-loop markers which go along with measurable peripheral physiological changes. Type II markers are the as-if-markers without measurable changes of the electrodermal activity.

3.3 Model Description

In the following, a model is presented that can both explain the empirical data and is in line with the above mentioned aspects of the somatic marker hypothesis (see Fig. 1). The X-axis represents the driving experience, the Y-axis indicates the strength of a somatic marker (both type I and type II).

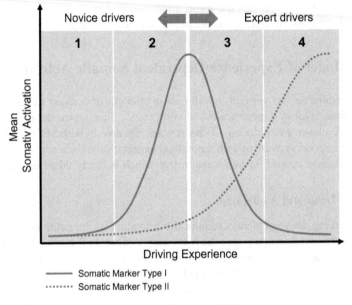

Fig. 1. The model of experience-dependent somatic activation. Note that only somatic markers type I are measurable by means of electrodermal activity.

Driving experience is divided into four phases. The first phase represents novice drivers at the beginning of their driving career. At this time, they have almost no previous experience when dealing with (hazardous) traffic situations. Consequently, neither somatic markers type I (solid line) nor somatic markers type II (dashed line) can be found. In phase 2, the novice driver continuously builds up experience. Therefore, this experience can be activated more and more in similar situations. The physiological expression for this activation is a stronger type I marker activation.

From the third phase on, drivers are considered as experts. They have gained a lot of driving experience and are able to use this experience to drive routinely and safely in many situations. Consequently, the proportion of type I markers decreases whereas the proportion of type II marker increases. Drivers in phase four have reached the highest level of experience and competence. They possess a great amount of driving experience which they can apply routinely in almost every situation. Accordingly, almost only type II markers are activated.

4 Implications

The model of experience-dependent somatic activation postulates an inverse U-shaped relationship between somatic activation and experience. Therefore, it provides the opportunity to assess the level of experience and associated skills and competencies by means of electrodermal activity. It is important to note that a loss of competence of a person can only be detected by a relative comparison of at least two measurement points. On an absolute scale, data from electrodermal activity have no diagnostic quality. If the two measurements differ significantly, this could be an indicator of a potentially critical change of experience and competence. Therefore, this could be a starting point for an intervention (e.g. redesign of the human-machine interface or user training). Detailed description of possible applications can be found in [8].

5 Summary

The model of experience-dependent somatic activation is consistent with both the empirical findings of our own experimental findings and the theoretical conception of somatic markers. At the present time, the model is to be regarded as a working hypothesis. Its validation requires further empirical studies. It should be noted that a recent paper from a different domain (surgical education) and without a theoretical reference to the somatic marker hypothesis also presented data which suggest an inverse U-shaped relationship between electrodermal data and competence (see [10]). This further supports the assumption that there is an actual relationship between these two variables.

References

1. DIN Deutsches Institut für Normung e.V. (2016) Grundsätze der Ergonomie für die Gestaltung von Arbeitssystemen (6385)
2. de Winter JCF, Hancock PA (2015) Reflections on the 1951 fitts list: do humans believe now that machines surpass them? Procedia Manuf 3:5334–5341. https://doi.org/10.1016/j.promfg.2015.07.641
3. SAE International (2014) Taxonomy and Definitions for Terms Related to On-Road Motor Vehicle Automated Driving Systems
4. Bainbridge L (1983) Ironies of automation. Automatica 19(6):775–779. https://doi.org/10.1016/0005-1098(83)90046-8
5. Damasio AR (1994) Descartes' error: emotion, reason, and the human brain. Putnam, New York
6. Bechara A, Damasio H, Tranel D, Damasio AR (1997) Deciding advantageously before knowing the advantageous strategy. Science 275(5304):1293–1295
7. Kinnear N, Kelly SW, Stradling S, Thomson J (2013) Understanding how drivers learn to anticipate risk on the road: a laboratory experiment of affective anticipation of road hazards. Accid Anal Prev 50:1025–1033

8. Heine T (2017) Vor-Sicht im Straßenverkehr - Experimentelle Untersuchung der somatischen Antizipation von Risiko. Dissertation, Karlsruher Institut für Technologie, Karlsruhe https://publikationen.bibliothek.kit.edu/1000075541
9. Bechara A (2004) The role of emotion in decision-making: evidence from neurological patients with orbitofrontal damage. Brain Cogn 55(1):30–40
10. Quick JA, Bukoski AD, Doty J, Bennett BJ, Crane M, Barnes SL (2017) Objective measurement of clinical competency in surgical education using electrodermal activity. J Surg. Educ. 74(4):674–680. https://doi.org/10.1016/j.jsurg.2017.01.007

Shared Ventilatory Drive as a Measure of Social Physiological Compliance During Team Decision Making

Robert A. Henning[✉] and Andrea M. Bizarro

University of Connecticut, Storrs, CT, USA
robert.henning@uconn.edu, andrea.bizarro@gmail.com

Abstract. According to the social cybernetic model of behavior that is rooted in control theory, the ability of individuals to establish mutual control over their behaviors is an essential part of teamwork, and this becomes even more important during periods of intense collaboration. We investigated the efficacy of using ventilatory drive to assess social physiological compliance (SPC) associated with mutual control of social behaviors during a joint decision-making task. SPC can be scored in numerous ways but generally involves calculating the extent that physiological changes among participants are shared or synchronized. SPC based on ventilatory drive was considered a good candidate for scoring the extent of mutual control during teamwork because ventilatory drive is known to closely track changes in body metabolism and an individual's bioenergetic state. Ventilatory drive, or the urge to breathe, can be estimated on a breath-by-breath basis by dividing the volume of one breath by the time it takes to inhale that breath.

Calibrated volumetric breathing signals were collected from dyadic teams (N = 54 teams) under laboratory conditions during a 15-min period of joint decision-making within a 40-min task period. SPC based on ventilatory drive was found to predict objective team performance outcomes, with more SPC associated with better performance outcomes. Although SPC showed no association with team satisfaction, more SPC was associated with lower team performance satisfaction. The results overall suggest that SPC based on ventilatory drive was reflective of mutual control during more intense periods of team collaboration, including those involving speech communications. SPC based on ventilatory drive may have potential uses in systems for augmented team cognition, team training, and for monitoring teams in high-demand environments because it can be measured continuously and non-invasively in applied settings.

Keywords: Teamwork · Breathing · Collaboration · Performance
Satisfaction · Dyads · Social cybernetics · Physiological compliance

1   Introduction

According to a social cybernetic model of behavior [1], social physiological compliance (SPC) is a phenomenon that is both an outcome and a determinant of effective social behaviors. As an outcome, SPC is understood to reflect shared physiological

© Springer Nature Switzerland AG 2019
S. Bagnara et al. (Eds.): IEA 2018, AISC 827, pp. 377–385, 2019.
https://doi.org/10.1007/978-3-319-96059-3_42

changes that result when behaviors become socially integrated in some way, such as during highly interdependent tasks. As an emergent systems phenomenon, attaining social physiological compliance is also understood to place individuals in a good position to exert a high degree of mutual control over their social behaviors going forward.

SPC based on heart rate variability, breathing rhythm, and electrodermal activity has been linked to various aspects of teamwork, including performance [2–4], stress resiliency [5], and cooperation [6]. This study introduces a novel measure of SPC based on shared levels of ventilatory drive between team members. Responding to a call for further validation of physio-behavioral measures of team dynamics [7], this paper reports on predictive relationships linking this new measure of SPC with team outcomes when two-member teams performed a joint decision-making task under laboratory conditions.

According to behavioral cybernetic theory that is akin to control theory, social behaviors depend on establishing dynamic motor-sensory control linkages that enable mutual control over each other's behaviors and bioenergetic states [1]. Effective forms of mutual control during teamwork depend on a combination of feedback control of behavior as well as feed-forward control of behavior as team members both react together to address current task demands while also taking proactive steps to jointly handle upcoming task demands. The integrated behaviors that emerge from this process reflect mutual control of behavior that can be distinguished from self-regulated behaviors by individuals functioning in isolation. In the present study, matched aspects of breathing were expected to reflect the extent of mutual control present when team members were actively collaborating and performing interdependent task activities.

A person's ventilatory control system for breathing is highly regulated as it acts to rid the body of excess carbon dioxide while also capturing the oxygen needed to support body metabolism. Ventilatory drive differs from conventional psychophysiological indices of breathing because it is more closely tied to changes in metabolic activity. For example, a person's breathing pattern may vary widely in terms of changes in rate (breaths/minute) or depth (tidal volume of each breath), and yet still ventilate the lungs by the amount needed to satisfy metabolic needs. This explains why matched ventilatory drive among team members may be superior to conventional measures of breathing when assessing mutual control of behavior during social interaction. While previous studies have reported a relationship between breathing pattern and high-level forms of teamwork [3, 6], to our knowledge this is the first test of ventilatory drive as a means to assess SPC during teamwork.

1.1 Hypothesized Relationships

Earlier research findings suggest that coordinated (i.e., predictable) behaviors during social interaction heighten an individual's ability to track (i.e., to both follow and anticipate) social behaviors, which enhances social learning [8] and also increases overall satisfaction with the interaction [9]. Therefore, SPC based on ventilatory drive in the present study was expected to predict team satisfaction (H1), performance satisfaction (H2), and objective team performance (H3).

2   Methods

2.1   Participants

Undergraduate students (N = 110) were recruited through an online participant pool to form 55 two-person teams. Participants were rejected if they were not strangers to each other at the beginning of the study in order to study aspects of team formation [10]. Gender information was lost for one team. The final sample consisted of 69 females and 39 males in 23 mixed- and 31 same-gender teams (N = 54). Participants had no previous experience with the team task.

2.2   Management Task

A simulated business decision-making task that revolved around movie production was chosen for this experiment because it was engaging and suitable for an undergraduate population. During the first 15 min of the 40-min experimental task period, each participant was asked to study a binder that contained unique information that they later could share when deciding as a team which screenplays to use from a list of available options, and which marketing strategies would be most profitable [11].

2.3   Measures

Ventilatory Drive. Inductive plethysmographs (Ambulatory Monitoring, Ardsley, NY) were used to non-invasively monitor changes in lung volume on a continuous basis. Each participant wore two elastic bands, one around the ribcage and one around the abdomen. The bands were imbedded with wires that created inductive magnetic fields. Changes in cross sectional area linked to breathing caused impedance changes in each of the elastic bands, and this was used to estimate changes in overall lung volume. In order to adjust the gain of each plethysmograph, participants performed an isovolume manoeuvre in which a held breath was actively shifted between their ribcage and abdomen, consistent with a two-compartment model of breathing. Additionally, the participant exhaled several times into a bellows spirometer (OMI Model PE91022; Houston, TX) so that the sum of both of their plethysmograph outputs could be calibrated to true volume.

The summed breathing signal from each individual was smoothed using two 8-pole Butterworth low-pass filters (Frequency Devices Model 901, 3 dB cutoff of 3 Hz) in series prior to 32 Hz analog-to-digital sampling. The resulting digital time series was smoothed prior to signal processing by a custom ventilatory drive algorithm. Estimated airway deadspace (150 ml), the volume of air in each breath that is not refreshed due to space in the mouth and throat, was used to set the breath detection threshold. Ventilatory drive was calculated on a breath-by-breath basis by dividing the size of each breath (i.e., tidal volume) by the time needed for that breath to be inhaled (i.e., inspiratory time). A time series for ventilatory drive was created for each participant that was normalized based on the first fifteen-minutes of the experimental period in which both team members were actively engaged in the task but not yet collaborating in a joint manner.

Social Physiological Compliance. Following Henning, et al. [3] in which a highly selective approach was used to score SPC in heart rate variability within long task periods, the present SPC scoring approach determined the threshold at which SPC was relatively high and presumably associated with more intense team collaboration and shared decision-making activity. This was accomplished by first creating a new time series record consisting of instantaneous root-mean-square (RMS) error values between the instantaneous normalized ventilatory drive values of each participant's time series. The scoring algorithm determined the cut-off threshold for the lowest 10 percent of all RMS error values in order for this SPC score to be immune to outliers caused by spurious events such as coughing and gross body movements. Only the first 15 min of the joint decision-making period was scored to avoid scoring a period of possible task inactivity near the end of the 25-min period allotted for joint decision-making, in case some team members completed the task early.

Objective Task Measures. A profit-to-cost ratio was calculated as the team-level objective performance score using spreadsheets provided by the simulation's first author [11].

Qualitative Measures. Stress, team satisfaction and performance satisfaction were assessed via a questionnaire. Sample items are provided in Table 1.

Table 1. Survey measures.

Construct	Sample item	Source
Team satisfaction	*"I would be willing to work with this team on another class project"*	Lancellotti and Boyd [16]
Performance satisfaction	*"I am not satisfied with the quality of the final recommendation"*	Lancellotti and Boyd [16]
Team stress	*"Before coming to the experiment today I was feeling irritated"*	Stanton et al. [17]

2.4 Procedure

Participants received instructions on how to don the breathing equipment, and then completed the pre-task stress assessment survey in isolation from each other. Participants were then seated in a common area, completed the equipment calibration, and received instructions for the task. Breathing patterns were monitored continuously throughout the 40-min experimental period. Participants completed a survey to assess team and performance satisfaction immediately following the task period and while remaining seated on either side of the task table.

3 Results

Descriptive statistics are provided in Table 2. Stress, team satisfaction, and performance satisfaction were aggregated to mean team-level scores in order to analyze the effects of SPC on team effectiveness outcomes using OLS regression at the team-level. Following analysis practices in related exploratory research on teamwork [11, 12], significance testing was conducted at $p < 0.10$.

Table 2. Descriptive statistics.

	M	SD
Individual-level		
Initial stress	6.83	1.39
Team satisfaction	5.89	.78
Performance satisfaction	5.43	1.10
Team-level		
Team stress	2.74	.81
SPC	.06	.01
Performance	2.09	.53

Tests of H1 revealed that SPC had no relation with team satisfaction ($ß = .03$, $p > .10$) but was predictive of lower performance satisfaction ($ß = .18$, $p = .061$; $\Delta R2 = .026$) after controlling for initial team stress levels (H2). Tests of H3 indicated that SPC predicted objective performance on the shared decision-making task after controlling for initial team stress levels ($ß = -.25$, $p = .071$; $\Delta R2 = .061$); higher levels of SPC (lower threshold scores) resulted in improved objective performance on the shared decision-making task (see Table 3).

Table 3. Regression results for SPC and effectiveness outcomes

	Team satisfaction	Performance satisfaction	Team performance
	ß	ß	ß
Step 1			
Team stress	0.03	0.32**	−0.12
Step 2			
Team stress	0.03	0.34**	−0.09
SPC	0.03	0.18*	−0.25*

Note: **p < .001, *p < .10

4 Discussion

This study provides some empirical evidence that social physiological compliance (SPC) based on ventilatory drive can be used to assess teamwork during joint decision-making tasks. More SPC of ventilatory drive was found to predict objective team performance but more SPC also predicted less satisfaction with one's overall team performance. These contrary findings are consistent with findings in our past work where SPC was positively correlated with objective team performance [4] but negatively correlated with subjective ratings of team performance [3]. Nonetheless, the results of the present study suggest that mutual control over social behaviors while collaborating, as reflected by more SPC, does benefit objective team decision-making performance.

SPC based on ventilatory drive offers a means to assess mutual control of social behaviors that is scientifically grounded in physiological control mechanisms, consistent with the social cybernetic model of behavior in which social interaction is understood to involve mutual control over behavior in parallel with control of the bioenergetic states of the participants [13]. As a result, SPC of ventilatory drive has the potential to support continuous assessment of team collaboration processes and social tracking effectiveness. Measurement of breathing needed to score SPC based on ventilatory drive is also relatively non-invasive, making scoring SPC feasible in applied settings and with multiple team members involved.

In terms of potential applications, it is possible that team training environments could be designed to help team members improve their skills and abilities to achieve higher levels of SPC during high-demand task periods. For example, training exercises could provide team members with opportunities to practice techniques for establishing high levels of SPC quickly. More specifically, it is possible that team learning could benefit if teams first functioned in each fundamental mode of social tracking (imitative, parallel-linked, series-linked; [1]) before engaging in joint decision making tasks that involve combinations of these fundamental social tracking behaviors, similar to part-task training used to gain proficiency in operating complex human-machine systems.

With a validated measure of SPC based on ventilatory drive, team tasks and their task environments could also be designed (or redesigned) to promote both the development and maintenance of SPC. For example, some team members may need to be located in closer proximity to each other to support the level of mutual visual-motor control necessary to establish and maintain adequate levels of SPC for the collaborative tasks at hand. Additionally, performing a series of preparatory task activities could become a standard operating procedure to help team members establish the level of mutual control and social-physiological compliance necessary to achieve high levels of performance proficiency, similar to what athletic teams do during team warm-up exercises prior to engaging in highly competitive events.

Social-physiological compliance based on ventilatory drive may also be particularly suitable for use in applications designed to augment team cognition which we have characterized in the past as dependent on periods of intense social tracking and mutual control of behavior in order for highly proficient team behavior to emerge [14]. Use of SPC as a means to assess emergent functional capacity at the team level differs from the conventional focus of augmented cognition systems which usually rely on psychophysiological indices to regulate the flow of information to humans to address presumed processing bottlenecks at the individual or team level. The focus here is instead on determining how best to promote a team's ability to establish and maintain mutual social control and social physiological compliance whenever needed as the main means to augment team cognition and performance. Intervention possibilities include providing the team with information about SPC that can be used to pace task activities.

Finally, SPC based on ventilatory drive could provide information regarding team member compatibility during the selection process for work teams in which collaboration is crucial to team effectiveness and where task failures have major consequences, such as in the selection of astronaut teams, emergency response teams, or military special forces.

4.1 Study Limitations

The positive findings in this laboratory study need to be replicated and also validated under field conditions with larger teams and for longer task durations. New wearable technologies developed for the continuous monitoring of individual physiological responses, coupled with synchronized telemetry systems that support scoring SPC, make such studies increasingly feasible. Inductive plethysmographs typically provide 10% volumetric accuracy, and so improvements in transducer technology could strengthen the size of experimental effects reported here.

The present study lacked a continuous process measure of task performance that would have enabled examination of the dynamic relationship between SPC and team task behaviors involving mutual control. This shortcoming was perhaps inevitable because of an inherent tradeoff between team tasks that are amenable to continuous process measurement, and team tasks that involve higher-level collaboration and decision making that is inherently more episodic in nature and consists of discrete non-linear steps rather than continuous outputs. Nonetheless, the present work complemented our past work on SPC and team behaviors where parallel-linked social tracking tasks performed over relatively brief task periods permitted continuous quantitative assessment of both team processes and performance outputs [13].

Lastly, the present study relied on only the single physiological response measure of ventilatory drive to score SPC rather than using multiple psychophysiological measures. It seems likely that SPC based on ventilatory drive has the potential to complement SPC measures based on heart rate variability, breathing rhythm, electrodermal activity, and a host of EEG measures.

5 Conclusions

Evidence revealed in this laboratory study suggests that social physiological compliance based on ventilatory drive offers some potential as a useful tool for researchers and practitioners interested in the objective assessment of team functioning during joint decision making tasks. More physiological compliance was predictive of objective performance outcomes on an interdependent decision-making task but was negatively associated with team members' satisfaction with their team's overall performance. Use of ventilatory drive as a measure of social physiological compliance may provide a more accurate measure of mutual control over interdependent task activity during teamwork than other breathing measures because of its close linkage with metabolic activity levels, including during speech behavior. Ventilatory drive also has the potential to provide a non-invasive means for researchers and practitioners to continuously assess mutual control of behavior among teammates in real time. Possible applications include systems to support augmented team cognition, selecting individuals who are more likely to collaborate successfully as a team, and in designing training or task environments to promote more effective teams and forms of teamwork.

References

1. Smith TJ, Smith KU (1987) Feedback-control mechanisms of human behavior. In: Salvendy G (ed) Handbook of human factors. Wiley, New York, pp 251–293
2. Elkins AN, Muth ER, Hoover AW, Walker AD, Carpenter TL, Switzer FS (2009) Physiological compliance and team performance. Appl Ergon 40:997–1003
3. Henning RA, Armstead AG, Ferris JK (2009) Social psychophysiological compliance in a four-person research team. Appl Ergon 40:1004–1010
4. Henning RA, Boucsein W, Gil MC (2001) Social-physiological compliance as a determinant of team performance. Int J Psychophysiol 40:221–232
5. Henning RA, Korbelak KT (2005) Social-physiological compliance as a predictor of future team performance. Psychologia 48:84–92
6. Chanel G, Kivikangas JM, Ravaja N (2012) Physiological compliance for social gaming analysis: cooperative versus competitive play. Interact Comput 24:306–316
7. Funke G et al (2012) Conceptualization and measurement of team workload: a critical need. Hum Factors 54(1):36–51
8. Smith KU (1971) Social feedback: determination of social learning. J Nerv Ment Dis 152 (4):289–297
9. Warner RM (1987) Rhythmic organization of social interaction and observer ratings of positive affect and involvement. J Nonverbal Behav 11(2):57–74
10. Bizarro AM (2013) The distinct roles of first impressions and physiological compliance in establishing effective teamwork (Unpublished Master's thesis). http://digitalcommons. uconn.edu/gs_theses/488
11. Devine DJ, Habig JK, Martin KE, Bott JP, Grayson AL (2004) Tinsel town: a top management simulation involving distributed expertise. Simul Gaming 35(1):94–134
12. Gorman JC, Cooke NJ, Amazeen PG (2010) Training adaptive teams. Hum Factors 52 (2):295–307
13. Dove-Steinkamp ML, Henning RA (2012) Training under imposed communication delays benefits performance effectiveness of distributed teams. In: Proceedings of the human factors and ergonomics society 56th annual meeting, pp 2432–2436
14. Henning RA, Bizarro A, Dove-Steinkamp M, Calabrese C (2014) Social cybernetics of team performance variability. In: Smith TJ, Henning RA, Wade MG, Fisher T (eds) Variability in cognitive performance. CRC Press, pp 193–210

15. Henning RA, Smith TJ, Korbelak K (2005) Social psychophysiological compliance as a gauge of the cognitive capacity of teams. In: Schmorrow DD (ed) Foundations of augmented cognition. CRC Press, New York, pp 1228–1238
16. Lancellotti MP, Boyd T (2008) The effects of team personality awareness exercises on team satisfaction and performance: the context of marketing course projects. J Mark Educ 30 (3):244–254
17. Stanton JM, Blazer WK, Smith PC, Parra LF, Ironson G (2001) A general measure of work stress: the stress in general scale. Educ Psychol Meas 61(5):866–888

The Influence of Room Size on Error Monitoring: Evidence from Event-Related Potential Responses

Chengwen Luo[✉], Georgios I. Christopoulos, Adam Roberts,
Arunika Pillay, and Chee Kiong Soh

School of Civil and Environmental Engineering, Nanyang Technological
University, 50 Nanyang Avenue, Singapore 639798, Singapore
cwluo@ntu.edu.sg

Abstract. The effect of environment on human behavior is a central topic for many disciplines. Many physical properties of the built environment, such as room size, modulate human perception and cognition: for instance, restricted physical space may lead to the perception of confinement, and potentially alter human cognitive functions, preferences, and performance. In the present study, we investigated the influence of room size on inhibition control with an established cognitive task (The Eriksen Flanker Task), while participant underwent EEG (electroencephalogram) recording. Specifically, we tested whether making error responses in a small room would lead to greater emotional disturbance as compared to a big room. Consistent with previous studies, reaction time was longer in trials with correct responses than those with error responses. Interestingly, participants in the small room, but not the big room, showed faster reaction time in error trials. For EEG results, we measured error positivity (Pe), a component reflecting motivation of error detection and emotional state after error responses. Our findings showed a comparable error positivity (Pe) responses regardless of room size. Possible implications are discussed.

Keywords: Room size · Inhibition control · Error positivity · Error detection
EEG

1 Introduction

The impact of the built environment on human performance is a critical topic for Ergonomics. Abundant evidence suggests that environmental factors modulate human perception and cognition. For instance, Levav and colleagues [1] reported that the perception of confinement in a shopping store altered customers' purchase behaviours. Participants in a smaller store tended to seek a greater variety of products compared to those in a larger store. Likewise, people tended to demonstrate greater self-regulation and less impulsive behaviours in confined spaces as compared to open spaces [2, 3].

There are a few hypotheses proposed to explain the influence of room size on cognitive performance and behavioural change. A possible explanation for the higher inhibition in constrained physical spaces is that the limited space restricts body

© Springer Nature Switzerland AG 2019
S. Bagnara et al. (Eds.): IEA 2018, AISC 827, pp. 386–391, 2019.
https://doi.org/10.1007/978-3-319-96059-3_43

movements to avoid colliding into objects/walls or to avoid the awkwardness of being too physically close to other people in social interactions. The increased bodily control would further affect behavioural and physiological responses, and thus influence human behaviour and performance [4]. Another possible explanation, not necessarily competitive, is that spatial constraints may be metaphorically linked to self-monitoring [5]. More restricted places are associated with increased inhibition control and therefore, reduce impulsive behaviours [3, 6].

To date, little physiological evidence has been reported to support this relationship between room size and error detection. Our study aims to see the influence of room size on inhibition control. We investigated whether spatial constraints influence cognitive function by employing a typical flanker task conducted in a small room versus a big room, and further examined the underlying mechanism by simultaneous EEG recording. Specifically we examined the Error-related Positivity (Pe), which is a response-locked positive deflection generated by the anterior cingulate cortex (ACC) [7]. Pe has been suggested to relate to additional processes that occurs after error detection, reflecting one's conscious appraisal of the mistake [8]. The Pe onset typically occurs around 150-200 ms after the commission of an error, although the latencies vary according to the difficulty levels of the task [7, 9].

2   Method

2.1   Participants

The protocols of the study were approved by the Internal Review Board at Nanyang Technological University. A total of 81 undergraduate students from Nanyang Technological University attended the experiment, with 3 of them excluded due to drowsiness/falling-off electrodes, and 6 excluded due to lack of interest (made more than 1/3 of mistakes in a simple orientation judgment task). Thus 72 participants were included in the final analyses, with the mean age of 21.34 ± 2.59.

2.2   Procedure

Participants were administered the Flanker task, in either a small room or a big room. The size of the rooms varied in one dimension, while the width and height kept constant (approximately 3.60 m), the length is 2.43 m for the small room and 3.53 m for the big room. This was achieved by manipulating the position of a Hollywood flat (portable wall). Experiments were conducted with Eprime 2.0. Details of the task are described as follows.

In each trial (Fig. 1), participants were instructed to indicate the direction of a central stimulus ('<' or '>') which was crowded either by congruent (same direction, i.e. '<<<<<' or '>>>>>') or incongruent arrows (different direction, i.e. '<<><<' or '>><>>') via a key press. They responded to the central target '<' with the index finger of one hand and the other '>' with the opposite hand. Each stimulus was presented for 100 ms, followed by a 600 ms response period. A fixation cross appeared in the centre throughout each intertrial interval (ITI).

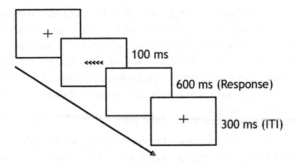

Fig. 1. The Eriksen Flanker Task. Participants were required to make a speeded button press by indicating the orientation of the central stimulus crowded by congruent or incongruent flankers.

Prior to the experiment, 40 practice trials were given to each participant to make sure they understand how the experiment works. The main task had a total of 600 trials, and participants were instructed to press the key correctly as fast as possible. No feedback was provided throughout. To enhance the difficulty of the task and ensure a sufficient number of error responses, 20% of trials were incongruent and 80% were congruent.

2.3   Data Acquisition and Analyses

2.3.1   Behavioural Measurements and Analyses

Behavioural measures were recorded in Eprime 2.0, including the sequence of trials, accuracy (correct/error) and reaction time of each trial. We calculated a linear mixed-effect model with reaction time as dependent variable (DV), and correctness (correct vs error: within-subject factor) and room size (big vs. small: between-subject factor) as independent variables (IV). The percentage of errors made in the big/small room for each participant was examined with Chi-square test.

2.3.2   EEG Recording and Processing

EEG was recorded from a 64-channel ANT EEG system (Netherlands), organized according to the international standard 10–20 system. EEG was digitized at 1024 Hz. All analyses were performed offline in Matlab with EEGLAB 14.1.1 toolbox (Delorme and Makeig, 2004). Datasets were down-sampled to 256 Hz and were filtered between 0.1 and 30 Hz with a basic FIR filter. The continuous EEG signal was then mastoid re-referenced and epoched in an interval of −250 ms to 800 ms with the onset of response at 0 ms; all epoched data were linear detrended. Following analysis suggested by Miyakoshi [10], we submitted the epoched data to automatic artefact rejection by a joint probability and kurtosis with the rejection threshold of 5 Standard Deviations of the signal mean(implemented in EEGLAB; for detail, Delorme, Sejnowski, & Makeig, 2007). The data were then submitted to Independent Component Analysis (ICA), using the Runica algorithm. ICA components carrying ocular and muscle movements arte-facts were identified and rejected. Finally, an automatic algorithm was applied to exclude remaining epoched data with amplitude larger than 50 μV.

2.3.3 EEG Analyses

The response-locked brain activity (i.e. Pe) was analysed following previous approaches [12]. First, we calculated the average amplitude between 150 ms and 450 ms after the onset of response, relative to a pre-response baseline from -250 ms to -50 ms for each trial [13–15]. Then the difference between trials with correct and error responses was compared using two-way Analysis of variance (ANOVA), with accuracy (correct vs error) and room size (big vs small) as factors.

2.4 Results

2.4.1 Behavioural Results

Mean reaction times and accuracy results are presented in Table 1. After controlling for within subject variance, the linear mixed-effect model showed that participants who completed the task in the small room tended to react faster compared to those in the big room ($t(70) = -1.97$, $p = .05$). The main effect of correctness was also significant ($t(70) = 14.10$, $p < .001$), RT was faster in error trials compared to correct trials. Although the interaction between correctness and room size was marginal ($p = .07$), we nevertheless performed planned post-hoc analyses. The findings showed that faster RT in small room was due to trials with error responses ($t(70) = 1.99$, $p = .05$), but not those with correct answers ($t(70) = -.65$, $p = .52$). For the percentage of error made in big versus small room, Chi-square test indicated that there was no difference across rooms ($p > .05$).

Table 1. Task performance measured by room sizes

Group	Reaction time (ms)		Accuracy	
	Error trials	Correct trials	Average number of error trials	Percentage correct (%)
Big room	271.34	374.70	75.31	87.45
Small room	246.81	382.12	70.94	88.18

2.4.2 EEG Results

Figure 2 presents the response-locked Pe for correct and error trials for participants in different rooms. Consistent with previous findings [16], we observed a typical Pe following error responses compared to correct responses - a larger positive deflection from 150 ms onwards to 400 ms, regardless of room sizes ($<.001$). However, the main effect of room and interaction between room and amplitude were not significant (both $ps > .05$)

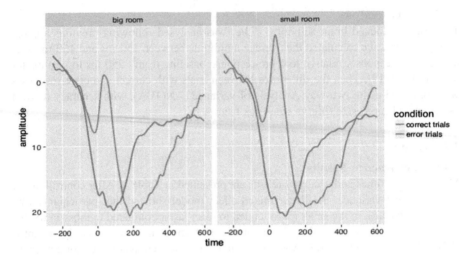

Fig. 2. Response-locked brain responses (Pe) to correct and error trials in big and small rooms. 0 ms marks the time of responses via a button-press. (a) ERP responses for participants who completed the task in a big room; (b) ERP responses (Pe) for participants who completed the task in a small room.

3 Discussion

The present study investigated the influence of physical space on inhibition control via a Flanker task and simultaneous EEG recording. Although the percentage of errors made across the two different room sizes was comparable, our findings showed a faster reaction time in small room compared to the big room, driven by error trials. We did not observe significant differences in Pe, a component suggested to measure conscious appraisal of errors and affect following error responses [17]. Therefore, our results suggest that room size does not influence emotional appraisal of errors made during the experiment.

A recent study by Paul and colleagues (2017) suggested that positive mood selectively decreased the amplitude of Pe following error responses, but not other error-related component. Their findings were consistent with an earlier EEG study, showing a smaller Pe in a more relaxed mood. A study examining participants with/without anxiety disorder showed comparable error-related Pe for both groups [18]. However, another error processing component – the error-related negativity (ERN) which occurs at much earlier stage (0–50 ms), reflecting error detection, demonstrated a significant difference in these studies. Thus, future analyses could be carried out on ERN to examine the influence of room size on cognitive processing.

Furthermore, it is well-known that people with claustrophobic disorders tend to feel stressful in a restricted space. Our study only includes undergraduate students without any claustrophobic symptoms, future studies could also be conducted in people with claustrophobic symptoms.

Acknowledgements. This material is based on research/work supported by the Land and Liveability National Innovation Challenge under L2 NIC Award No. L2NICCFP1-2013-2. Any opinions, findings, and conclusions or recommendations expressed in this material are those of the authors and do not necessarily reflect the views of the L2 NIC.

References

1. Levav J, Zhu R (2009) Seeking freedom through variety. J Consum Res 36(4):600–610
2. Scholer AA, Higgins ET (2011) Promotion and prevention systems: regulatory focus dynamics within self-regulatory hierarchies
3. Xu AJ, Albarracín D (2016) Constrained physical space constrains hedonism. J Assoc Consum Res 1(4):557–568
4. Mozrall JR et al (2000) The effects of whole-body restriction on task performance. Ergonomics 43(11):1805–1823
5. Zhong C-B, Liljenquist K (2006) Washing away your sins: threatened morality and physical cleansing. Science 313:1451–1452
6. Zhong C-B, Leonardelli GJ (2008) Cold and lonely does social exclusion literally feel cold? Psychol Sci 19:838–842
7. Larson MJ et al (2010) Temporal stability of the error-related negativity (ERN) and post-error positivity (Pe): the role of number of trials. Psychophysiology 47(6):1167–1171
8. Nieuwenhuis S et al (2001) Error-related brain potentials are differentially related to awareness of response errors: evidence from an antisaccade task. Psychophysiology 38 (5):752–760
9. Paul K et al (2017) Modulatory effects of happy mood on performance monitoring: insights from error-related brain potentials. Cogn Affect Behav Neurosci 17(1):106–123
10. Miyakoshi M et al (2010) EEG evidence of face-specific visual self-representation. Neuroimage 50(4):1666–1675
11. Delorme A, Sejnowski T, Makeig S (2007) Enhanced detection of artifacts in EEG data using higher-order statistics and independent component analysis. Neuroimage 34(4):1443–1449
12. Colosio M et al (2017) Neural mechanisms of cognitive dissonance (revised): an EEG study. J Neurosci 37(20):5074–5083
13. Hanna GL et al (2012) Error-related negativity and tic history in pediatric obsessive-compulsive disorder. J Am Acad Child Adolesc Psychiatr 51(9):902–910
14. Carrasco M et al (2013) Increased error-related brain activity in youth with obsessive-compulsive disorder and other anxiety disorders. Neurosc Lett 541:214–218
15. Hall JR, Bernat EM, Patrick CJ (2007) Externalizing psychopathology and the error-related negativity. Psychol Sci 18(4):326–333
16. Yeung N, Botvinick MM, Cohen JD (2004) The neural basis of error detection: conflict monitoring and the error-related negativity. Psychol Rev 111(4):931
17. Hajcak G, McDonald N, Simons RF (2003) To err is autonomic: error-related brain potentials, ANS activity, and post-error compensatory behavior. Psychophysiology 40 (6):895–903
18. Ladouceur CD et al (2006) Increased error-related negativity (ERN) in childhood anxiety disorders: ERP and source localization. J Child Psychol Psychiatr 47(10):1073–1082

Development of Chairs for Nonintrusive Measurement of Heart Rate and Respiration and Its Application

Mieko Ohsuga[(✉)]

Faculty of Robotics and Design, Osaka Institute of Technology,
Osaka 530-8568, Japan
mieko.ohsuga@oit.ac.jp

Abstract. This report describes the development of "a vital sensing chair" and "a vital sensing cushion" to obtain data on heart rate using capacity-coupled electrodes instead of electrodes attached on the skin and respiration pressure sensors embedded in the sheet. The measurement test for the vital chair was conducted with five participants during two events: watching TV programs or movies projected on a wall and operating a smartphone. The results did not yield sufficient performance; however, there is the possibility to apply the developed chair in the assessment of a human state in a relaxed posture. The evaluation of the developed cushion has not been performed yet and it will be a part of future work. A game for a pair of participants applying nonintrusive respiration measurement was developed to facilitate respiration synchronization. Sixty-eight participants experienced it, and it was well received.

Keywords: Non-intrusive measurement · Heart rate · Respiration
Human-robot interaction

1 Introduction

For an affective interaction between humans and robots, the adaptive control of robot's behaviors depending on the human's states is important. Our previous study, we aim to detect a decrease in concentration when people carry out tasks, not by alarming the individual but by waiting for an opportunity to interrupt without being a nuisance. Two situations where a robot approaches a human are assumed; one is a short break recommendation during a long period of computer task and the other is a recommendation to carry out light exercise while watching television in the living room [1]. Several studies have demonstrated the availability of non-invasive physiological indices, particularly autonomic indices to recognize emotions in the space of arousal and valence [2, 3]. In our previous study, degraded concentration were detected using the measures derived from electrocardiogram (ECG) and respiration [4].

The present study focuses on the non-intrusive measurement of ECG and respiration. Capacity-coupled electrodes attached on the backrest and cushion of a seat are introduced to measure ECG and obtain heart rate with clothes on to eliminate the

© Springer Nature Switzerland AG 2019
S. Bagnara et al. (Eds.): IEA 2018, AISC 827, pp. 392–404, 2019.
https://doi.org/10.1007/978-3-319-96059-3_44

burden of attaching sensors for measurement [5–9]. Pressure sensors embedded in the seat of the chair were used to obtain respiration data [10].

2 Previous Works

2.1 Capacity-Coupled Electrodes for Driver Assessment

The original aim of developing the capacitive electrodes is to measure heart rate (HR) during driving a car to assess their arousal and readiness state (Fig. 1 [8]). Two electrodes were placed on the backrest of the chair and two on the seat cushion (Fig. 1 left), and the combinations of the electrodes were examined. The electrodes and the circuit were also improved several times (Fig. 2 [6], Fig. 3 [8]). In the preliminary experiments, the performance of obtaining HR information from ECG were better for pairs of one of the electrodes placed on the backrest and another on the cushion. However, since it is easier to incorporate the circuit using electrode pairs on the backrest (BR–BL) and cushion (CR–CL), these pairs were thoroughly investigated (Fig. 4 [7]).

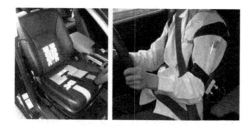

Fig. 1. Electrodes placed on car seat (left) and actual scene of the vehicle experiment (right) [8]

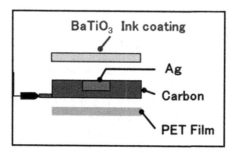

Fig. 2. Electrode materials [6]

An improved method was applied in an experiment using ten volunteers (five males and five females) aged 24–51 years, who provided written informed consent. The participants were required to drive on an expressway and a rural road for 15 min each.

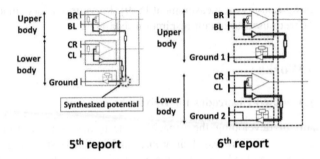

5th report   **6th report**

Fig. 3. Circuit improvement [8]

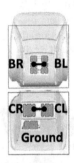

Fig. 4. Pair of back electrodes (BR–BL), cushion electrodes (CR–CL), and a ground electrode [7]

False positive (FP) and false negative (FN) R-wave detections were evaluated by comparing the results with those detected from direct chest ECG (d-ECG). The results in the BR–RL derivation significantly improved; five males and three females wearing stockings exhibited good performance (less than 10% for both FP and FN). The R-wave detections obtained using CR–CL derivation were poor [8].

2.2   Measurement During PC Work

Original Method [4]. Chips of conductive tape were placed on the surface of airbags placed on the backrest of the chair and a ground electrode was placed on the seat cushion (Fig. 5). A pair of back electrodes was used for the differential input of the amplifier. Respiration was measured by detecting variations in the back pressure using pressure sensors attached to the air bags on the back of the chair. Five healthy adults who gave written informed consent participated in the experiment. Respiration was measured successfully during most periods of the computer task in three of the five participants, while ECG measurement was not successful during most periods in one participant who moved or changed postures frequently. R-waves were measured during 60–90% of the period for other participants.

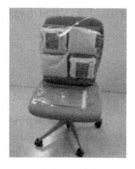

Fig. 5. Chair equipped with air bags and chips of conductive tapes (left) and a scene of the experiment during PC task (right) [5]

Introduction of Capacity-Coupled Electrodes Developed for Driver Sensing. The electrodes developed for driver sensing were introduced during measurements in PC tasks using an office chair with a spring-loaded seatback [9]. The electrodes were placed on the surfaces of the airbags attached to the backrest (Fig. 6 left). Four healthy paid volunteers (4 males) aged 21–23 years who gave written informed consent participated in the experiment (Fig. 6 right). The Kraepelin test was carried out using a display and a keyboard as a simulated computer task. After 3-min training and 3-min rest, the participants were required to execute a 40-min task as a simulated computer task. The results of FP and FN using ch.4 (CR–CL) derivation were poor even in this experiment. However, four participants (less than 5% for both FP and FN during the period without body movement) for ch.2 (BR–CL), three participants for ch.3 (BL–CL), and two for ch.1 (BR-BL) yielded good performance.

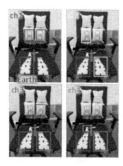

Fig. 6. Chair equipped with air bags and ECG electrodes developed for driver sensing (Left), and a scene of experiment during a simulated PC task (Right)

3 Development of a Vital Sensing Chair and a Vital Sensing Cushion

3.1 Vital Sensing Chair

We developed a vital sensing chair that can be used in a living room or hall, which incorporates the electrode developed for the driver into a relaxing chair. The aim is to adapt the robot's approach based on the degree of concentration during the task of watching TV in the living room and change the contents to be provided depending on the state of the participants during the event. We also aim to carry out measurements quickly just by sitting so that measurements can be carried out on several persons simultaneously.

A commercially available reclinable and high backrest chair (Bulb Chair made by SWITCH, Fig. 7) was adopted. A sensor unit with a pair of ECG electrodes (Fig. 8) and a preamplifier on the backrest and the seat, respectively, as well as an amplifier circuit were embedded in the chair. The respiration sensor was also developed by TS Tech and consists of four pressure sensors, and the position of the sensor was determined based on our previous collaborative research with Nippon University [10]. The surface of the electrode part is covered with cloth containing a conductive thread.

Fig. 7. Appearance of vital sensing chair (Color figure online)

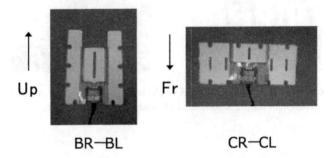

Fig. 8. Capacity-coupled electrodes (L: back, R: cushion)

3.2   Vital Sensing Cushion

To carry out measurements during the desk task, we developed a cushion that can be used by placing it on a chair when needed instead of incorporating the sensor in an office chair. The posture correction cushion (Back joy Posture Plus) was adopted, and the backrest was custom-made (Fig. 10). The sensor unit is the same as that of the vital chair. In addition, an air lumbar for the car seat was placed on the backrest so that the contact condition can be maintained.

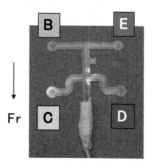

RESPIRATION SENSOR

Fig. 9. Pressure sensors for respiration (4 channels)

Fig. 10. Appearance of vital sensing cushion

4 Experiments

4.1 Methods

Participants. Five healthy paid volunteers (4 males and 1 female) aged 21–23 years who gave written informed consent participated in the experiment (Approval No. 2016-35).

Experimental Conditions. The experiment was conducted in a semi-soundproof room under two conditions: watching a favorite movie on a projection screen equivalent to 40 in. ahead and operating a smart phone by hand (contents are free). Each task was carried out with light clothing (T-shirt only) and thick clothing (T-shirt and trainer, etc.) for 15 min each.

Measurement. In addition to the 2 channels of capacitive ECG (backrest BR–BL and seating surface CR–CL) and 4 channels of respiration by the vital sensing chair, ECG and respiration were measured using sensors attached to the participants for comparison. ECG was measured using disposal electrodes attached on the chest and a commercially available bio-amplifier (BA 1008, Nihon Suntech). The two types of respiration were measured using a respiratory pickup sensor (TR - 751T, Nihon Kohden) and DC respiration sensor (MaP2290 DRS (G), Nihon Suntech). These data were sampled at 1000 Hz and accumulated in a PC.

Analysis. The filtering procedure for R wave detection from c-ECG is as follows: (1) high frequency noise elimination, (2) 13–20 Hz band pass filter (2nd order Butterworth, phase compensation), (3) conversion to absolute value, and (4) 3 Hz low pass filter (2nd order Butterworth, phase compensation). Steps (3) and (4) correspond to full-wave rectification (Fig. 11, upper). The capacity-coupled ECG after filtering is called c-ECG (capacitive-ECG). R wave was sequentially detected from the largest peak to satisfy the condition that the contacting peak is separated by 0.5 s or more (Fig. 11, lower).
The evaluation of R wave detection was carried out on two types of errors: False-Positive (FP) and False-Negative (FN) (Fig. 12). FP is the detection of R wave peak in c-ECG without detection chest ECG (d-ECG (direct-ECG)). On the other hand, FN is R wave peak detection in d-ECG without that in c-ECG. The allowable range of the time lag from each peak was $\pm 0.1 + dt$ [s]. dt was set to the optimum value of each data taking the difference between c-ECG and d-ECG peaks into consideration.

The respiration waveform measured by the four pressure sensors (B to E in Fig. 9) was subjected to anti-aliasing filter, then resampled at 50 Hz, and a 0.05–0.6 Hz band pass filter (2nd order Butterworth, phase compensation) was applied. The maximum of the absolute value of the cross-correlation coefficient between the above mentioned filtered respiration and the respiration measured by the DC sensor was obtained for each 1-min interval. The average and standard deviation of 13 sections of 1 min from 2 to 14 min were obtained.

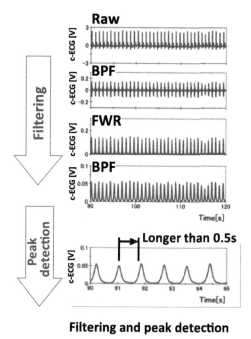

Filtering and peak detection

Fig. 11. Filtering and peak detection procedure for capacitive-coupled ECG

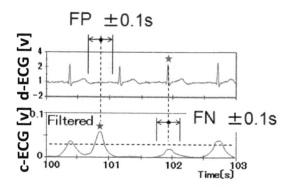

Fig. 12. Definition of False Positive (FP) and False Negative (FN)

5   Results

Figure 13 shows an example of C-ECG analysis when the measurement condition is good.

The results of FP and FN are given in Table 1. The green background is less than 5%, yellow is less than 10%, gray is less than 20%, 20–40% is white, and 40% or more is not measurable and designated as x. Only two people's results showed the condition that both FP and FN became 5% or less. The condition was when looking at the front

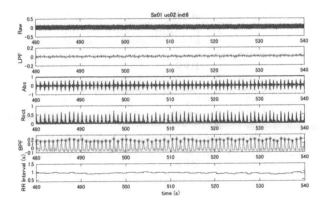

Fig. 13. Example of C-ECG analysis

Table 1. Rate of FP and FN for each channel and each condition [%]

		BR–BL				CR–CL			
		smartphone		display		smartphone		display	
Participant	clothes	FP	FN	FP	FN	FP	FN	FP	FN
1	heavy	×	×	35.4	32.4	22.6	32.2	0.4	0.5
	light	11.2	18.0	29.3	×	0.7	0.9	0.3	0.7
2	heavy	10.7	9.6	×	34.2	9.8	19.5	8.7	25.1
	light	14.1	10.6	×	×	×	31.7	2.1	1.0
3	heavy	×	×	6.4	8.7	×	×	1.7	5.6
	light	9.7	12.9	13.3	32.0	×	×	12.5	17.5
4	heavy	×	×	×	×	×	×	33.0	32.0
	light	30.7	30.1	35.2	36.9	16.2	20.5	5.8	8.5
5	heavy	×	×	9.3	7.6	×	×	×	31.4
	light	×	×	22.1	18.2	×	×	35.8	23.1

screen, and the measurement site was CR-CL (seating surface). The difference due to clothing was not clear. Even in cases with many FPs and FNs, good results may be obtained in the section with less body movement, and an analysis taking account of the presence or absence of body movement is necessary.

Figure 14 shows examples of waveforms of the pressure sensors and the DC respiration sensor when the measurement condition is good. The result of calculation of the correlation coefficient is shown in Fig. 15. There are individual differences in the value of the cross-correlation, but the measurement situation in front of the seating surface (CR–CL) is good in all five. Respiration is also greatly affected by body movements.

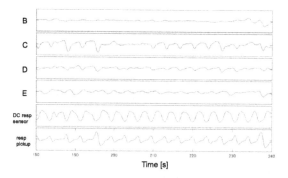

Fig. 14. Example of respiration analysis

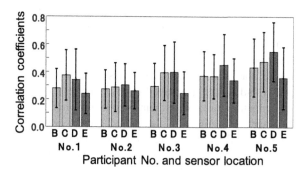

Fig. 15. Average values and standard deviations of the correlation coefficients between respiration waves measured by pressure sensors and DC respiration sensor

6 Application of Respiration Measurement

6.1 Development of a Game Facilitating Respiration Synchronization

A game for a pair of participants applying nonintrusive respiration was developed to facilitate respiration synchronization. It aims at the fact that two people's feelings also match when they make an effort to match their breathing. The environment of the game is set in the sea; the character is a sunfish. When the participants breathe in, the sunfish rises, while it falls if they exhale. The angle of the sunfish is determined by the difference value (change rate) of the respiration waveform. When the difference between the position and angle of two sunfish (corresponding to the size and phase of the breath) is within the set range, the sunfish coalesce into a special version of sunfish. This informs the participants that their breaths are synchronizing. The performance of the game is evaluated by the ratio of time in which the special sunfish has emerged (breath is synchronizing), and it is scored and indicated by the number of starfish on the ocean floor.

The flow of the experience is as follows. First, in order to decide which of the four pressure sensors' signals are to be selected for each individual, participants are required to breath according to the target waveform presented on the display, which consists of a 2-s inspiration and 3-s expiration, and have them breathe accordingly. The cross-correlation coefficient between the target waveform and the sensor output signals are obtained, and a sensor showing the largest coefficient is selected. The maximum and the minimum values of the output of this sensor are reflected in the display range of the next game. Next, for each person, a game to pass through the rings was prepared as an aid to understanding the relationship between the sunfish movement and breathing and a practice for getting used to fitting the breath. After these two steps, the participants experience the breathing game.

6.2   Demonstration and Assessment

At "The Lab" (Grand Front Osaka North Building 3rd Floor, a place for publication of research results), the demonstration for assessment was executed with visitors from the general public (Fig. 16). Oral explanations, calibration, games for practice, and respiration games were performed in this order. Questionnaires, attributes, and impressions were obtained from those who gave informed consent (parents also in the case of juveniles) (approval number 2016-23). Many people experienced it, and it was so well received that a queue appeared on holidays. Questionnaires were obtained from 68 participants (28 males and 40 females) aged 6–66. Over 90% of the people answered that they found the sunfish to move with their own breath. Half of the people answered that they felt that their breathing with the paired person was synchronizing while playing a breathing game. Although the reaction varied depending on the pair, it could be said that breathing games are promising as entertainment. In addition, middle-aged and elderly participants commented that this game could also be used for breathing exercises.

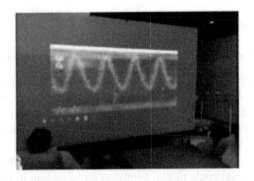

Fig. 16. Experiment at "The Lab"

7 Discussion and Future Work

The evaluation of the developed cushion has not been performed yet and it will be a part of future work.

The possibility of heart rate measurement with the developed chair was shown; however, when compared with the results during driving and during computer work in the past experiments, the rate of measurement failure is high. For the reason of better measurement during driver monitoring, it is conceivable that the body is fixed to the seat by the seat belt and the movement of the body is relatively restricted. As to the previous computer work experiment, the working time was short and the airbag and the restorative backrest were able to absorb the movement of the body. In this way, it is necessary to devise measures to improve the maintenance of contact on the back, but it is impossible for the user's back to touch the backrest of the chair always, under the condition that the user can freely move. Therefore, it is realistic to improve the measurement further only on the seating surface.

Respiratory measurement is also greatly affected by body movements and its use is limited to the condition that the user is sitting still. It is also necessary to consider combining the non-contact measurement method using cameras for both heartbeat and respiration. Since measurement failure will not disappear, the evaluation of the reliability of the measurement result and outputting it simultaneously should be considered. It is conceivable to evaluate the validity of measured values by the time series change of measurement results and statistical properties. It is also a practical method to detect the body motion from the output of the pressure sensors for respiration measurement and ignore the measured value during body movement.

Regarding the application, it is worthy of evaluation that breaths of multiple people can be measured easily, and something that is enjoyable can be provided simply by sitting. It can be expected to be applied to multi-person measurements at concert venues and theaters.

Acknowledgement. A part of this work was supported by JSPS KAKENHI Grant Number JP15K00385. All experiments referred to in this paper were executed with the permission of the president of Osaka Institute of Technology in accordance with the report of the Ethics Committee on Life Sciences of Osaka Institute of Technology. The author thanks individuals who worked in the driver sensing research groups, especially Mr. Shinji Sugiyama and Dr. Yukiyo Kuriyagawa and other researchers and students who worked and discussed for this research. The author also appreciates all the experiment participants.

References

1. Yamada E, Ohsuga M, Hashimoto W, Inoue Y, Nakaizumi F (2011) Proposal of system which promotes physical activity at home and effects of promotion using a low-cost home robot. In: Proceedings of the fourth international conference on human-environment system, pp 79–84
2. Pattyn N, Neyt X, Henderickx D, Soetens E (2008) Psychophysiological investigations of vigilance decrement: boredom or cognitive fatigue? Physiol Behav 93:369–378

3. Valenza G, Lanata A, Scilingo EP (2012) The role of nonlinear dynamics in affective valence and arousal recognition. IEEE Trans Affect Comput 3(2):237–249

4. Osuga M, Boutani H (2014) Detection of a decrease in concentration using indices derived from heart rate and respiration toward affective human-robot interaction. In: Proceedings of the 5th international conference on applied human factors and ergonomics, Kraków, pp 1462–1467

5. Ohsuga M, Boutani H (2014) Preliminary study on nonintrusive measurement of ECG and respiration while sitting on a chair for human states estimation. In: Proceedings of the 1st Asian conference on ergonomics and design, Jeju

6. Ohsuga M, Sugiyama S (2015) Obtaining heart rate information from a driver using capacity coupled electrodes (4th report). In: Proceedings of JSAE annual congress in spring of society of automotive engineers of Japan

7. Ohsuga M, Sugiyama S (2016) Obtaining heart rate information from a driver using capacity coupled electrodes (5th report). In: Proceedings of JSAE annual congress in spring of society of automotive engineers of Japan

8. Ohsuga M, Sugiyama S (2016) Obtaining heart rate information from a driver using capacity coupled electrodes (6th report). In: Proceedings of JSAE annual congress in autumn of society of automotive engineers of Japan

9. Ohsuga M (2016) Non-intrusive measurement of heart rate during computer work. In: The proceedings of the 18th international conference on human-computer interaction, Toronto

10. Iida S, Kuriyagawa K, Kageyama I, Kobayashi H, Ohsuga M, Itoh T, Sugiyama S (2013) Obtaining of respiratory information from a driver using seat pressure. In: Proceedings of JSAE annual congress in spring of society of automotive engineers of Japan

Development of Intention Inference Algorithm Based on EMG Signals at Judging Directional of Arrow Cues

Yuzo Takahashi[✉]

Graduate School of Information Sciences, Hiroshima City University, 3-4-1,
Ozuka-higashi, Asaminami-ku, Hiroshima City, Hiroshima 7313194, Japan
y-taka@hiroshima-cu.ac.jp

Abstract. Inferring a device wearer's intention is important for supporting devices that exert force, dexterity and sustainability. In this research, we considered a simple ON and OFF function. If one can discriminate whether an order to move a specific muscle is voluntary or involuntary when the command reaches the periphery, estimating the direction in which the supporting device assists the movement is easy. Therefore, in this study, we measured voluntary biceps brachii muscle contraction in response to a visual stimulus, with various stimulus intensities. We found that the preparation start time and the lifting start time were earlier during the condition ("right directional arrow" and "left directional arrow") where there was high confidence in the choice to execute a forearm lift, among other available stimuli. In contrast, the center frequency of the biceps brachii muscles, at the time of preparing the lifting motion, tended to be higher. Finally, by using an algorithm to infer movement intention, it was possible to identify if the next movement comprised flexion or extension, before muscle torque generation.

Keywords: Intention inference algorithm · EMG
Muscle voluntary movement

1 Introduction

In this research, focused on a simple ON or OFF command to a supporting device. Inferring the intention of someone wearing a supporting device is important while using a device that exerts force, dexterity and sustainability. The wearer's intention is generated in the brain, which then forms an anticipatory plan to initiate intentional motion in not only one specific part but also various parts of the body that participate in movement execution. For this reason, algorithms that inference a body part to execute a voluntary movement, from an anticipatory plan that is generated in the brain, and extracting a command to target body parts is very complicated. Therefore, we investigated an algorithm involving the detection of subsequent extension or flexion, prior to muscle torque generation. If a voluntary movement (extension or flexion) requiring device assistance can be known before movement execution, then the supporting device can generate torque in conjunction with the commencement of movement.

© Springer Nature Switzerland AG 2019
S. Bagnara et al. (Eds.): IEA 2018, AISC 827, pp. 405–412, 2019.
https://doi.org/10.1007/978-3-319-96059-3_45

Accomplishing this necessitates that two problems be solved. First, the lengthy lead time, beginning when the support equipment receives information on the wearer's intention to when the equipment gets activated [1, 2]. Second, in specific muscles, the command is returned to the spinal cord after travelling from the central nervous system to the periphery. Therefore, it is difficult to distinguish whether if specific muscle activation on electromyogram occurs because of the first command or the command that occurs after the spinal reflex [3]. Muscle agonists move, particularly when the muscle load is several seconds to several tens of seconds, and this movement potentially changes the control of a specific action [4]. In order to infer operative intention, multiple muscles must be simultaneously measured, complicating the muscle control mechanism.

If it is possible to discriminate whether a specific muscle command is voluntary or involuntary when the command reaches the periphery, then estimating the direction of support device assistance is easy. Therefore, we measured voluntary biceps brachii muscle contraction in response to a visual stimulus with various stimulus intensities. Then we report the results of our attempts to develop an algorithm for separating the efferent single carrying movement intention from the signal controlling the motion preparation state, on an EMG (electromyogram) signal.

2   Method

2.1   Participants

Participants were eight healthy university students and graduate students (19–23 years, males = 4 and females = 4). All participants were right hand and right foot dominant. After entering the laboratory, all participants provided written informed consent following a comprehensive introduction to the study protocol and objectives.

2.2   Apparatus

Stimulus Equipment. The visual stimuli presented to the participants consisted of black arrows (Fig. 1). Stimuli were presented on one nineteen-inch screen (Flexcan S1901-B, iiyama, Japan) placed one meter in front of the participants. At the same time as visual stimulus presentation, a one-volt rectangular wave (stimulation trigger) was created using a DA converter CSI-360112 (Interface, Japan) installed in a personal computer (Lavie PC-LG17FDNGM, NEC, Japan) used for control of stimulation presentation.

Electromyogram (EMG) Measures. We amplified the electromyogram of the biceps brachialis muscle using EMG amplifiers (PTS-137, DKH and K 800, Biometrics) connected to the bioelectrode SX - 230 - 1000 (Biometrics). The EMG signals were sampled using the TRIAS System (DKH) + personal computer (Dynabook PS L21 220C/W, TOSHIBA) using a sampling interval of 1 ms synchronised with the stimulus trigger. The experimental setting is shown in Fig. 2.

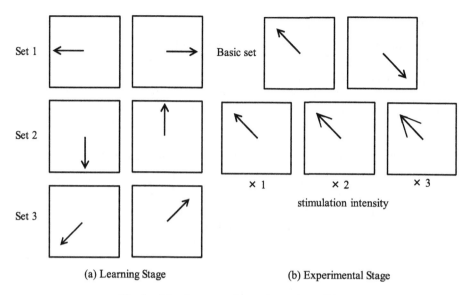

(a) Learning Stage   (b) Experimental Stage

Fig. 1. Stimulus set and experimental conditions

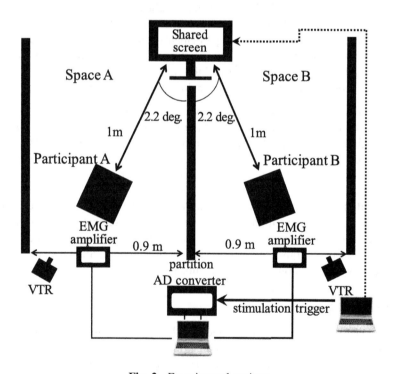

Fig. 2. Experimental settings

2.3 Task Sequences

Experimental Task. In the experiment, visual stimuli (arrows) were presented on the shared screen and participants were asked to lift their forearms as fast as possible. Participants had dumbbells on both forearms weighing 250 g to measure discriminative myoelectric potential signals. In order to strengthen the reaction to the stimulus, the *Learning Stage* was carried with the goal of strengthening the stimuli responses (elevation of the right or left forearm). Subsequently, from the reaction acquired at the *Learning stage*, at the *Experimental Stage*, each participant was asked to decide which forearm to lift and then fulfill that movement intention with movement execution.

Learning Stage. Three sets of stimuli were presented during the learning stage. During the first set, we asked the participants to lift their right forearm when they saw the right-pointing arrow; Fig. 1(a) on the Set 1 was presented on the screen. Alternatively, they were asked to lift their left forearm when the left-pointing arrow was presented. In the second set, we asked participants to lift their right forearm in response to the upper-pointing arrow; Fig. 1(a) on the Set 2 was presented on the screen. Alternatively, they were asked to lift their left forearm in response to the lower-pointing arrow. In the third set, we applied the contents of the previous two sets. When an arrow was presented in the upper right quadrant [the first quadrant: the arrow in the right-side box on the set 3, in Fig. 1(a)], participants were asked to lift their right forearm. Alternatively, when an arrow was presented within the lower left quadrant [the third quadrant: the arrow in the left-side box on the set 3, in Fig. 1(a)], participants were asked to lift their left forearm. All subjects progressed through the learning stages in the same order, with 40 forearm lifts completed at each stage (the left and the right 20 times, respectively). The arrow presentation sequence at each stage was randomised for each subject.

Experimental Stage. In this stage, arrows were presented within the upper left quadrant [second quadrant: the arrow in the right-side box on the Basic set, in Fig. 1 (b)] or within the lower right quadrant [fourth quadrant: the arrow in the left-side box on the Basic set, in Fig. 1(b)]. During the experimental stage, we did not give participants instructions as to which forearm to lift; rather, participants independently decided which forearm to lift. To examine the influence of stimulus intensity [retinal size of the flight: "stimulation intensity" on Fig. 1(b)], the same flight stimulus (by 1), twice the flight length and triple the flight length were presented. Experiments were conducted for each stimulus intensity, and the number of repetitions in each stimulus condition was set at 80 times (40 repetitions in each direction). Stimulus intensity order and directions were randomised for each subject.

2.4 Collected Data

EMG signals were taken from the left and right biceps brachii muscles using the device system shown in Fig. 2. We determined the lifted forearm by analysing the VTR image in order to be accurate. In addition, after completing each stimulus condition during the learning stage, we asked each participant why he or she decided to lift that forearm.

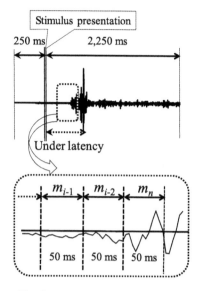

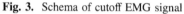

Fig. 3. Schema of cutoff EMG signal

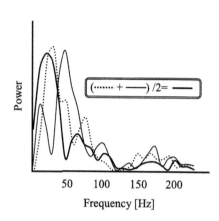

Fig. 4. Average waveform of FFT result

3   Intention Inference Algorithm

3.1   EMG Signal Processing

Initially, trigger signals and EMG signals for the visual stimuli, presented simultaneously during each forearm lifting were extracted from the polygraphical time series data. The cutout section was from 250 ms before the stimulus presentation to 2,250 ms after the stimulus presentation (Fig. 3). In this study, we defined the time at which the M1 wave [3] was observed as the preparation start time, and the time at which the M2 or M3 [3] waves were observed was defined as the lifting start time.

Preparation Start Time. Estimated time (the preparation start time) before occurrence of the agonist's myotatic reflex determined which forearm responded to the visual stimulus. Specifically, FFT processing was applied to 128 points of the myoelectric potential data every 50 ms after the arrow was presented, then the average waveform was calculated between two successive sections (Fig. 4). Next, we calculated the center frequency for each average waveform. We defined preparation start time as the first time the difference between the center frequency of the previous section and the center frequency of the current section exceeded the threshold value of 7.9 Hz.

Lifting Start Time. To calculate lifting start time, the EMG signal in the section 2,500 ms from the time of visual stimulus presentation (Fig. 3) was rectified [Fig. 5(a)]. Next, the cumulative added waveform was derived from the rectified waveform [Fig. 5(b)], and we calculated the difference value every 50 ms. The lifting start time was defined as the first time the difference value between the previous section and the current section exceeded double [Fig. 5(c)].

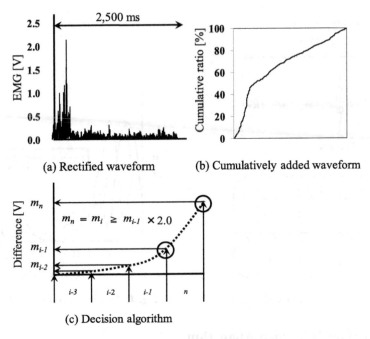

(a) Rectified waveform   (b) Cumulatively added waveform

$$m_n = m_i \geq m_{i-1} \times 2.0$$

(c) Decision algorithm

Fig. 5. Judging algorithm for lifting start time

4   Results

4.1   Time Domain Aspects of Intention

Figure 6 shows the preparation time results. (a) shows the pooled results of all the experimental conditions, and (b) shows the results for each individual experimental condition. Both Fig. 6(a) and (b) showed a tendency to retreat in the preparation time in

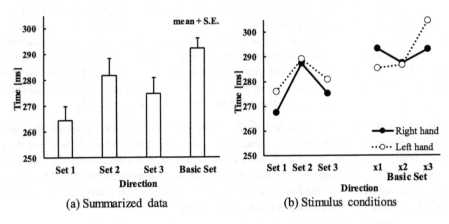

(a) Summarized data   (b) Stimulus conditions

Fig. 6. Preparation start time between experimental and stimulus conditions

(up-pointing and down-pointing arrows). In (a), the longest preparation time was observed during the Basic Set. When examining the influence of stimulus intensity, we observed no significant relationship between stimulus intensity and preparation time. Similar results were observed at the lifting start time [Fig. 7(a) and (b)]. In other words, there was no significant relationship between the regression of the lifting start time and the stimulus intensity in the Set 2 condition in the learning stage.

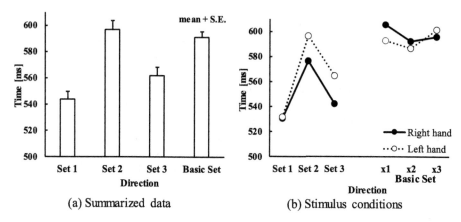

(a) Summarized data   (b) Stimulus conditions

Fig. 7. Lifting start time between experimental and stimulus conditions

4.2   Frequency Domain Aspects of Intention

Figure 8 shows the difference in the center frequency between the preparation time and the stimulus presentation time. As shown in Fig. 8, during the learning stage, the degree of coincidence with the stereotype was low, and when learning was required, the difference value tended to increase. In addition, as the stimulus intensity increased during the experimental stage, the difference in the center frequency between when the stimulus was presented tended to be larger.

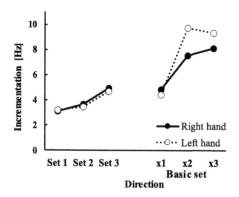

Fig. 8. Incrimination of center frequency during preparation start time

5 Concluding Remarks

As a result of applying the algorithm to infer intention from the EMG signal generated in response to each type of visual stimulus, the preparation time and the lifting start time clearly showed the nature of the stimulation. The learning stage and the experimental stage was not indicated a conclusory relationship between the experimental condition or arrow-pointing direction. However, the center frequency of the preparation time (which is considered to correspond to the M1 wave [3] of the agonist's myotatic reflex) markedly differed from the center frequency at the time of visual stimulus presentation; (1) the central frequency increased when learning was required, (2) the center frequency also increased as the stimulation intensity increased. Therefore, intentional action may accompany increases in center frequency of the M1 wave. In other words, the center frequency of the biceps brachii muscles, when preparing a lifting motion, tended to be higher. Finally, we used an algorithm to infer intention. It seems possible to identify whether the next movement would consist of flexion or extension, before torque generation, in the muscles controlling the movement. Regarding the stimulus strength, we need to consider not only the length of the flight but also the influence of ambiguity due to the way the flights are coupled.

References

1. Kosaki T, Atsuumi K, Takahashi Y, Li S (2017) A pneumatic arm power-assist system prototype with EMG-based muscle activity detection. In: Proceedings of 2017 IEEE International Conference on Mechatronics and Automation, pp 793–798
2. Kitai J, Kosaki T, Atsuumi K, Takahashi Y, Sano M (2014) Development of an EMG-based motion detector for a pneumatic arm assistive device. In: Proceedings of the 9th JFPS International Symposium on Fluid Power, pp 679–683
3. Takahashi K, Fukuda O, Iisaka H, Suzuki S (1988) Supraspinal modulations of stretch reflex by movement preparation: Input-Output properties. Ann Rep Coll Med Technol Hokkaido Univ 1:45–59
4. Takahashi Y, Kumashiro M (2007) Time series analysis of postural coordination during quiet stance. In: Proceedings of the 8th Pan–Pacific Conference on Occupational Ergonomics, Conference CD –ROM, p 5

Adverse Workstyle and Its Correlation with Other Ergonomic Risk Factors in Work Related Musculoskeletal Disorders

Deepak Sharan[✉]

Department of Orthopedics and Rehabilitation, RECOUP Neuromusculoskeletal
Rehabilitation Centre, 312, 10th Block, Anjanapura,
Bangalore 560108, KA, India
deepak.sharan@recoup.in

Abstract. Workstyle has been reported as a mediating factor in the relation between job demands and Work Related Musculoskeletal Disorders (WRMSD). A retrospective report analysis of 9500 IT professionals in an Industrially Developing Country was conducted. The participants' data was extracted from the database from 2006 to 2017. The average age of participants was 32.4 ± 9.2 years and 78% of respondents were males. 68% of participants worked between 8 to 12 h on a computer and 55% used a desktop computer. Neck and upper back pain were the more prevalent, followed by wrist, lower back and shoulder pain. 28% of participants were reported to have a high risk of an adverse workstyle (score ≥ 28). Lack of breaks, deadlines/pressure and social reactivity subscales of workstyle questionnaire were the highest predictors of pain and loss of productivity. Regression analyses revealed that workstyle factors and duration of computer use per day were significant predictors of pain.

Keywords: Work-related musculoskeletal disorder · Information Technology
Risk factors

1 Introduction

Workstyle has been reported as a mediating factor in the relation between job demands and Work Related Musculoskeletal Disorders (WRMSD). The concept "Workstyle" is defined as a psychosocial, physiological, and behavioural response that occurs in an individual due to high work demands (Feuerstein and Nicholas 2006). These modifying responses generated by self (e.g. fear of losing a job, work ethic) or environment (e.g. expectation of supervisor, work culture) further exacerbate the demands placed on workers. The aim of this study was to identify workstyle related and other ergonomic risk factors that may be associated with the onset or exacerbation of WRMSD among Information Technology (IT) professionals and to look for the correlation between the two.

© Springer Nature Switzerland AG 2019
S. Bagnara et al. (Eds.): IEA 2018, AISC 827, pp. 413–414, 2019.
https://doi.org/10.1007/978-3-319-96059-3_46

2 Methodology

A retrospective report analysis of 9500 IT professionals in an Industrially Developing Country was conducted. The participants' data was extracted from the database from 2006 to 2017. Ergonomic workplace analysis and demographic data (age, gender etc.), workstation information, working posture information, perceived pain and discomfort, and workstyle questionnaire addressing psychosocial factors (Feuerstein and Nicholas 2006) were analysed.

3 Results

The average age of participants was 32.4 ± 9.2 years and 78% of respondents were males. 68% of participants worked between 8 to 12 h on a computer and 55% used a desktop computer. Neck and upper back pain were the more prevalent, followed by wrist, lower back and shoulder pain. 28% of participants were reported to have a high risk of an adverse workstyle (score ≥ 28). 72% of participants reported pain symptoms during or shortly after they finish work on the computer. 38% of participants experienced numbness/tingling sensation in their fingers after working on the computer. Loss of strength in hands was reported by 24% of participants. 42% of participants indicated a loss in productivity due to the symptoms of pain and discomfort. 12% of participants indicated that days were taken off work due to the pain symptoms. Lack of breaks, deadlines/pressure and social reactivity subscales of workstyle questionnaire were the highest predictors of pain and loss of productivity. Regression analyses revealed that workstyle factors and duration of computer use per day were significant predictors of pain. Correlation coefficient analyses indicated a significant positive correlation between workstyle score and pain, and posture and regional pain.

4 Conclusions

Correlations were observed between ergonomic risk factors, psychosocial risk factors and musculoskeletal pain symptoms. Adverse workstyle appears to be a mediating factor for musculoskeletal pain, discomfort, and loss of productivity. It is recommended that intervention efforts directed towards prevention of WRMSD should address psychosocial work factors such as adverse workstyle in addition to biomechanical and environmental risk factors.

Ergonomics in Advanced Imaging

Ergonomic Guidance for Virtual Reality Content Creation

Takashi Kawai[1(✉)] and Jukka Häkkinen[2]

[1] Department of Intermedia Art and Science, Waseda University,
3-4-1 Okubo, Shinjuku, Tokyo 169-8555, Japan
tkawai@waseda.jp
[2] Institute of Behavioral Sciences, University of Helsinki,
Siltavuorenpenger 1-5, P.O. Box 9, 00014 Helsinki, Finland

Abstract. Virtual reality (VR) content aims to elicit a "sense of presence", defined as quality of immersive experience. In this paper, the authors suggest six pieces of guidance to follow to create and improve this sense of presence in VR from an ergonomic viewpoint.

– Analysis of VR content
– Parametric framework for VR
– Evaluation of VR content
– Usage environment for VR
– Individual differences in VR discomfort
– Sensory conflict in VR

Keywords: Ergonomic guidance · Virtual Reality · Content creation

1 Introduction

Recently, virtual reality (VR) has begun to be used in entertainment and many other fields. Although head-mounted display (HMD) used in VR has been advanced in function and quality, problems such as visual fatigue and cyber sickness remain. This is because these factors involve complicated interactions such as elements in content and attributes of users, and cannot be solved only by the performance of the HMD.

The authors have been working on elucidating user experience by immersive media including stereoscopic content [1, 2]. VR content aims to elicit a "sense of presence", defined as the quality of immersive experience. To that end, various creative trial and error attempts, including storytelling, are being done actively day by day. To support the creation of VR content, the authors are trying to form ergonomic guidance based on past knowledge and experimental data. In this paper, six pieces of guidance that are currently being examined are described.

© Springer Nature Switzerland AG 2019
S. Bagnara et al. (Eds.): IEA 2018, AISC 827, pp. 417–422, 2019.
https://doi.org/10.1007/978-3-319-96059-3_47

2   Analysis of VR Content

The first piece of the guidance is analyzing VR content. In VR, final user experience is formed by the complex interplay of various hardware devices and design techniques. Among others, our experiments have suggested that how users receive visual information and experience discomfort is affected by gaze motion during viewing, and that their emotional responses are affected by the structure and layout of the visual targets in virtual space [3]. In the experiment, the participants were 20 adults with normal visual function, and five 360° videos lasting 90 s each were presented. A HMD with eye tracking function (SMI Mobile Eye Tracking HMD) was used for presenting 360° videos and for measuring eye movements. The results are described on a spherical VR space. In Fig. 1, the results of two types of content having different characteristics are shown. In the case of content with many camera movements, the gaze concentrates in the front direction, and in the case of few camera movements, there is a tendency to look around at surroundings.

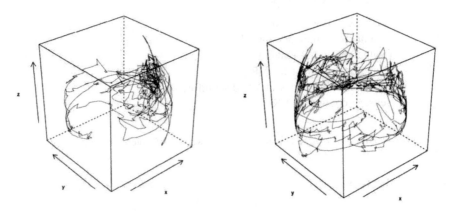

Fig. 1. Gaze path on spherical VR space (left: a 360° video with many camera motions, right: a 360° video with few motions).

Therefore, VR content creators should pay attention to the characteristics of the devices and techniques they plan to use to create their VR contents and experiences.

3   Parametric Framework for VR

The second piece of guidance regards handling of a parametric framework. Varied research has been done to determine human perceptual threshold parameters related to VR. One such parameter is angular velocity, a component of rotational motion contained in visual information; others include field of view (FOV), refresh rate, depth magnitude, and image quality. Together, these form a parametric framework that indicates the technical requirements for VR that is comfortable and elicits sense of presence. However, one must always bear in mind how one parameter could interact with others. In other words, improvement of a certain parameter may not lead to the improvement of user experience. For example, FOV is deeply related with the induction of sense of presence, but expansion of FOV can also be a cause of cyber sickness.

Furthermore, previous studies have usually focused on one or two parameters and a general parametric framework was difficult to form. Therefore, creators of VR content need to exercise caution in how they handle previous results. Further research is needed to integrate existing scientific literature concerning critical parameters for VR.

4   Evaluation of VR Content

The third piece of guidance is evaluating the produced content. While no standardized or consensus techniques for evaluating VR content in terms of user experience have yet been established, a few trends are apparent. Some researchers use both subjective indicators (such as questionnaires) and objective indicators (such as physiological measurements), and/or apply frequently used questionnaires such as the Simulator Sickness Questionnaire (SSQ) [4]. The SSQ evaluates motion sickness by total score based on three factors: nausea, oculomotor, and disorientation.

Figure 2 shows change of SSQ after viewing the 360° videos introduced in Fig. 1, with the baseline before viewing. In each factor and the total score of SSQ, the increase by content with many camera movements is significant. From these results, the extent of sickness level was found to be related to camera movements.

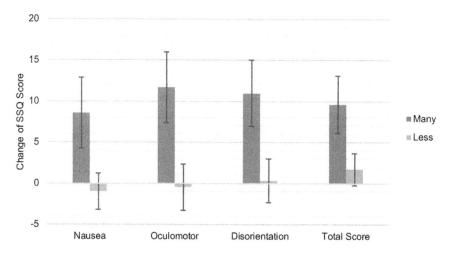

Fig. 2. Results of SSQ for the 360° videos described in Fig. 1.

The most effective approach would be for creators to use the same evaluation items, preferably ones with high reliability and that are convenient to measure.

5   Usage Environment for VR

The fourth piece of guidance is to pay attention to the environment in which VR is used. The devices and techniques used to create a sense of presence are affected by various environmental factors. One such factor, chair swiveling, is generally regarded as a way of implementing the act of "looking around" a 360° video. However, results of the abovementioned experiment suggest that excessively amplifying chair swiveling can fatigue or burden users. Figure 3 compares the results between with and without chair swiveling from Fig. 2. The condition with chair swiveling is observed to have an increasing tendency to change after viewing.

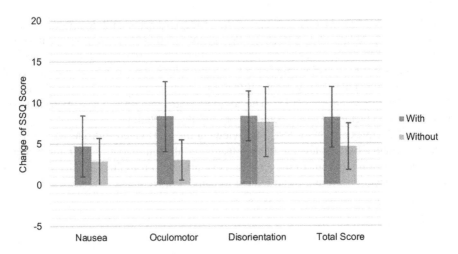

Fig. 3. Results of SSQ with/without chair swiveling conditions

When considering what environmental factors to incorporate in VR, creators must be aware of discrepancies between the anticipated effectiveness of a factor and the actual effects that it induces, especially for any factor that directly impacts the user's posture and actions.

6   Individual Differences in VR Discomfort

Feelings of discomfort sometimes induced by VR are known to vary not only according to attributes such as age and gender, but also greatly between individuals.

Figure 4 shows the results of Fig. 2 divided into two groups of high (7 persons) and low (13 persons) scores by cluster analysis of the Total Score. In other words, the responses by the high score group were considered to be significantly reflected in the results shown in Figs. 2 and 3.

Susceptibility of users to discomfort cannot be classified simply as "high" or "low": the sensitivity may be specific to certain VR devices or techniques, or depend on

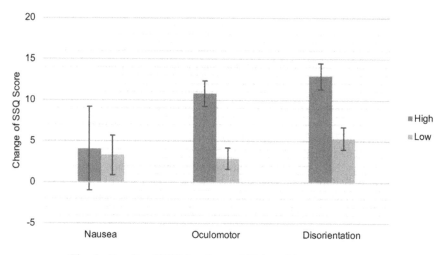

Fig. 4. Results of SSQ for clustered high and low score groups

familiarity with them. Most users probably lack self-awareness of their own sensitivity to VR-induced discomfort, and much less about the specific feelings accompanying it, meaning that creators must anticipate and take into account a variety of susceptibilities. At the same time, efforts would be necessary to raise users' self-awareness by various means, such as through periodic "status checks" and adjusting usage time or environment as necessary.

7   Sensory Conflict in VR

Visual information can induce sensation of motion. This illusion, called "vection", is deeply involved in generating a sense of presence in VR. However, such visual information causing sense of motion can differ from the proprioceptive sensation of remaining at rest. This gap is termed "sensory conflict" [5], a term that refers to the conflict between the sensory input due to VR devices and content and the sensory input that users encounter in their everyday lives; this conflict is one factor responsible for VR-induced discomfort. Sensory conflict causes both positive and negative user experiences, making it well-suited for dealing with VR comfort (and safety) in a single framework. In other words, the optimization of sensory conflict could lead to VR creation frameworks that preserve and improve sense of presence.

Acknowledgment. The pieces of guidance in this paper were described based on the research results sponsored by The Mechanical Social Systems Foundation (28-03).

References

1. Häkkinen J, Kawai T, Takatalo J, Leisti T, Radun J, Hirsaho A, Nyman G (2008) Measuring stereoscopic image quality experience with interpretation based quality methodology. In: SPIE, vol 6808, pp 6808B1–6808B12 (2008)
2. Kawai T, Hirahara M, Tomiyama Y, Atsuta D, Häkkinen J (2013) Disparity analysis of 3D movies and emotional representations. In: SPIE, vol 8648, pp 8648Z1–8648Z9
3. Banchi Y, Tsukada S, Yoshikawa K, Kawai T (2017) Behavioral and psychological effects by short time viewing 360 videos using a HMD. In: Proceedings of the 2nd Asian conference on ergonomics and design, pp 600–603
4. Kennedy RS, Lane NE, Berbaum KS, Lilienthal MG (1993) Simulator sickness questionnaire: an enhanced method for quantifying simulator sickness. Int J Aviat Psychol 3(3):203–220
5. Oman CM (1991) Sensory conflict in motion sickness: an observer theory approach, pictorial communication in virtual and real environments. Taylor & Francis, Bristol, pp 62–376

Virtual Reality in Education: How Schools Use VR in Classrooms

Takashi Shibata$^{(\boxtimes)}$

School of Education, Tokyo University of Social Welfare, Tokyo, Gunma, Japan
tashibat@ed.tokyo-fukushi.ac.jp

Abstract. In the present paper, I discuss several attempts of conducting experimental classes in schools to evaluate the educational effects and advantages of using stereoscopic 3D images and virtual reality (VR) techniques in schools. In comparison with 2D educational material, 3D material can help students focus on specific parts in images as well as to understand 3D spaces and concavo–convex shapes. The use of stereoscopic 3D images was also helpful, leading to an inquiry-based learning approach. Furthermore, the classes that made good usage of the advantages of VR techniques could encourage collaborative learning. The important thing to use VR in education is to support students to achieve learning objectives in class by their active attitude and discussions.

Keywords: Stereoscopic 3D images · Virtual Reality · Education
School · ICT

1 Introduction

Recently, substantial developments in display technology have been achieved. In addition to being incorporated into daily activities, display technologies promote the use of information and communication technology (ICT) in education. For example, tablets are increasingly popular in school classrooms as an approach to improve educational quality.

The applications of functionalities of new technologies such as stereoscopic 3D images and virtual reality (VR) also attract attention from the field of education because of their potential to provide an interactive learning environment for students. For an example, Bamford explored the efficiency of 3D images in classrooms and suggested that the motivation and engagement of students can be improved [1]. Shibata reported that the usage of stereo 3D movies and other similar educational material could provide students with a strong sense of depth in their learning school education [2]. Moreover, the Google Expeditions program provides students with virtual field trips with a head-mounted display (HMD) such as Google Cardboard [3]. Students can locate themselves in the center of 360° images and 3D scenes to learn the historical importance of locations that are represented by such images.

© Springer Nature Switzerland AG 2019
S. Bagnara et al. (Eds.): IEA 2018, AISC 827, pp. 423–425, 2019.
https://doi.org/10.1007/978-3-319-96059-3_48

2   Utilization of Stereoscopic Viewing in School

The advantages of using stereoscopic 3D images in education were previously examined by conducting experimental classes in an elementary school [4, 5]. A unit of the Tumulus period in Japan for sixth-grade students was selected as the source of 3D educational materials. The unit represents part of the coursework for the topic of Japanese history. The educational materials used in the study included stereoscopic 3D images to examine the stone chambers and terracotta clay figures of the Tumulus period.

The results of the experimental classes showed that 3D educational materials helped students to focus on specific parts in images as well as to understand 3D spaces and concavo–convex shapes. In addition, 3D educational materials also helped students in experimental classes to ask novel questions regarding the attached objects of terracotta clay figures as well as the spatial balance and alignment. The educational use of stereoscopic 3D images is helpful for an inquiry-based learning approach to history (Fig. 1).

Fig. 1. Scenes from experimental 3D classes in an elementary school; left: autostereoscopic mobile device; center: 3D laptop; and right: 3D television.

3   Use of Virtual Reality Techniques in School

Experimental classes using a desktop-style VR system were conducted in a school. The applications of VR techniques, such as viewing stereoscopic 3D images, interactivity, and sharing 3D objects in a space, in education were examined [6]. By attending the classes, students learned about plate tectonics and the mechanism of earthquakes.

The usage of VR education system helped the students to focus on the details of various overlapping plates and also to understand 3D structures. In addition, sharing 3D educational materials could promote discussions related to the learning topic in a group. These studies indicate that the classes that use the advantages of VR techniques can facilitate collaborative learning. However, it is important to consider the possibility of visual fatigue and motion sickness related to 3D and VR in the class design (Fig. 2).

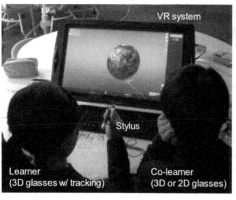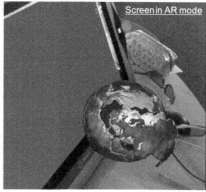

Fig. 2. The desktop VR system (zSpace Inc.) and an augmented reality (AR) presentation tool allowed students and the teacher in classroom to observe the objects that were displayed in space on the VR system.

4   Conclusion

Several works were discussed herein regarding the educational effects of using stereoscopic 3D images and VR techniques in conducting experimental classes in schools. Although VR techniques hold great potential for users, stereoscopic 3D representation is a key technology to present a realistic and immersive environment that could be helpful for students' learning. It is important for students to gain an interest in learning, which can be facilitated by the effective use of 3D and VR. Furthermore, using VR in education aims to support students to achieve learning objectives in class by their active attitude and discussions.

References

1. Bamford A (2011) The 3D in education white paper. Technical report, International Research Agency
2. Shibata T (2014) Utilization of stereoscopic 3D images for social studies class in elementary school. In: Proceedings of EdMedia 2014, pp. 2575–2580
3. Google Inc. https://edu.google.com/expeditions/. Accessed 20 May 2018
4. Shibata T, Ishihara Y, Sato K, Ikejiri R (2017) Utilization of stereoscopic 3D images in elementary school social studies classes. Electron. Imaging 2017:167–172
5. Shibata T, Sato K, Ikejiri R (2017) Generating questions for inquiry-based learning of history in elementary schools by using stereoscopic 3D images. IEICE Trans Electron 100 (C11):1012–1020
6. Shibata T, Drago E, Araki T, Horita T (in press) Encouraging collaborative learning in classrooms using virtual reality techniques. In: Proceedings of EdMedia 2018

Some Novel Applications of VR in the Domain of Health

David Grogna[1]([✉]), Céline Stassart[2], Jean-Christophe Servotte[3],
Isabelle Bragard[3], Anne-Marie Etienne[2], and Jacques G. Verly[1]

[1] Faculty of Applied Sciences, University of Liège, Liège, Belgium
dgrogna@uliege.be
[2] Faculty of Psychology, University of Liège, Liège, Belgium
[3] Faculty of Medicine, University of Liège, Liège, Belgium

Abstract. Recent progress in virtual reality (VR) technologies make immersion more accessible to everyone, and, in particular, developments aimed at the entertainment industry are being brought into to the domain of health.

The main uses of VR in health are of two forms. First, it is a new method to diagnose and to treat patients; second, it is a new method to train and/or teach healthcare and emergency-response professionals.

There are several reasons for using VR in healthcare. First, virtual environments (VE) are fully under control, so that the user (patient or professional) is then safe from any harm and the session can be interrupted if necessary. Second, there are many instances where placing the user in a real environment would be very hard to do and/or very costly. A major advantage of VR is that this user can instead be immerged in an equivalent artificial/virtual environment through the use of immersive technologies. Third, with regard to teaching, a significant advantage of VR is that it allows one "to bring the body to learning", thereby effectively embedding new knowledge into the muscles.

Below, we describe several uses of VR at our university in the domain of health.

Keywords: Virtual reality · Virtual reality applications

1 Treatment of Phobias

One way to treat phobias is to incite the patient to be confronted gradually with the object of his/her fear. The benefit of VR is that it allows the therapist to be fully in control of the exposition of the patient to his phobia.

For example, to treat the fear of spiders (arachnophobia), the Psychological and Speech Therapy Consultation Center (CPLU) of the university of Liège uses a VE representing a generic apartment with several rooms. In each room, there is one or more spiders, of various sizes, either moving or not. Depending on the degree of comfort of the patient, the therapist may guide the patient to enter or, on the contrary, to avoid some rooms. The same can be done for the fear of snakes and heights.

© Springer Nature Switzerland AG 2019
S. Bagnara et al. (Eds.): IEA 2018, AISC 827, pp. 426–427, 2019.
https://doi.org/10.1007/978-3-319-96059-3_49

2 Training of Healthcare and Emergency-Response Professionals

The recent increase in the frequency of major disasters – whether natural disasters, vehicle crashes, or terrorist attacks - forces healthcare professionals to transition the emergency protocol from an individual model to a collective model. This transition implies learning new skills that are specific to these situations. VR allows the learner to evolve in a realistic environment that is nearly impossible to setup in real life.

The Faculty of Medicine of the university of Liège offers the possibility to train emergency-response personnel to develop skills such as accessing victims, triage, and first aid in an EV replicating the 2012 bus crash in the tunnel in Sierre, France.

3 Interaction with the Environment

Usually, interactions with EV are made either through the headset and/or the controllers, but for some applications, these tools are not sufficient. The addition of eye tracking to the headset allows a precise knowledge of what the subject is looking at. Having this information is capital for some diagnostics and treatments, such as for phobias or alcoholism. One should note that most of the controllers that are well suited for entertainment interactions don't allow realistic interactions in other settings, such as those described above, which may constitute a significant obstacle to effective immersion. For example, an adequate combination of hand tracking and of matching between real and virtual objects is required to allow for the realistic grabbing of the virtual objects with the user's real hand.

Visual Ergonomics in Radiology

S. Hoffmann[1]([✉]) [iD] and N. Berger[2] [iD]

[1] MAS Work + Health, University of Zurich, Hirschengraben 82, 8001 Zurich,
Switzerland
sven.hoffmann@uzh.ch
[2] Institute of Diagnostic and Interventional Radiology, University Hospital of
Zurich, Raemistrasse 100, 8091 Zurich, Switzerland
nicole.berger@usz.ch

Abstract. In the past two decades, diagnostic radiology has changed in technology and frequency of imaging per patient. Further, the change from film to
soft-based image diagnostics as well as further technological developments in
cross-sectional and dynamic imaging have introduced new opportunities but
also health hazards for radiologists, too.

Keywords: Radiology developments · Radiologists' health hazards
Radiology workplace design

1 Aim

The aim of this workshop is, to review the current technological and procedural
developments in image diagnostics. Furthermore, the work environment and work-
related health hazards of radiologists shall be reviewed, addressing e.g. lighting design,
required attentiveness, evoked physical activity and postures.

2 Methods

2.1 Literature Review

The current literature findings, senior radiologist statements and own workplace
assessments will be presented and discussed with the participants.

2.2 Interactive Development of Radiology Workplace Recommendations

Interactive discussions and small group work on individual adaptation and development of radiology work environments. Especially lighting conditions, sit-stand workplaces, distractions and future technologies, e.g. augmented reality (AR) will be
discussed.

© Springer Nature Switzerland AG 2019
S. Bagnara et al. (Eds.): IEA 2018, AISC 827, pp. 428–429, 2019.
https://doi.org/10.1007/978-3-319-96059-3_50

3   Literature Results

3.1   Developments in Image Diagnostics

Several studies report of major changes in image diagnostics, e.g. change from film-based images to soft images, increase use and request of radiology images, especially increase demand for cross-sectional and merged images.

Imaging
Image diagnosis on a film image attached to a light box required at least part-time standing during the work hours, whereas soft-based image diagnostics is mostly performed as sedentary work. Further, using a light box required at light to get dimmed or even the room darkened, whereas scanning a soft image on the computer screen would allow normal office lighting. However, practice shows that even young radiologists still adopt working in darkened rooms, which may be associated to fatigue, eye strain and decreased attentiveness.

Use of Radiology Imaging
New technology with soft-based imaging led to an increased capacity in performing radiology imaging on patients. Furthermore, in the past decade, the request for radiology imaging has increased dramatically. Literature reports of a ten-fold increase of imaging, whereas the individual radiologist workload has increased five-fold. Besides the traditional X-ray images, the use of cross-sectional imaging and fusion of different types of imaging has increased, too.

3.2   Design of Radiology Work Environments

First, the question shall be raised what are the fundamental human needs on a healthy work environment in radiology. What are the major tasks and how may the work environment contribute e.g. to ease the strain of the workload, avoid distraction and support attentiveness and mood. Furthermore, what are the foreseeable technological developments and how may those affect work, work-related health hazards and occupational well-being of radiologists.

4   Conclusion

The workshop shall conclude in deriving recent major ergonomic hazards and challenges from the literature, as well as in recommendations on current and future design of radiology work environments.

Ergonomic Challenges and Interventions in Radiology Digital Imaging

S. Hoffmann[1]([envelope]) [iD] and N. Berger[2] [iD]

[1] MAS Work+Health, University of Zurich, Hirschengraben 82, 8001 Zurich,
Switzerland
sven.hoffmann@uzh.ch
[2] Institute of Diagnostic and Interventional Radiology, University Hospital of
Zurich, Raemistrasse 100, 8091 Zurich, Switzerland
nicole.berger@usz.ch

Abstract. This presentations aims to present the current literature and own field experience on clinical radiology workplaces. The presentation divides into work-related health problems and ergonomics challenges.

Special attention shall be payed to the occupational environment, including physical and organizational factors, e.g. impact of light exposure and workplace lighting design on attentiveness, sleep quality and wellbeing at work.

Keywords: Work-related health problems · Occupational environment
Work organization · New technologies

1 Work-Related Health Problems in Radiologists

Frequent work-related health problems in radiologists are visual fatigue and eye strain, musculoskeletal disorders and mental problems, frequently associated to long working hours and sleep deprivation. The latter are coming more into focus, since radiologist show a high prevalence of depression and burnout, being the highest among physicians.

2 Ergonomics Challenges

Three major ergonomic challenges were identified, coming from the (a) occupational environment, (b) work organization and (c) technology.

2.1 Occupational Environment

From an ergonomics perspective, reading and interpreting radiology images covers many aspects of PC office work. This refers to the physical work environment, e.g. space, location and materials used; further environmental factors, e.g. noise, lighting and indoor air quality, etc. Furthermore, postures and movements come into focus, which on the one hand side refer to the physical environment and furniture, on the other

© Springer Nature Switzerland AG 2019
S. Bagnara et al. (Eds.): IEA 2018, AISC 827, pp. 430–431, 2019.
https://doi.org/10.1007/978-3-319-96059-3_51

hand to workplace exposures, e.g. noise and odors, as well as workload and associated fatigue and distress.

2.2 Work Organization

From an ergonomics perspective, working schedule, tasks & responsibilities as well as distraction and interruptions come into focus. Many studies report of an more than dramatic increase in diagnostic workload for radiologists in the last decade. Besides interpreting radiology images and 3d reconstructions, senior radiologists have a large variety of responsibilities, e.g. monitoring resident radiologists, attending staff meetings, responding to phone calls or technician's requests. Furthermore, interruptions of diagnostic work was reported, e.g. by staff requests or telephone calls.

2.3 Technology

The introduction of digital systems in healthcare has brought numerous advantages, e.g. reduced radiation exposure to patients and staff, availability and accessibility of patient information, previous examinations of a patient, as well as computer-aided diagnostics. However, those advantages come alongside with health hazards and ergonomic challenges.

By cross-sectional imaging becoming wide-spread available as well as cheaper, the number of classical and cross-sectional images per patient has increased dramatically. In the future, additional challenges may apply by further developments in augmented 4d-reality imaging, computer-aided automatized image diagnostics as well as developments in radiology telemedicine.

Immersive Visualization of 3D Protein Structures for Bioscience Students

Tetsuri Inoue$^{(\boxtimes)}$, Kazutake Uehira, and Ayumi Koike

Kanagawa Institute of Technology, Atsugi, Kanagawa 243-0292, Japan
inoue@nw.kanagawa-it.ac.jp

Abstract. We have developed a visualization software to display protein structures in 3D on high-performance head-mounted displays (HMDs). This software aims to support students studying bioscience in understanding protein structures easily and rapidly. An experiment was conducted to examine how the immersive HMD visualization could enhance students' understanding of protein structures and raise their interest and motivation for learning. In the experiment, we compared 2D visualization of the 3D structures designed for desktop computers with 3D visualization designed for an immersive VR environment using high-performance HMDs. The results showed that students understood the protein structures after participating in the HMD observation better than after the observation on the desktop computers. When participating in the immersive VR environment, students had the feeling that they were actually inside the protein molecules. They viewed the 3D molecules as if they were real objects in front of them and tried to grab them. These results indicate that immersive visualization of 3D protein structures is effective for improving students' understanding.

Keywords: Immersive visualization · HMD · Bioscience education

1 Introduction

Recent advances in computer and visual display technologies produce high-performance displays such as multiscreen stereoscopic displays and wide field-of-view head-mounted displays. These displays can provide users with highly immersive virtual reality (VR) environments. VR environments offer us experiences that cannot be easily obtained in the real world. VR visualizations that use VR technologies to create visual imagery have been used in various fields such as medicine, engineering, design, training, and education.

Education is an important area in which VR visualizations can be effectively applied for teaching and learning. One example of VR visualization application in education can be found in medical schools. VR visualization can be used in 3D virtual models of the human body that students can explore [1]. VR astronomy is another example of an educational scientific visualization. Students can learn about the solar system by using 3D display systems in which they can move planets, look at stars, and track the progress of a comet in a virtual space simulator. These 3D virtual models have many advantages over conventional 2D images on a computer desktop. For example, 3D models can be observed from different angles, whereas 2D images are limited to one view [2].

© Springer Nature Switzerland AG 2019
S. Bagnara et al. (Eds.): IEA 2018, AISC 827, pp. 432–439, 2019.
https://doi.org/10.1007/978-3-319-96059-3_52

In a previous study in our laboratory, we created a VR simulation of sea fish for a science education program [3, 4]. The simulation provided a virtual undersea environment in which students could observe various types of 3D virtual fish models. We used a multi-screen display to present the VR simulation. The VR simulation provided actual-size objects and produced stereoscopic 3D images of fish located close to the observer (see Fig. 1). The idea is that students can access and interact with the 3D virtual fish electronically to improve their understanding.

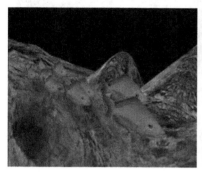 

Fig. 1. Example scenes of educational VR simulation of sea fish developed in a previous study. Stereoscopic 3D actual-size images of fish are located close to the user in the simulator.

In this study, we have developed VR visualization software to display protein structures on high-performance HMDs. This software aims to support bioscience students in understanding protein structures easily and rapidly. We investigated how VR visualization could enhance students' understanding of protein structures and raise their interest and motivation for learning.

2 Immersive Virtual Reality Display

Immersion in virtual reality is the perception of actually being in a simulated virtual environment. It is known that this perception can be enhanced by surrounding the user in the VR system with high resolution, stereoscopic wide field of view images. One example of a popular immersion technology is multi-projection display systems such as the CAVE-like display. Another example is head-mounted display (HMD) systems with a large field of view, such as Oculus Rift^{TM} (Oculus VR of Facebook Inc.) and HTC VIVE^{TM} (HTC Corp.).

The CAVE-like display is a room-type multi-projection display comprising four large screens (front, right, left, and floor) that surround the user [5]. Each screen is projected by stereoscopic 3D projectors that users can view using 3D glasses (see Fig. 2(a)). The screens around the participants display wide field of view images and provide them with an immersive and interactive virtual reality environment. An advantage of the CAVE-like display over other VR displays is that multiple viewers can observe the same scene at the same time and place. The disadvantage is that the equipment requires large room space and is quite expensive. It is not easy for an educational institution to provide such equipment.

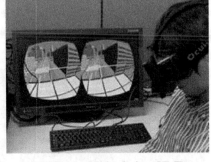

(a) CAVE-like display   (b) Wide field-of-view HMD

Fig. 2. Examples of popular immersive display technologies that provide the user with high resolution, stereoscopic wide field of view images.

HMDs have a small display optic in front of the user's eyes. With typical HMD devices, users are shielded from their physical surroundings and then immersed into a virtual environment produced by displayed images (see Fig. 2(b)). In addition, most HMDs have a head tracker device so that the system can respond to head movements of the user. That means that if users move their head to the left, the images in the display will change to make it seem as if they are actually looking at the left-hand part of the scene [6]. Recent HMDs feature improved performance in terms of resolution, refresh rate, field-of-view and compactness. On the other hand, their price has gone down. This is impractical for situations in which there are multiple students because only one person at a time can experience the VR environments via the HMD. For reasons of performance and price, we chose HTC VIVE^{TM}, a high-performance HMD, as the immersive display equipment in this study.

3   Bio-molecular Structure Education

Students of bioscience study proteins and other biomolecules in order to understand how life functions on the molecular level. Learning about protein structures is one of the most important elements in bioscience education because a wide variety of different functions by different proteins is directly related to the 3D structure of the proteins [7].

Large-scale datasets of 3D protein structures describe highly complex biomolecular entities that often consist of thousands of atoms and residues in large 3D strands of amino acids. Various types of 3D models are used to display these data sets, ranging from conventional ball-and-stick models to feature-presenting ribbon models (see Fig. 3). Comparative visualization of various structures in any model of representation is helpful for understanding the relationship between the function and structure of proteins. Due to the complexity of protein structures, these features are difficult to recognize using the 2D visualizations that are found in most software for PC desktops and textbooks. Therefore, different levels of representation are used to reduce the complexity of the geometric representation to be rendered. VR environments allow a detailed inspection and comparison of related molecular structures and offer a superior quality [8–10].

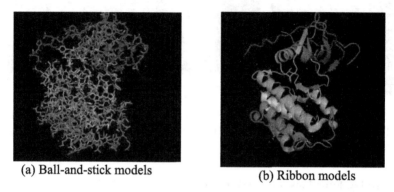

(a) Ball-and-stick models   (b) Ribbon models

Fig. 3. Examples of 3D models to display 3D protein structure datasets. Ball-and-stick models display both the three-dimensional position of the atoms and the bonds between them. Ribbon models are 3D schematic representations of protein structure.

If 3D models of protein structures are displayed with an immersive VR display in life-sized environments, participants can view the structure of 3D protein models from the inside and interact with the models. Such experiences are expected to be enormously helpful for participants to understand the structure.

4 Immersive Visualization Software of 3D Protein Structures

4.1 Overview

We have developed visualization software to display protein structures in 3D on high-performance HMDs. This software aims to help students studying bioscience to understand protein structures easily and rapidly. The software can read the molecular structure data of protein obtained from the Worldwide Protein Databank (PDB: http://www.pdb.org/) and display them in an immersive 3D VR environment. Students wear HMD headsets to look at the 3D models representing proteins. This software enables students to interactively move, rotate, and scale molecular models freely using hand-held controllers (see Fig. 4).

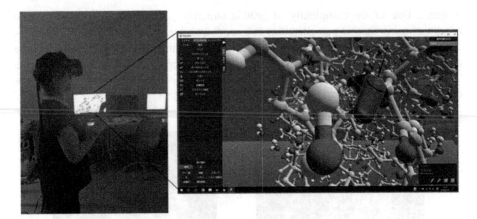

Fig. 4. Immersive visualization software of 3D protein structures developed for the HTC Vive HMD. Left: a user wearing the HMD to view 3D models. Right: Desktop image of the software.

4.2 Display Functions and User Interface

This software provides several 3D models to represent protein structure, including ball, ball-stick, ribbon and cartoon (see Fig. 5). Users manipulate a controller to translate, rotate, and scale the 3D models. They can also walk through the environment and enter the structure. They can highlight a specific atom or acid-base by pointing at them using the beam line from the controller, and the name and weight are then displayed (Fig. 6).

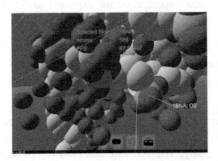
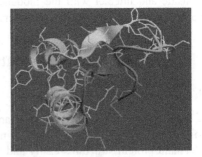

Fig. 5. Examples of desktop screen shots of the developed software. Users view the same images in wide field of view 3D images via the HMD.

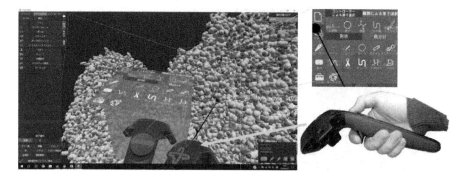

Fig. 6. Virtual controller that manipulates images in the virtual environments. Menu icons are displayed near the controller.

5   Investigation

5.1   Experiment

We conducted an experiment to examine how the immersive HMD visualization enhances students' understanding of protein structures and raises their interest and motivation for learning. In the experiment, we compared 2D visualization of the 3D structures designed for desktop computers with 3D visualization designed for an immersive VR environment using high-performance HMDs. We used the PyMOL software (https://pymol.org/2/) for the 2D visualization. PyMOL is widely used in bioscience education as a standard molecule viewing tool. To provide immersive 3D visualization of protein structures, we used our developed software.

Ten bioscience students from our university participated in the experiment. The students viewed 3D models representing proteins displayed on desktop computer screens generated by a 3D computer graphic software and then viewed the 3D models generated by the developed visualization software on HMD devices. At the end of each observation, the students answered chemistry-related questions, and at the end of the experiment, they filled out a questionnaire and expressed their impressions of the VR visualization.

5.2   Results and Discussion

The results showed that students understood the protein structures after participating in the HMD observation better than after the observation on the desktop computers. When participating in the immersive VR environment, students had the feeling that they were actually inside the protein molecules, and they could observe the molecules from different angles. They viewed the 3D molecules as if they were real objects in front of them and tried to grab them. These results indicate that immersive visualization of 3D protein structures is effective for improving students' understanding (Fig. 7).

Fig. 7. Actual images of participants in the experiment. The participants observed protein structures from both the outside and the inside in the virtual environments. They could view the protein models as if they were real objects in front of them and tried to grab them.

6 Summary

We have developed visualization software to display protein structures in 3D on high-performance HMDs. This software aims to support bioscience students in understanding protein structures easily and rapidly. The software displays molecular protein structures in a fully immersive VR environment. Students wear the HMD headsets and interactively move, rotate, and scale molecular models freely using handheld controllers. They can also observe protein structures from both the outside and the inside. Experimental results indicate that immersive visualization of 3D protein structures is effective for improving students' understanding.

Acknowledgements. This research is supported by the MEXT (the Ministry of Education, Culture, Sports, Science and Technology of Japan) Supported Program for the Strategic Research Foundation at Private Universities (No. S1511019L).

References

1. Zajtchuk R, Satava RM (1997) Medical Applications of Virtual Reality. Commun ACM 40 (9):63–64
2. Al-khalifah AH, McCrindle RJ, Sharkey PM, Alexandrov VN (2006) Using virtual reality for medical diagnosis, training and education. In: Proceedings of 6th International Conference on Disability, Virtual Reality & Assoc. Tech, pp 193–200
3. Shibata T, Lee J, Inoue T (2014) Ergonomic approaches to designing educational materials for immersive multi-projection system. In: Proceedings of SPIE 9012, Engineering Reality of Virtual Reality 2014: 90120Q1-8
4. Inoue T, Shibata T (2015) Evaluation of visual fatigue and sense of presence for CAVE-like multi-projection display. In: Lindgaard G, Moore D (eds) Proceedings of the 19th Triennial Congress of the IEA, No 670

5. Cruz-Neira C, Sandin DJ, DeFanti TA (1993) Surround-screen projection-based virtual reality: the design and implementation of the CAVE. In: Proceedings of the SIGGRAPH 1993, pp 135–142
6. Slater M, Usoh M, Steed A (1994) Depth of presence in immersive virtual environments. Presence: Teleoper Virtual Environ 3:130–144
7. Akkiraju N, Edelsbrunner H, Fu P, Qian J (1996) Viewing geometric protein structures from inside a CAVE. IEEE Comput Graph Appl 16(4):58–61
8. Limniou M, Roberts D, Papadopoulos N (2008) Full immersive virtual environment CAVETM in chemistry education. J Comput Educ 51(2):584–593
9. Moritz E, Meyer J (2004) Interactive 3D protein structure visualization using virtual reality. In: Proceedings of the 4th IEEE Symposium on Bioinformatics and Bioengineering 2004
10. Salvadori A, Gianluca FD, Pagliai M, Mancini G, Barone V (2016) Immersive virtual reality in computational chemistry: Applications to the analysis of QM and MM data. Quantum Chem 116(22):1731–1746

Author Index

© Springer Nature Switzerland AG 2019
S. Bagnara et al. (Eds.): IEA 2018, AISC 827, pp. 441–443, 2019.
https://doi.org/10.1007/978-3-319-96059-3

Printed in the United States
By Bookmasters